Nanoscale Semiconductor Memories

Technology and Applications

Devices, Circuits, and Systems

Series Editor

Krzysztof Iniewski
CMOS Emerging Technologies Research Inc.,
Vancouver, British Columbia, Canada

FORTHCOMING TITLES:

High-Speed Devices and Circuits with THz Applications
Jung Han Choi and Krzysztof Iniewski

Labs-on-Chip: Physics, Design and Technology
Eugenio Iannone

Laser-Based Optical Detection of Explosives
Paul M. Pellegrino, Ellen L. Holthoff, and Mikella E. Farrell

Metallic Spintronic Devices
Xiaobin Wang

Microfluidics and Nanotechnology: Biosensing to the Single Molecule Limit
Eric Lagally and Krzysztof Iniewski

MIMO Power Line Communications: Narrow and Broadband Standards, EMC, and Advanced Processing
Lars Torsten Berger, Andreas Schwager, Pascal Pagani, and Daniel Schneider

Mobile Point-of-Care Monitors and Diagnostic Device Design
Walter Karlen and Krzysztof Iniewski

Nanoelectronics: Devices, Circuits, and Systems
Nikos Konofaos

Nanomaterials: A Guide to Fabrication and Applications
Gordon Harling and Krzysztof Iniewski

Nanopatterning and Nanoscale Devices for Biological Applications
Krzysztof Iniewski and Seila Selimovic

Optical Fiber Sensors and Applications
Ginu Rajan and Krzysztof Iniewski

Organic Solar Cells: Materials, Devices, Interfaces, and Modeling
Qiquan Qiao and Krzysztof Iniewski

Power Management Integrated Circuits and Technologies
Mona M. Hella and Patrick Mercier

Radio Frequency Integrated Circuit Design
Sebastian Magierowski

Semiconductor Device Technology: Silicon and Materials
Tomasz Brozek and Krzysztof Iniewski

Smart Grids: Clouds, Communications, Open Source, and Automation
David Bakken and Krzysztof Iniewski

Soft Errors: From Particles to Circuits
Jean-Luc Autran and Daniela Munteanu

Technologies for Smart Sensors and Sensor Fusion
Kevin Yallup and Krzysztof Iniewski

Edited by **Santosh K. Kurinec** • **Krzysztof Iniewski**

Nanoscale Semiconductor Memories

Technology and Applications

CRC Press
Taylor & Francis Group
Boca Raton London New York

CRC Press is an imprint of the
Taylor & Francis Group, an **informa** business

MATLAB® is a trademark of The MathWorks, Inc. and is used with permission. The MathWorks does not warrant the accuracy of the text or exercises in this book. This book's use or discussion of MAT-LAB® software or related products does not constitute endorsement or sponsorship by The MathWorks of a particular pedagogical approach or particular use of the MATLAB® software.

CRC Press
Taylor & Francis Group
6000 Broken Sound Parkway NW, Suite 300
Boca Raton, FL 33487-2742

First issued in paperback 2017

© 2014 by Taylor & Francis Group, LLC
CRC Press is an imprint of Taylor & Francis Group, an Informa business

No claim to original U.S. Government works

Version Date: 20131017

ISBN 13: 978-1-138-07264-0 (pbk)
ISBN 13: 978-1-4665-6060-4 (hbk)

Visit the Taylor & Francis Web site at
http://www.taylorandfrancis.com

and the CRC Press Web site at
http://www.crcpress.com

Contents

PART I Static Random Access Memory

PART II Dynamic Random Access Memory

PART III   Novel Flash Memory

PART IV   Magnetic Memory

PART V   Phase-Change Memory

PART VI   Resistive Random Access Memory

Preface

At no time in the history of the semiconductor industry has memory technology assumed such a pivotal position. The last decade has seen a remarkable shift in usage and value of semiconductor memory technologies. These changes have been driven by the elevation of three particular target applications for the development of memory technology performance attributes.

The first and most obvious shift is that mobile multimedia applications such as tablets and advanced cell phones have now replaced desktop data processing as the primary target for many new semiconductor technologies. The significance of this shift is that the smaller form factor and smaller semiconductor content automatically increases the percentage of value contributed by the analog wireless and the memory components. The second trend is driven by the explosive growth in the sheer volume of data that is being created and stored. The continuing growth in digital information is heavily driven by mobile multimedia access to cloud storage on the Internet as well as the astounding increase in image data storage and manipulation. The third trend is the shift of emphasis from the individual components to the ability to configure some high-volume elements in subsystems and multidie packages rather than as discrete components on a motherboard.

Over the past three decades, numerous memory technologies have been brought to market with varying degrees of commercial success, such as static random-access memory (SRAM), pseudostatic RAM, NOR flash, erasable programmable read-only memory (EPROM), electrically erasable programmable read-only memory (EEPROM), dynamic RAM (DRAM), and NAND flash. Generally speaking, these "memory" technologies can be split into two categories: volatile and nonvolatile. Volatile memory does not retain data when power is turned off; conversely, nonvolatile memory retains data once power is turned off. The dominating memory technologies in the industry today are SRAM, DRAM (volatile), and NAND flash (nonvolatile). Storage class memory (SCM) describes a device category that combines the benefits of solid-state memory, such as high performance and robustness, with the archival capabilities and low cost per bit of conventional hard disk magnetic storage. Such a device requires a high areal density nonvolatile memory technology that can be manufactured at a very low cost per bit.

The general technology requirements of memories are compatibility and integration with complementary metal oxide semiconductor (CMOS) platform, high functional bit density, high speed, low power dissipation, and low cost. The major technology barriers are stability, reliability, data retention, on–off ratio, and endurance. There is a significant interplay between requirements and barriers, and optimized trade-offs between them are expected. The current memory technologies have entered the nanoscale regime and are encountering very difficult issues related to their continued scaling to and beyond the 16 nm generation.

SRAM is typically constructed from core CMOS technology; all issues associated with MOSFET scaling apply to scaling of SRAM. In addition, research is ongoing to find a dense SRAM replacement that can substantially reduce the area occupied by the traditional 6T SRAM bit cell. Discovery of such a bit cell would have profound implications on the die cost of integrated circuits given the ever-increasing area ratio occupied by this type of memory. In addition to area scaling, there is also a need to develop alternate architecture that maintains stability while operating at lower voltages, thus allowing the industry to substantially reduce standby power consumption in the memory arrays.

Embedded SRAM continues to be a critical technology enabler for a wide range of applications from high-performance computing to mobile applications. It is important for SRAM to reduce both leakage and dynamic power, keeping products within the same power envelope at the next technology node. Redundancy and error correction code (ECC) protection are also keys to ensure yield and reliability when embedded SRAM products go to production.

In the DRAM, the one-transistor/one-capacitor cell, which can be trench or stack capacitor, requires photolithographic processes of very high aspect ratio. To meet retention and refresh requirements, the transistor has to control both subthreshold leakage and junction leakage. The transistor structure is becoming nonplanar such as recessed channel and FinFET. This is especially challenging for the extra requirement of nonplanar surface for the capacitor in order to get adequate capacitance with minimal layout area. To minimize the capacitor area, higher permittivity dielectrics are a natural path. Embedding DRAM into a CMOS process flow has become more popular over the last decade.

Flash memory is composed of one transistor with two stacked gates, a floating gate underneath a control gate. The threshold voltage of this transistor depends on whether the floating gate is charged or not. These transistors are then arranged in either a NOR or a NAND configuration to create the memory device. In a NOR flash memory configuration, the gate is connected to a word line, while the drain and source are connected to a bit line and to ground. NAND flash memory connects transistors that compose the memory device in series. These blocks of memory are further linked together in a NOR style configuration. The series bitcell string of NAND flash eliminates contacts between the cells compared with the NOR type and results in a smaller cell size, which reduces manufacturing costs. The NAND configuration is more prevalent due to its capacity to achieve higher density. Possible nearterm scenarios include 3D stacked NAND vertical gates as a solution to further increase NAND density. New strategies for using nanocrystals or quantum dots as charge trapping locations are underway.

The floating gate technology has extended its process span by employing SONOS configuration, which consists of a stack of oxide (SiO_2), nitride (Si_3N_4), and oxide. Charge is stored in the electron traps in the nitride film. Since the electron traps are discrete, the leakage path affects a very small fraction of the stored charge. To improve the blocking performance further, a high work function metal gate is introduced (TANOS). Scaling of charge-based storage to these dimensions had been deemed questionable in past decades due to reliability concerns, and this had sparked investigations into alternative technologies.

In the past decade, there has been significant focus on the emerging memories field to find possible contenders to displace either or both NAND flash and DRAM.

Some of these newer emerging technologies include magnetic RAM (MRAM), spin-transfer torque RAM (STT-RAM), ferroelectric RAM (FeRAM), phase change RAM (PCRAM), resistive RAM (RRAM), and memristor-based RRAM. Innovations in fabrication processes and devices are continuing to fuel all competing technologies. Increased storage capability with reduced costs, significantly higher speed random access, and light weight have pushed flash memory into competition with hard drives for notebook computers and high-performance systems and has been one of the enabling technologies for lightweight, low-power tablet PCs. Currently, any competing solid-state memory technology has to either outperform flash memory in its own memory segment, which is difficult in terms of density unless multibit per cell operation is achieved or has to offer higher performance.

In MRAM, the most common basic cell is composed of one n-channel metal oxide semiconductor (NMOS) transistor as the access device and one magnetic tunnel junction (MTJ) as the storage element (1T1J structure). The MTJ constitutes a pinned magnetic layer (e.g., CoFe or NiFe/CoFe) and a free magnetic layer (e.g., CoFe or NiFe/CoFe) separated by an insulating barrier (e.g., MgO). Information is stored in the magnetization direction of the free layer. By employing a magnetic field, the orientation of the free magnetic layer can be flipped in order to make the two layers parallel or antiparallel with each other. These two conditions correspond to high or low barrier conductance, respectively, and thus define the state of the memory bit. The latest MRAM technology is STT-RAM. In STT-RAM, the direction of magnetization of the free layer is changed by directly passing spin-polarized currents through MTJs. STT-RAM has the advantage of scalability, which means that the threshold current to make the state reversal will scale down as the size of the MTJ becomes smaller. FeRAM utilizes the permanent polarization of a ferroelectric material such as PZT (lead–zirconate–titanate), SBT (strontium–bismuth–tantalate), or BLT (La-substituted bismuth tantalate) as the storing mechanism. It has a DRAM-like cell structure for a 1-transistor, 1-capacitor cell.

PCRAM is one of the leading candidates among alternatives to flash and DRAM. This memory works based on the thermally induced reversible phase transition in phase change materials that exhibit two stable material phases: a low-resistance crystalline phase and a high–resistance, short-range-ordered amorphous phase. The most commonly used material is a chalcogenide, $Ge_2Sb_2Te_2$ (GST), which is widely used in optical storage devices such as compact discs and digital video discs wherein heating by a laser beam enables the GST layer to switch between the two states. These two states have a distinct difference in optical reflectivity. A basic PCRAM cell consists of the phase change material layer sandwiched between two electrodes. The device is driven by a bipolar or field-effect transistor in a 1 transistor/1 resistor (1T1R) configuration or by a diode in a 1 diode/1 resistor (1D1R) configuration. The two states of the PCM are known as SET (low resistance) and RESET (high resistance) states. The RESET state is achieved by applying a pulse to heat the PCM above its melting point and rapidly quenching it to its high-resistance short-range-order state. To return to SET state, a longer pulse is applied to heat the PCM above its crystallization temperature but below its melting point, allowing it to crystallize to its low-resistance state. Some commercial applications, such as cellular phones, have recently started to use PCRAM, demonstrating that reliability and cost competitiveness in emerging memories is becoming a reality. Fast write speed and low read-access time are being achieved in many of these emerging memories.

RRAM is a type of nonvolatile memory that shares some similarities with PCRAM as both are considered to be types of memristor technologies—a passive two-terminal electronic device that is designed to express only the property of an electronic component that lets it recall the last resistance it had before being shut off ("memristance"). In the case of RRAM, the memory cell is a metal–insulator–metal (MIM) structure. Resistance switching is accomplished by changing the conductivity of the insulator layer. Resistance switching is observed in a wide range of transition metal oxides, including NiO, TiO_2, and HfO_2. Based upon the types of switching mechanisms, RRAM cells can be further classified as filament-based, interface–based, or programmable metallization- based cells (PMC). The redox-based nanoionic memory operation is based on a change in resistance of a MIM structure caused by ion (cation or anion) migration combined with redox processes involving the electrode material, the insulator material, or both. The material class for redox memory is comprised of oxides, chalcogenides (including glasses), semiconductors, as well as organic compounds including polymers. Another form of RRAM is the Mott memory, where charge injection induces a transition from strongly correlated to weakly correlated electrons, resulting in an insulator–metal transition (IMT) or Mott transition. Electronic switches and memory elements based on the Mott transition (sometimes referred to as CeRAM—correlated electron random access memory) have been explored using several materials systems such as VO_2, $SmNiO_3$, NiO, and others.

Other emerging areas of memory that are not discussed in this book include molecular memory, using individual molecules or small clusters of molecules as building blocks, and nanoelectromechanical memory (NEMM). NEMM is based on a bistable nanoelectromechanical switch (NEMS). In this concept, mechanical digital signals are represented by displacements of solid nanoelements (e.g., nanowires, nanorods, or nanoparticles), which result in closing or opening of an electrical circuit. Several different modifications of suspended beam/cantilever NEMMs are currently being explored using different materials, including Si Ge, TiN, carbon nanotubes (CNT), and others. A difficult challenge of the cantilever NEMM is scalability as the cantilever spring constant and therefore the pull-in voltage increases as the beam's length decreases.

In the quest for a universal memory, engineers hope to find a memory system that fits all the requirements of an "ideal memory" capable of high-density storage, low-power operation, unparalleled speed, high endurance, and low cost. Today's memory technologies cannot satisfy all these criteria and are thus oriented toward specific categories. In the future, memory systems may have most of the desired features and may be able to provide broad based application's currently served separately by conventional memory types.

This brief review will be incomplete without providing a future vision towards 3D integration. In a typical 2D architecture, memory arrays and peripheral logic devices are generally located on the same plane above the Si substrate since both devices use single crystalline Si as the channel material. These 2D chips have a cell-area efficiency of approximately 60% and in other words, peripheral logic devices use 40% of the chip area. In order to increase the cell-area efficiency, the 3D vertical-chip architecture is preferred to have the memory and logic cells stacked vertically. Trends towards 3D heterogeneous integration of memory with logic are emerging. The Hybrid Memory Cube (HMC), envisioned by Micron blends the best of logic and DRAM processes into a heterogeneous 3D package. At its foundation is a small logic layer that sits below vertical stacks of DRAM die connected by through-silicon -vias (TSVs). An energy-optimized DRAM array provides access to memory bits via the internal logic layer and TSV – resulting in an intelligent memory device, optimized for performance and efficiency. By placing intelligent memory on the same substrate as the processing unit, each system can be more efficiently than previous technologies. Specifically, processors can make use of all of their computational capability without being limited by the memory channel. A radically new technology like HMC requires a broad ecosystem of support for mainstream adoption. To address this challenge, Micron, Samsung, Altera, Open-Silicon, and Xilinx, collaborated to form the HMC Consortium (HMCC), in 2011. This architectural breakthrough will lead to stack multiple memories onto one chip.

This introduction provides a basic overview of various memory technologies presented in this book. The readers are directed to an excellent review "Nanoscale memory devices," written by A. Chung, J. Deen, Jeong-SooLee, and M. Meyyappan, published in *Nanotechnology* 21 (2010) 412001, for further understanding of different memory technologies.

The book is divided into six parts dedicated to current and prototypical memory technologies. Part I consists of three chapters on SRAM. The first chapter addresses the design challenges as the technology scales, followed by two chapters on explaining and designing strategies to mitigate radiation induced upsets in SRAM.

Part II consists of three chapters. Chapter 4 discusses the state of the art in DRAM technology and the need to develop high-performance sense amplifier circuitry. Chapters 5 and 6 are devoted to the novel concept of capacitorless 1T DRAM known as advanced-RAM or A-RAM.

Part III consists of a single chapter. Chapter 7 covers quantum dot–based flash memory, describing the advantages of using self-organized quantum dots created with heterostructures made out of III–V semiconductors in which charge carriers are confined.

Part IV consists of two chapters that focus on emerging magnetic memories. Chapter 8 describes STT-RAM with an emphasis on scalable embedded STT-RAM. Chapter 9 discusses the physics and engineering of magnetic domain wall "racetrack" memory. Racetrack memory, envisioned by IBM researchers, promises a novel storage-class memory combining characteristics of low cost per bit of magnetic disk drives and the high performance and reliability of conventional solid state memories.

Part V is dedicated to state-of-the-art modeling applied to phase change memory devices. Chapters 10 and 11 present the work by leading groups in the area of nanoscale PCM modeling and simulations, which are extremely important in designing future PCRAM.

Part VI provides an extensive review and discusses the latest updates in RRAM. Chapters 12 and 13 cover the physics of operation of RRAM and provide an in-depth analysis of different materials systems currently under investigation.

Santosh K. Kurinec
Rochester, New York

Krzysztof (Kris) Iniewski
Vancouver, British Columbia, Canada

MATLAB® is a registered trademark of The MathWorks, Inc. For product information, please contact:

The MathWorks, Inc.
3 Apple Hill Drive
Natick, MA 01760-2098 USA
Tel: 508 647 7000
Fax: 508-647-7001
E-mail: info@mathworks.com
Web: www.mathworks.com

Editors

Santosh K. Kurinec received her PhD in physics from the University of Delhi, India, and later became a scientist at the National Physical Laboratory, New Delhi. She worked as a postdoctoral research associate in the Department of Materials Science and Engineering at the University of Florida, Gainesville, Florida, where she conducted research on thin metal film composites. She then served as an assistant professor of electrical engineering at the Florida State University/Florida A&M University College of Engineering in Tallahassee, Florida, where she researched on light emission from silicon and superconducting thin films. She joined the Rochester Institute of Technology (RIT) in 1988. She served as the head of the Department of Microelectronic Engineering at RIT from 2001 to 2009. She is currently a professor of electrical and microelectronic engineering at RIT. Dr. Kurinec has been actively engaged in outreach for promoting engineering education. She received the RIT Trustee Scholarship Award in 2008 and was honored as the Engineer with Distinction by the Rochester Engineering Society in 2013. She is a fellow of the IEEE, a member of the American Physical Society (APS) and the New York State Academy of Sciences (NYAS), associate editor of *IEEE Transactions on Education*, and an IEEE EDS Distinguished Lecturer. She received the 2012 IEEE Technical Field Award for Outstanding Undergraduate Teaching. Her current research activities include nonvolatile memory, photovoltaics, and advanced integrated circuit materials and processes. She has over 100 publications in research journals and conference proceedings. She can be reached at santosh.kurinec@rit.edu.

Krzysztof (Kris) Iniewski is managing R&D at Redlen Technologies Inc., a start-up company in Vancouver, Canada. Redlen's revolutionary production process for advanced semiconductor materials enables a new generation of more accurate, all-digital, radiation-based imaging solutions. Kris is also a president of CMOS Emerging Technologies Research Inc. (www.cmosetr.com), an organization of high-tech events covering communications, microsystems, optoelectronics, and sensors. In his career, Dr. Iniewski held numerous faculty and management positions at the University of Toronto, University of Alberta, Simon Fraser University (SFU), and PMC-Sierra Inc. He has published over 100 research papers in international journals and conferences. He holds 18 international patents granted in the United States, Canada, France, Germany, and Japan. He is a frequent invited speaker and has consulted for multiple organizations internationally. He has written and edited several books for CRC Press, Cambridge University Press, IEEE Press, Wiley, McGraw-Hill, Artech House, and Springer. His personal goal is to contribute to healthy living and sustainability through innovative engineering solutions. In his leisurely time, Kris can be found hiking, sailing, skiing, or biking in beautiful British Columbia. He can be reached at kris.iniewski@gmail.com.

Contributors

Maryline Bawedin
Institut d'Electronique du Sud
Université de Montpellier II
Montpellier, France

Dieter Bimberg
Institute of Solid State Physics
Berlin Institute of Technology
Berlin, Germany

Mansun Chan
Department of Electronic and Computer
    Engineering
The Hong Kong University of Science
    and Technology
Kowloon, Hong Kong

Lin Chen
School of Microelectronics
Fudan University
Shanghai, People's Republic of China

Lawrence T. Clark
School of Electrical, Computer and
    Energy Engineering
Arizona State University
Phoenix, Arizona

Sorin Cristoloveanu
IMEP-LAHC
Grenoble INP Minatec
Grenoble, France

Azer Faraclas
Department of Electrical and Computer
    Engineering
University of Connecticut
Storrs, Connecticut

Michael C. Gaidis
Physical Sciences Division
IBM T.J. Watson Research Center
Yorktown Heights, New York

Francisco Gamiz
Department of Electronics and
    Computer Technology
University of Granada
Granada, Spain

Gilles Gasiot
Technology R&D
Central CAD and Design Solutions
STMicroelectronics
Crolles, France

Martin Geller
Experimental Physics and CeNIDE
University of Duisburg-Essen
Duisburg, Germany

Ali Gokirmak
Department of Electrical and Computer
    Engineering
University of Connecticut
Storrs, Connecticut

Jin He
Shenzhen SOC Key Laboratory
PKU HKUST Shenzhen-Hong Kong
    Institution
Peking University
Shenzhen, Guangdong, People's
    Republic of China

and

The Hong Kong University of Science
    and Technology
Kowloon, Hong Kong

Kangho Lee
Advanced Technology
Qualcomm Inc.,
San Diego, California

Myoung Jin Lee
R&D Division
Hynix Semiconductor Inc.
Kyoungki-do, South Korea

Qingqing Liang
Institute of Microelectronics
Chinese Academic of Sciences
Xicheng, Beijing, People's Republic
    of China

Hangbing Lv
School of Microelectronics
Fudan University
Shanghai, People's Republic of China

Andreas Marent
Institute of Solid State Physics
Berlin Institute of Technology
Berlin, Germany

Tobias Nowozin
Institute of Solid State Physics
Berlin Institute of Technology
Berlin, Germany

Philippe Roche
Technology R&D
Central CAD and Design Solutions
STMicroelectronics
Crolles, France

Noel Rodriguez
Department of Electronics and
    Computer Technology
University of Granada
Granada, Spain

Helena Silva
Department of Electrical and Computer
    Engineering
University of Connecticut
Storrs, Connecticut

Qingqing Sun
School of Microelectronics
Fudan University
Shanghai, People's Republic of China

Luc Thomas
TDK-Headway Technologies, Inc.
Milpitas, California

Haijun Wan
School of Microelectronics
Fudan University
Shanghai, People's Republic of China

Yujun Wei
Shenzhen SOC Key Laboratory
PKU HKUST Shenzhen-Hong Kong
    Institution
Peking University
Shenzhen, Guangdong, People's
    Republic of China

and

School of Electronic Engineering and
    Computer Science
Peking University
Haidian, Beijing, People's Republic of
    China

Hong Yu Yu
Department of Electrical and Electronic
    Engineering
South University of Science and
    Technology of China
Shenzhen, Guangdong, People's
    Republic of China

Peng Zhou
School of Microelectronics
Fudan University
Yangpu, Shangai, People's Republic
    of China

Part I

Static Random Access Memory

1 SRAM
The Benchmark of VLSI Technology

Qingqing Liang

CONTENTS

In all semiconductor memories, static random access memory (SRAM) is with the closest device structure to the conventional devices and is most representative of the VLSI technology. There are both N- and P-type transistors in the memory cell, and usually only a few implant differences between the devices in the cell are used in standard logic units. Due to this similarity, it is well recognized as one of the key technology benchmark. Compared to other types of memory cells, the design and optimization of SRAM cell is an unavoidable task and strongly associated to more general issues in the technology development.

Since the early 1960s, we have witnessed a continuous, exponential growth through each technology generations: the device area shrinks down by half with better performance or power consumption in every 18–24 months [1]. Among various obstacles during the technology evolution, the fluctuation of device electrical behavior is emerging as one of the most fundamental limits to the yield of small devices such as SRAM

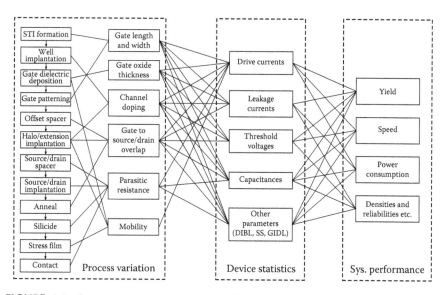

FIGURE 1.1 Standard process flow of sub-65 nm CMOS technology and correlations among each process step, process-induced variables, device electric behavior, and circuit/system performance.

cells [2–11]. First, as the area shrinks, the fluctuation inevitably increases by nature [12–14]. Second, as the process steps of the state-of-the-art technology keep increasing, more variations are introduced and complicate impacts on device behavior are expected [15–21]. Moreover, as the applied voltage is decreased to achieve lower power consumption—from 3.3 V in sub-micron node down to 0.9 V in sub-32 nm node—issues like the threshold voltage fluctuation become more problematic even its magnitude keeps the same, since the normalized sigma or proportional fluctuation increases.

Obviously, an accurate characterization of device variations is the key to evaluate and optimize the advanced VLSI technology. As shown in Figure 1.1, comprehensive statistics analysis (including the sigmas and correlations) is involved to link the process modules, device behavior, and circuit performance, and should be conducted on either the bottom-up or the top-down design approaches. More specifically, the statistics study should provide not only the guidelines to process engineers such as which module dominates device fluctuation (hence the yield), but also the information to circuit/system designers such as performance-power corners to reserve adequate redundancies. In this chapter, we will investigate these issues from the following aspects: the origins of device variations in the advanced VLSI technology, the methodology for accurate characterization on the device statistics, and the design and optimization of the technology benchmark: SRAM cell.

1.1 ORIGINS OF DEVICE VARIATION

The left side of Figure 1.1 shows a typical process flow of conventional sub-65 nm CMOS technology [22–29]. In general, every single step is more or less a variation source. Moreover, recent technologies adopt lots of new material and process modules to keep device scaling (e.g., stress film liner, stress memory technique, embedded SiGe source/drain, laser anneal, and high-K metal-gate), which cause additional variations. There could be hundreds of independent process variation sources in a standard CMOS technology flow. Even though monitoring the variation of each step in the flow is important for process development, it is more feasible to group them into fewer categories for characterization. Indeed, as shown in early studies [2–11], the effects of these process steps on electrical behavior are linked to just a few primary responses (i.e., many process-induced variations can be lumped into one or more key categories) from the electrical data. It has been demonstrated that about six or seven primary responses [8,9] are enough to represent statistics of device electrical characteristics. Then one can correlate the primary electrical behavior responses to process variables, which are detailed in the next section.

1.1.1 GATE LENGTH AND WIDTH

Among all process-induced variables, gate length is dominant. Besides physical gate length edge roughness (LER) caused by litho resist and RIE, variation of effective gate length (L_{eff}) is also caused by spacer, extension and source/drain implant, and rapid thermal anneals. Measuring the sigma of either physical gate length L_{gate} or L_{eff} is rather difficult. Electrical measurement of L_{gate} requires large arrays of MOS capacitors, which is not representative to the sigma of a single FET. The scanning/transmission electron microscope (SEM/TEM) measurements offer only a small population of data. However, the average gate length can be adjusted by simply changing the layout. Hence, here we denote it as an explicit variable because the impact of changing gate length can be clearly characterized. Similarly, gate width is also an explicit variable, and its variation is associated with various process modules: divot in the formation of shallow trench insulation [30], fringing dopant segregation [31], stress proximity [32], etc. These effects are generally negligible in wide devices (e.g., W > 1 μm), thus they can be decoupled through wide-to-narrow width average comparison.

1.1.2 GATE OXIDE THICKNESS

The variation of gate oxide thickness (T_{inv}) is not only due to gate dielectric deposition but also due to doping fluctuation in the polysilicon gate, where a portion of T_{inv} (effective gate oxide thickness in the inverse-biased condition) comes from poly-gate depletion. Moreover, if high-K gate dielectric is used [33,34], the thermal process later on may introduce regrowth of interface oxide, which causes additional variation. Like gate length, the average T_{inv} value can be measured in large arrays of capacitors, whereas the measurement of the sigma of single FETs requires advanced testing techniques (e.g., charge-based capacitance measurement [35]). It is an explicit variable since the impact of T_{inv} variation can be monitored by measuring data from wafers that only change the gate deposition process.

1.1.3 CHANNEL DOPING

Channel doping is another dominant variable in small devices and is mainly driven by well and halo implantation. The major outcome is random doping fluctuations of threshold voltages and drive currents, which is inversely proportional to the effective transistor channel area [12–14]. It is also an explicit variable that can be characterized using different well or halo implant conditions. As different gate stacks are used in CMOS technology development [25–29], different sources for V_t variation are introduced. Dipole, density of interface traps (DIT), and metal work function cause different impacts on threshold voltage (e.g., temperature dependency [36]). Further studies on these effects need more process experiments and are still ongoing. Here, for simplicity, we still lump these process-induced variables into channel doping (as an example, one can assume these are δ-function doping profiles located at the interface). However, if temperature is varying, this portion will become an additional variable since it plays a different role to the carrier mobility compared with normal doping.

1.1.4 GATE TO SOURCE/DRAIN OVERLAP

As illustrated in Figure 1.2, whereas the overlap distance of gate-to-source/gate-to-drain can be estimated as $(L_{gate} - L_{eff}) = 2$, the impact of the overlap should also account for the thickness and doping level in this region. The overlap region not only contributes parasitic resistance and capacitance but also influences electrostatics and leakage currents. Combined with the channel doping, it can be used to approximate the 2D profile dependency of threshold voltages and drive currents. The associated process steps are spacer thickness, extension (or LDD) implantation, and thermal anneals thereafter. If using extension implant conditions (e.g., dose, energy, and tilted angle) as the primary driving factor, the impacts of the overlap region on electrical behavior can be distinguished from other variables.

1.1.5 MOBILITY

Since the introduction of 90 nm technology node [37–39], mobility has become a knob in device design. The commonly used approaches to apply stress on CMOS

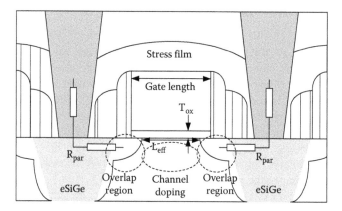

FIGURE 1.2 Cross-sectional view of a standard MOSFET device structure and corresponding process-induced variables.

devices use a stress liner (or contact etch stop liner) covering the FET [38], and/or use an embedded SiGe source/drain [39], as shown in Figure 1.2. In any case, the effective stress applied on the intrinsic channel depends on the device structures (e.g., stress liner thickness, gate pitch, and e-SiGe proximity), hence the variation of mobility is unavoidable.

On the characterization side, how to accurately extract the mobility of a short channel FET is still a well-known issue, since it is hard to decouple the impact of mobility from other variables (e.g., parasitic resistance in the overlap region) due to the distributive nature of the device profile. Therefore, it is denoted as an implicit variable, which requires additional information to derive the trend of the impacts on electrical characteristics.

1.1.6 PARASITIC RESISTANCE

Whereas the parasitic source/drain resistance strongly depends on the overlap region, the additional parts such as silicide and metal contact are not correlated to the intrinsic device behavior. The fluctuations of these parts are due to source/drain implantation, thermal anneals, silicidation, and metal contact. The impact of these components on device behavior is different from the influence of the overlap region (e.g., different trends on parasitic resistance and parasitic capacitance) and is not negligible (especially in sub-65 nm devices where source/drain and silicide resistance significantly degrade the performance [40]). However, the former is usually overwhelmed by the latter and is hard to be distinguished. Therefore, it is an implicit variable that needs to be considered in the analysis besides the overlap region.

These variables cover most of primary device responses for the whole process flow. As mentioned earlier, each process step may induce one or more variables in the list. Therefore, any of the variables is more or less correlated to each other. This raises more difficulties in statistical analysis, which will be discussed in the following sections. Moreover, all the process variation can be either pure random or systematic. Since the systematic variation is easy to be characterized, only the random portion is studied here.

1.2 ACCURATE CHARACTERIZATION OF STATISTICS

1.2.1 GENERAL REPRESENTATION OF VARIATIONS

According to Figure 1.1, one needs to get the primary responses of device electrical behavior before linking them to the key process-induced variables. If the primary responses and their statistics are accurately extracted, the device electrical behavior should be fully represented and the model construction is then straightforward: one can either use conventional compact models (e.g., BSIM and PSP) or behavioral models as long as those responses can be fitted well.

The question is how to obtain the primary responses from scores of electrical measured points, especially in devices with strong nonlinear characteristics that principal component analysis (PCA) is no longer valid (since the correlation coefficients directly extracted from non-Gaussian distributions are skewed). Considering that, we established a parameter transferring methodology to "Gaussify" all the measured parameters:

$$\int_{-\infty}^{+\infty} P(x)dx = \int_{0}^{1} \frac{P(x)}{y'(x)} d[y(x)] = \int_{0}^{1} dy$$

$$\Rightarrow y(x) = \int_{-\infty}^{x} P(x_1)dx_1. \tag{1.1}$$

$$\int_{-\infty}^{+\infty} F(z)dz = \int_{0}^{1} \frac{F(z)}{y'(z)} d[y(z)] = \int_{0}^{1} dy$$

$$\Rightarrow y(z) = \int_{-\infty}^{z} F(z_1)dz_1. \tag{1.2}$$

where
    x is the original parameter with probability distribution equals P(x)
    y is the "normalized" parameter from x, and its probability distribution is a box
        function
    F(z) is a Gaussian function

Hence,

$$z = \mathrm{erf}^{-1}(2y-1) = \mathrm{erf}^{-1}\left(\int_{-\infty}^{+\infty} 2P(x)dx - 1 \right) \tag{1.3}$$

is a transferred parameter with a normal distribution. Figure 1.3 shows an example of a random parameter transferred to parameters with box and Gaussian function as probability distribution, respectively.

If the space constructed by the distribution of original parameters is connected and convex, then one can apply PCA or linear decomposition on the transferred parameters (otherwise, one needs to split the space and apply the same technique on the subspaces). The goal for decomposition is to separate the dependent and independent parameters (V_d and V_i), which should satisfy the following equations:

$$V_a = \begin{bmatrix} V_d \\ V_i \end{bmatrix}, \quad C_{aa} = \begin{bmatrix} C_{dd} & C_{di} \\ C_{id} & C_{ii} \end{bmatrix} \tag{1.4}$$

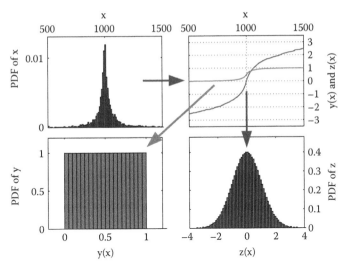

FIGURE 1.3 Plots of how to "normalize" or "Gaussify" a random parameter x using Equations 1.1 and 1.3, respectively.

and

$$C_{dd} \approx C_{di}C_{ii}^{-1}C_{id} = C_{di}C_{ii}^{-1}C_{di}^* \qquad (1.5)$$

where

C_{aa}, C_{dd}, and C_{ii} are the self-correlation matrices of V_a, V_d, and V_i, respectively
C_{di} is the cross-correlation matrix between V_d and V_i

If these equations are satisfied, all dependent parameters V_d can be written as a linear combination of the independent ones, that is, $V_d = C_{di}C_{ii}^{-1} V_i$. The statistics of all parameters can then be separated to correlation matrices C_{ii} and C_{di}, and transfer functions of V_d and V_i. Independent parameters V_i can be used as principal drivers or primary responses of the overall randomness in device electrical behavior.

In early studies [8], the number of independent parameters in a 65 nm SOI technology is six. This is coincidently consistent with the number of process-induced variables discussed in the previous section, whereas the two numbers are not necessarily the same. If the number of process-induced variables is larger than the number of primary responses, then there must be at least one variable whose impact on the parameters is equivalent to (or a function of) that of other variables.

Thus, one cannot distinguish this variable from others just using the measured electrical behavior. However, it may offer more flexibility in device design, for example, trading the requirements of one process step to that of the others. On the other hand, if the number of primary responses is larger than the number of process-induced variables, then there should be some impact neglected when lumping the process impacts (e.g., 2D/3D distributive nature of doping profile). Additional physical variables are needed to account for this effect. Therefore, the analysis of the links between primary parameters and process-induced variables will shed light on device design and optimization of a given technology. The next study is to establish the correlations/trends between them.

1.2.2 LINKS BETWEEN PROCESS AND DEVICE

To extract the correlations between each primary response and process-induced variables, one has to decouple the process-induced variables. Unlike the electrical responses that can be measured on individual FETs, these variables are generally not measurable (or not practical to measure) on each sample. As discussed in the previous section, only average values can be obtained on explicit variables, whereas little information can be obtained on implicit variables. It is rather difficult to directly derive the correlation functions of the mentioned variables. As each explicit variable is mainly driven by one or more process step(s), a commonly used method is measuring design-of-experiments (DOE) that adjust the variable (through the specific process) on a large scale, and with numerous FET samples in each case. The random components are then minimized using the average values, and the impact of the variable is singled out.

For the implicit variables, even averages values are not accessible, since they are hard to exclusively control. To decouple their impact from other variables, one would think to use a screening technique to reduce fluctuations caused by others. The basic theory of the screening technique is shown in the following equation:

$$\begin{bmatrix} R_1 \\ R_2 \end{bmatrix} = \begin{bmatrix} f_1(V_1, V_2) \\ f_2(V_1, V_2) \end{bmatrix} \tag{1.6}$$

where
  R_1 and R_2 are two measured parameters
  V_1 and V_2 are two process-induced variables

$$\text{If} \quad \left| \frac{\partial f_1}{\partial V_2} \right| \left| \frac{\partial f_2}{\partial V_1} \right| << \left| \frac{\partial f_1}{\partial V_1} \right| \left| \frac{\partial f_2}{\partial V_2} \right| \tag{1.7}$$

$$\Rightarrow dR_1 \big|_{R_2 = C_2} \approx \frac{\partial f_1}{\partial V_1} dV_1, \quad \text{and} \quad dR_2 \big|_{R_1 = C_1} \approx \frac{\partial f_2}{\partial V_2} dV_2. \tag{1.8}$$

Then one can decouple the impact of V_1 on R_1 from V_2, or the impact of V_2 on R_2 from V_1. More specifically, the first step is to find the measured parameter R_2 (either independent or dependent) that is a strong function of the variable V_2 (so that Equation 1.7 is satisfied) to be screened. Then screen the data of R_1 so that R_2 equals a constant C_2, and the impact of V_1 on R_1 is derived.

As an example, to extract mobility's influence on drive currents, we need to separate other variables such as gate length, gate oxide thickness, channel doping,

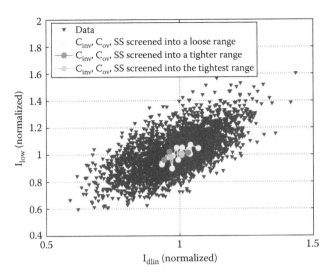

FIGURE 1.4   Scattering plot using conventional screening technique.

and parasitic resistance. According to basic device physics, the gate capacitance at inversion bias C_{inv}, overlap capacitance C_{ov}, and subthreshold slope SS are strong functions of these variables and very weak functions of mobility. To distinguish the mobility impact, one would think to screen the data by these parameters since by this way, in the selected sample, fluctuations of other variables are greatly reduced.

The conventional screening strategy is simply to find the data that specified parameters (i.e., C_{inv}, C_{ov}, and SS in this case) lie in a small target range such as ±5%. This approach requires lots of samples located in this range, which is not feasible due to the limits on time and cost. Figure 1.4 shows a typical I_{low} (drive current at $V_{ds} = V_{dd}$ and $V_{gs} = V_{dd} = 2$) vs. I_{dlin} (drive current at $V_{ds} = 0{:}05$ V and $V_{gs} = V_{dd}$) data and trend with screened C_{inv}, C_{ov}, and subthreshold slope parameters. The reason to choose I_{low} and I_{dlin} is that these two parameters are known to show different responses to mobility variation. The sample size is 3000, which is decently large for statistic analysis. One can see that the extracted curve using a loose range is too noisy, whereas the curve with a tighter range ends up with fewer points. One can hardly derive a valid trend on these screened data. Therefore, a more practical technique is needed.

After comparing several screening approaches, we found that the Delaunay triangulation method [41] offers elegant tessellation and high accuracy in the prediction on multidimensional interpolation. Applying this technique, sparser data population is still feasible for screening. As shown in Figure 1.5, if parameter values at $R_1 = 0{:}4$ and $R_2 = 0{:}3$ are needed, one can find the triangle (tetrahedron if screening three parameters) enclosed in the point and calculate the values using the interpolation of the vertices of the triangle. This method is used in the following analysis and verifications.

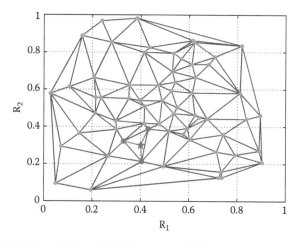

FIGURE 1.5   Data interpolation using Delaunay tessellation.

1.3 EXTRACT THE SIGMAS AND CORRELATIONS

TCAD and MATLAB® simulations can be used to prove the accuracy of the decoupling technique discussed earlier. The advantages of using simulations are that one can tune the variations to cover most of the scenarios to study and can selectively turn on/off individual variations for decoupling verification. A commercial TCAD tool (*Sentaurus* [42]) with a calibrated 2D device structure is used here to mimic 45 nm node CMOS technology [25]. Figure 1.6 shows the simulation flow. For simplicity, only wide NFET (narrow channel effect is neglected) is analyzed here, whereas PFET can be studied in a similar manner.

Measured parameters I_{off}, I_{dlin}, I_{dsat}, I_{low} are the drain current at off region ($V_{ds} = V_{dd}$, $V_{gs} = 0$), at linear region ($V_{ds} = 0.05$, $V_{gs} = V_{dd}$), at saturation region ($V_{ds} = V_{dd}$, $V_{gs} = V_{dd}$), and at $V_{ds} = V_{dd}$, $V_{gs} = V_{dd}$ 2, respectively. V_{tlin} and V_{tsat} are threshold voltages when $V_{ds} = 0.05$ and $V_{ds} = V_{dd}$, respectively. SS is the subthreshold slope. C_{inv} and C_{ov} are the inversion and overlap capacitance, respectively (note that these capacitors should be measured on individual FETs using the test technique as in [35]). The first six parameters are primary responses according to previous studies [8,9], and C_{inv}, C_{ov}, and SS are the parameters used as screening. The process-induced explicit variables are gate length (L_{gate}), gate oxide thickness (T_{inv}), overlap region influenced by extension implant (Ext), and channel doping influenced by halo/well implant (Halo). Implicit variables are mobility (Mob) and parasitic resistance (R_{par}, which includes resistance from source/drain, silicide, and metal contact).

Figure 1.7 shows the "spider" charts of the correlation coefficients between a set of measured parameters and each of the six process-induced variables. Each axle represents a process variable, and the scalar on the axle represents the correlation coefficient between the parameter and the variable, with the outer limit equaling 1. The purpose of plotting "spider" charts is to qualitatively demonstrate the impacts of each process variable on electrical parameters. The coefficients could be extracted directly from hardware measurement with sufficient sampling points or from

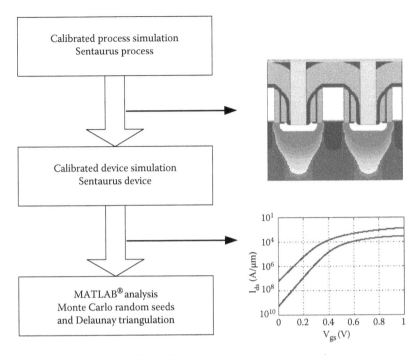

FIGURE 1.6   TCAD and MATLAB® simulation flow for demonstration.

carefully calibrated TCAD simulations. Note that actual values of the coefficients vary with different process tools and recipes, while the first-order dependencies are similar.

As expected, according to basic physics in CMOS devices, SS is a strong function of L_{gate}, T_{inv}, Ext, and Halo; C_{inv} is a strong function of L_{gate} and T_{inv} only; C_{ov} is a strong function of R_{par}, Ext, and T_{inv}.

Figure 1.8 shows the simulated V_{ts} and subthreshold slope as functions of L_{gate}. In this simulation, following commonly known process centering strategy, we tuned the nominal halo implant so that the maximum of V_{tlin} locates near the 40 nm gate length. Then at this gate length, the variation of V_{tlin} induced by L_{gate} is minimized.

According to the technique described in the previous section, one can extract the functions of explicit variables by intentionally changing them in large scales in the DOEs. As shown in Figure 1.9, I_{off} and V_{tlin}, which are different functions of the four explicit variables, are analyzed. The I_{off}–V_{tlin} trend driven by explicit variables (i.e., L_{gate}, T_{inv}, Halo, and Ext) can be extracted from the medians of the DOEs with decent agreement to the "theoretical" trend. Here the "theoretical" trend comes from the Monte Carlo simulation with just one of the variables (labeled in the figure) turned on. It is the perfect reference but can only be extracted in an ideal simulation.

For implicit variables (i.e., mobility and R_{par}), we adopt the previously mentioned Delaunay triangulation technique. Figure 1.10 shows the theoretical (in solid lines) and the extracted (in solid symbols) I_{low}–I_{dlin} trend driven by mobility and Rpar. Using this new interpolation technique, excellent agreement between the theoretical and extraction trends is achieved. This proves that the device designer can now—from

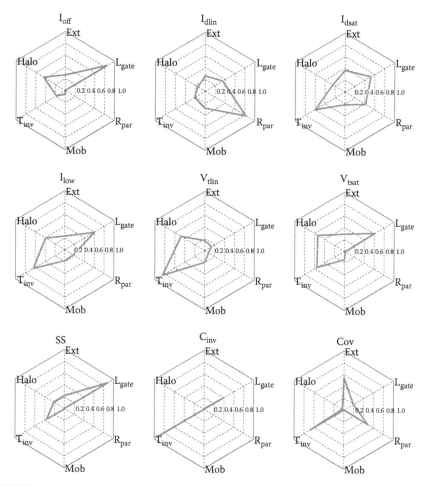

FIGURE 1.7 Spider charts of the correlation coefficients between the measured parameters and six process-induced variations obtained by TCAD simulations.

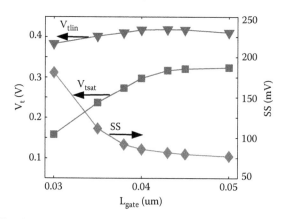

FIGURE 1.8 Simulated V_{tlin}, V_{tsat}, and SS as functions of L_{gate}.

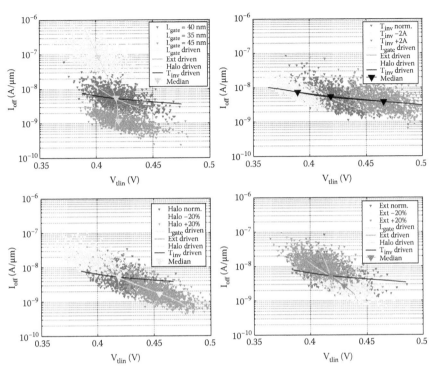

FIGURE 1.9 Simulated I_{off} vs. V_{tlin} data with different gate lengths, T_{inv}, Halo, and Ext, respectively. Theoretical trends (in solid lines) that are driven by different variables, and extracted trend (large triangles) from the medians are plotted as well. The sample size is 3000 points per DOE.

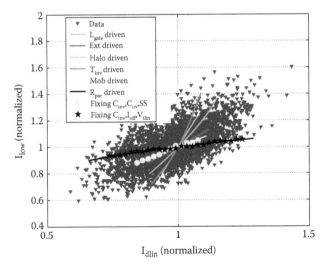

FIGURE 1.10 Simulated I_{low} vs. I_{dlin} data at $L_{gate} = 40$ nm, with theoretical trends (solid lines) driven by different variables and extracted trend (dots and stars) using the presented screening technique. The sample size is 3000 points.

measured data, with almost no assumptions—conclude what is the main driver of performance shifting and extract the relative mobility changes. On the other hand, one can predict the values of all measured parameters if only mobility changes.

Moreover, since the mobility impacts on C_{inv}, C_{ov}, or SS are negligible, fixing these parameters will not reduce the mobility varying range. This is a key feature because one can directly back-calculate the sigma values of mobility without additional DOEs to fully extract all trends.

In addition, one can derive all variation trends and then calculate the sigma values and intracorrelation coefficients of process-induced variables, following the next equation:

$$
\begin{bmatrix} R_1 \\ \vdots \\ R_n \end{bmatrix} = \begin{bmatrix} f_1(V_1,\cdots,V_m) \\ \vdots \\ f_n(V_1,\cdots,V_m) \end{bmatrix}
$$

$$
\approx \begin{bmatrix} \partial f_1/\partial V_1 & \cdots & \partial f_1/\partial V_m \\ \vdots & \ddots & \vdots \\ \partial f_n/\partial V_1 & \cdots & \partial f_n/\partial V_m \end{bmatrix} \begin{bmatrix} V_1 \\ \vdots \\ V_m \end{bmatrix} \tag{1.9}
$$

The statistics of primary responses R_i are extracted from the measured data using the approach mentioned in Section 1.3. The correlations ($\partial f_i = \partial V_j$) between primary response R_i and process-induced variable V_j are extracted using the screening technique. The sigmas of V_j can then be derived. Table 1.1 lists the input and extracted sigma values of process variables. An excellent agreement is achieved, proving the validity of the approach.

So far, an accurate model on the device and process statistics (including the sigmas and correlations) is clearly established. The next work is to check the impacts of the variations on the circuit level and how to optimize these impacts, which leads to our final goal—the SRAM design.

TABLE 1.1

Simulation Input and Extracted Sigmas of Different Process-Induced Variables

	Input Sigma (%)	Extracted Sigma (%)
L_{gate}	3.34	3.33
Ext	3.85	3.79
Halo	3.17	3.25
T_{inv}	5.96	5.98
Mob	6.67	6.56
R_{par}	25 (Ω)	25 (Ω)

1.4 DESIGN OF SRAM

The yield of the integrated circuit/system is an ultimate criterion for the success of this technology. It becomes more and more challenging to maintain the yield with the device area shrink-down and performance step-up. On the one hand, the fluctuation of device characteristics becomes more significant. On the other hand, the complicated process flow and the prolonged design rules in the advanced technology reduce the degree of freedom in circuit design. Both these facts shift the focus of system optimization to the process level (e.g., implantation and gate dielectrics formation). Nevertheless, if the links between the process variation and device characteristics are accurately developed, the SRAM optimization is quite straightforward as discussed in the following section.

1.4.1 BASICS OF SRAM

A commonly used SRAM cell in the industry is a 6-transistor (6-T) structure, as shown in Figure 1.11. The SRAM cell is with the closest device structure to standard logic FETs: the 6-T cell includes two pass-gate (PG), pull-up (PU), and pull-down (PD) devices. The two pull-up and pull-down FETs construct a two-inverter loop to hold the data. The two pass-gate FETs are used to control the access from bit lines (denoted as VBL and VBR in the figure) to internal nodes (denoted as VL and VR), by setting the voltage level of the word line (denoted as VWL).

Like other memories, there are three operation modes for SRAM cell: standby (or hold), read, and write modes. In the standby mode, the word line is set to a low-voltage level and both the internal nodes are isolated from the bit lines. In a large SRAM array (e.g., >1 MB), most of the cells are in the standby state, which dominates the overall power consumption. In the read mode, both the bit lines are usually precharged to a high-voltage level before the PG FETs are turned on, the charges in the bit lines will disturb the charges stored in the internal nodes, and if the inverters are not "strong" enough (i.e., the static noise margin is too small, as shown in Figure 1.12a), the bit lines may not be sufficiently discharged to the expected values,

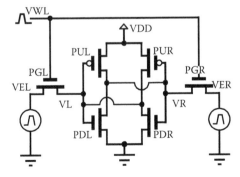
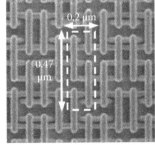

FIGURE 1.11 A standard 6-T SRAM cell structure used in the VLSI technology. The right image is a typical top-down SEM picture of SRAM array. (From Basker, V.S. et al., *IEEE Symp. VLSI Tech. Dig.*, 19, 2010.)

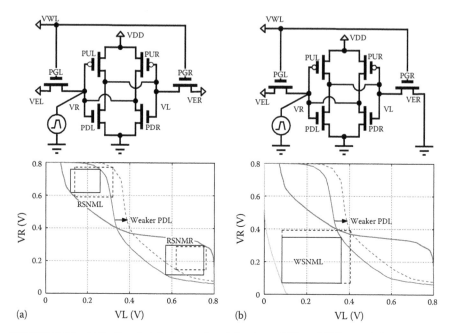

FIGURE 1.12 (a) Left and right internal nodal voltage (VL and VR) trends in read mode and extracted left/right read-static-noise margin (RSNM). (b) Left and right internal nodal voltage trends in write mode and extracted left write-static-noise margin (WSNM). Note that a weaker pull-down NFET in the left side (e.g., higher V_{th}) increases RSNM on the right node and WSNM on left node, while decreases RSNM on left node.

or even overwrite the original data; this is referred to as "read fail." In the write mode, the two bit lines are set at the two complimentary voltage levels (shown in Figure 1.12b), if the PGs are too "weak," the internal nodes are overwhelmed by the stored charge and cannot be switched by the external bit lines; this is referred to as write fail.

Both the read and write fails determine the "soft" yield of the SRAM, which is partially fixable by adjusting the biasing voltages. Besides that, the standby power consumption, overall cell size, and access speed are the other factors to be considered and are converged with the general requirements of device technology development. For a given technology, there is not much design space for the latter three factors, which are strongly associated to the tuning of process recipes.

Here we will focus on the "soft" yield optimization in this chapter. Note that the optimization highly relies on not only a precise representation of the device statistics as demonstrated previously but also on an accurate calculation of the "soft" yield dependency as discussed in the following sections.

1.4.2 SNM and Butterfly Curves

Since Jan Lohstroh proposed the methodology in 1979 [44], static noise margin (SNM) is widely used as an index for yield analysis. The virtue of SNM is that it quantitatively measures the yield probability in one cell. The definition of SNM is illustrated in Figure 1.12. The voltage dependencies between the two internal nodes are used to extract the SNM.

In the read mode, both of the bit lines are biased at a high-voltage level (usually the bit lines in read mode are with a high effective resistance; however, pure voltage source is used here to consider the worst-case scenario). If the voltage of one internal node is disturbed, the voltage of the other node should be changed correspondingly. As shown in the left plot, the blue curve is the voltage of the right internal node (VR) responding to the voltage of the left internal node (VL), and the red curve is vice versa. The two curves shape the well-known "butterfly" trajectory, and the dimensions of the largest squares inscribed in the two "eyes" of the "butterfly" trajectory are the read SNMs (denoted as RSNMR and RSNML). These dimensions measure the disturbing voltage that the SRAM cell can sustain without losing the original data, assuming that two disturbing sources are simultaneously applied on both internal nodes with the same magnitude but different polarities. If the disturbing voltages are higher than the read SNM, one of the "eyes" disappears in the shifted trajectory and there is only one stable state for the internal nodes, which overwrites the original data. The larger the dimension of the square, the higher the voltage required to disturb the read operation, and the higher the read yield.

In the write mode, one of the bit lines (VR in this case) is biased at a high-voltage level, and the other (VL) is biased at a low-voltage level. Unlike the red curve in the read mode, the green curve of the right plot is the VL responding to VR. One can define the write SNMs in a similar manner as the read SNMs, that the dimension of the largest square inscribed in the "write-safe" zone is the WSNM. This is also assuming that two disturbing sources are applied on both nodes. If the disturbing voltages are higher than the write SNM, there will be additional cross points between the green and the blue curve beside the upper-left one (i.e., VR $\approx$ 0.8 V and VL $\approx$ 0 V). The internal nodes may stay at some of the additional cross points since those are stable states and will not reach the expected upper-left one. This leads to the write fail described earlier. Note that like the read SNM, the larger the dimension, the higher the voltage required to disturb the write operation, and the higher the write yield. Also note that one can define the other WSNM at the inverted bit line bias condition: that is, VR is low and VL is high. Hence, there are two WSNM values (denoted as WSNMR and WSNML) like the read SNMs.

For either read or write mode, SNM > 0 ensures the cell is unsusceptible to each fail. The SNM value shifts as the characteristics of each device change, as shown in Figure 1.12. For example, if in one cell the left PD FET is weaker than nominal due to fluctuation (e.g., a higher threshold voltage, smaller width, longer L_{gate}, thicker T_{inv}, higher R_{par}, and lower mobility), the blue curve will shift to the right. This results in a lower right RSNM and left WSNM, and a higher left RSNM. Furthermore, one needs to consider the impact of the fluctuations of all six FETs. Figure 1.13 shows the simulated left RSNM and WSNM as functions of threshold voltage (V_{th}) and parasitic resistance (R_{par}) variations. One can see that different FETs exhibit different impacts on the SNMs. Note that as discussed in previous sections, there are six independent variables that represent the statistics of one device. Then we need to include

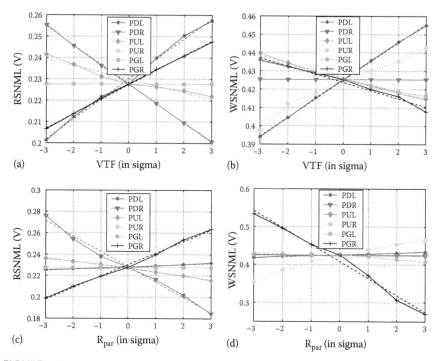

FIGURE 1.13   Read (a and c) and write (b and d) static noise margin on the left side (RSNML and WSNML) as functions of V_t and R_{par} fluctuations on each device of the 6-T cell; VDD and VWL are both 0.8 V. Negative sigma in the x-axis represents a lower V_t (i.e., stronger FET) or a lower parasitic resistance. Dashed lines are linear fits of the trend.

all these variables to estimate the overall SNM trends. Another observation is that the SNM values are approximately linear functions of the fluctuations in sigmas; this characteristic is very useful in the overall yield calculations and optimizations.

1.4.3   YIELD ESTIMATION, V_{min}, AND OPTIMIZATION

By definition, one can estimate the yield by calculating the probability of SNM of one cell drop to 0. In the advanced VLSI technology, there are six transistors, each transistor includes six independent variables. One needs to integrate the probability function in these $6 \times 6 = 36$ dimensions in theory. Usually, a direct integration over 36 variables is time consuming and practically impossible. The Monte Carlo simulation is then adopted and becomes a reliable method for the yield estimation [45,46]. However, as the sizes of current SRAMs increase to multimillion- or giga-bits, the Monte Carlo approach is still not fast enough to conduct a comprehensive optimization. For example, the designers need to check the bias dependency of the yield (i.e., the Schmoo chart) and locate the minimum operational voltage (i.e., the V_{min}). Furthermore, the designers need to check the impacts on yield by adjusting the process or device structures. These introduce more design variables in the optimization. Therefore, it is critical to find an even faster method to calculate the yield.

The simulated SNM trends show that they are approximately linear functions of device fluctuations such as in threshold voltage or in parasitic resistor (observe that the fitted lines are pretty close to the trends in Figure 1.13). Measured data from [47] also proved these characteristics. Based on this linear assumption, the yield can be calculated as the linear combination of the sigmas of uncorrelated variables as

$$\text{SNM} = \text{SNM}_0 + \sum_{i=1}^{36} s_i \sigma_i \Rightarrow \sigma_{all} = \frac{\text{SNM}_0}{\sqrt{\sum_{i=1}^{36} s_i^2}} \qquad (1.10)$$

Here, the s_i is the slope of the studied SNM changes as the ith variable changes (in sigma). Usually, the 36 variables are correlated, and then the uncorrelated linear combinations of these 36 variables are used in Equation 1.10 instead. To extract the combinations, one has to apply the singular-value decomposition on the correlation matrix that is obtained using the methodology previously introduced. The σ_{all} is the yield probability of one of the studied fail modes (e.g., left or right, read or write). For simplicity, the sigma of Gaussian distribution is quoted here: sigma = 4.89 is equivalent to 1 fail in 1×10^6 cells, sigma = 6.11 is equivalent to 1 fail in 1×10^9 cells. The sum of the four fails (RSNMR, RSNML, WSNMR, and WSNML) is considered as the overall fail count, assuming a worst-case scenario.

A technique to further speed up the SNM calculation is to adopt a behavioral look-up table to replace the compact model in simulation. Since the I–V curves of numerous devices can be in-line measured, it is straightforward to build a look-up table including the statistics of the technology (e.g., the sigmas and correlations of different I–V points). The butterfly curve can be simulated using the table with linear interpolations. This approach not only dramatically increases the speed calculating the SNM but also greatly reduces the delay in constructing fully calibrated compact models such as BSIMs or PSP.

Using this algorithm, one can calculate the Schmoo chart (e.g., yield vs. bit and word-line voltage sources). As in Figure 1.14, the Schmoo chart shows the impacts of the biasing word line (VWL) and bit line (VDD) voltages. The plot determines the minimum voltage (i.e., V_{min}) at which the SRAM is functional. Note that the write fail dominates when bit-line voltage source is higher than word-line voltage source because of weaker pass-gate, and read fail dominates vice versa because of weaker inverters. Therefore, yield is decent only when voltage sources are biased in the diagonal canyon region. One can read the V_{min} of a 1 Mb SRAM array (i.e., sigma = 4.89) is about 0.6 V on the word line supply voltage and 0.55 V on the bit line supply voltage.

Furthermore, we use two device parameters—gate length (L_{gate}) and difference between the threshold voltages (NV_{th}–PV_{th}) of NMOS and PMOS—as design variables at fixed bias voltage sources (e.g., both VWL and VDD are 0.5 V) to find the optimum device/process configurations. Figure 1.15 shows the yield contours on the two variables. Observe that an optimum N-PMOS V_{th} delta (i.e., −120 mV) exists at L_{gate} = 25 nm for yield higher than 4.89 sigma (i.e., 1 fail in 1 Mb array), implying that the minimum gate length of the SRAM cell is restricted by the yield.

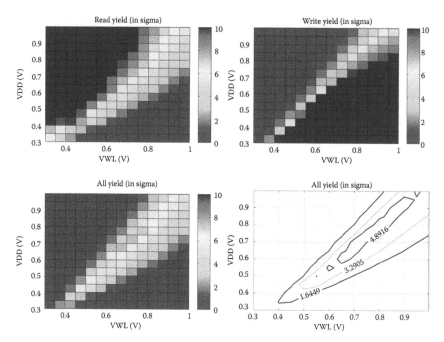

FIGURE 1.14 Soft yield (read, write, and combined) Schmoo chart, where bit line voltage (VDD) and word line voltage (VWL) are the sweeping variables.

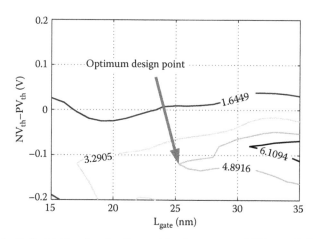

FIGURE 1.15 Soft yield (read and write combined) contours of different N–P V_t offset and L_{gate} design, with VDD and VWL biasing at 0.5 V.

Figure 1.15 is just one demonstration of how to optimize SRAM on the device and process levels. One can also calculate the contours on other design variables such as the width, implant difference between the PG and PD FETs, mobility, etc. This approach offers a detailed analysis on the technology limit and a clear solution to achieve a high-yield SRAM design.

1.5 CONCLUSION

SRAM cell is a typical circuit block that represents the advanced CMOS technology. The design and optimization should start from the basic statistic analysis on standard devices. A set of models that accurately captures the sigmas and correlations of the device variations is required for further circuit-level study. Yield is the top concern in the SRAM cell design and can be estimated by extracting the static-noise margins. A simple technique is introduced here to quickly calculate the yield. Applying this technique can help us conduct a comprehensive optimization of the SRAM cell and extract the limits (e.g., minimum device size, implant level, and maximum device number) that best characterize a given technology.

REFERENCES

1. http://www.itrs.net/Links/2011ITRS/2011Chapters/2011ExecSum.pdf
2. J. Power et al., *IEEE Intl. Conf. Microelec. Test Struc.*, 1993, 63.
3. H. Sato et al., *IEEE Trans. Semicon. Manufac.*, 1998, 575.
4. P.A. Stolk et al., *IEEE Trans. Elec. Dev.*, 1998, 1960.
5. A. Asenov et al., *IEEE Trans. Elec. Dev.*, 1999, 1718.
6. Z. Krivokapic et al., *IEEE Trans. Semicon. Manufac.*, 1999, 437.
7. M. Orshansky et al., *IEEE Trans. Semicon. Manufac.*, 1999, 403.
8. Q. Liang et al., *IEEE Intl. Semicon. Dev. Res. Symp.*, 2007, 1.
9. C. Cho et al., *IEEE Trans. Semicon. Manufac.*, 2008, 55.
10. S. Shedabale et al., *IET Cir. Dev. Syst.*, 2008, 451.
11. S.K. Saha, *IEEE Des. Test Comp.*, 2010, 8.
12. M.J.M. Pelgrom et al., *IEEE J. Solid State Circ.*, 1989, 1433.
13. M. Steyaert et al., *IET Elec. Lett.*, 1994, 1546.
14. T. Mizuno et al., *IEEE Trans. Elec. Dev.*, 1994, 2216.
15. E. Felt et al., *IEEE/ACM Intl. Conf. CAD*, 1996, 374.
16. J.-O. Plouchart et al., *IEEE ISSCC*, 2006, 2142.
17. S.K. Springer et al., *IEEE Trans. Elec. Dev.*, 2006, 2168.
18. D. Kim et al., *IEEE Custom IC Conf.*, 2006, p-41-1.
19. A.B. Kahng, *IEEE Intl. Elec. Dev. Meet.*, 2007, 644.
20. K. Agarwal et al., *IEEE Trans. VLSI Syst.*, 2008, 86.
21. B.P. Harish et al., *IEEE Intl. Symp. Circ. Syst.*, 2009, 2309.
22. F. Lallement et al., *IEEE Symp. VLSI Tech. Dig.*, 2004, 178.
23. P. Bai et al., *IEEE Intl. Elec. Dev. Meet.*, 2004, 657.
24. E. Leobandung et al., *IEEE Symp. VLSI Tech. Dig.*, 2005, 126.
25. S. Narasimha et al., *IEEE Intl. Elec. Dev. Meet.*, 2006, 1.
26. M. Chudzik et al., *IEEE Symp. VLSI Tech. Dig.*, 2007, 194.
27. C. Auth et al., *IEEE Symp. VLSI Tech. Dig.*, 2008, 128.
28. B. Greene et al., *IEEE Symp. VLSI Tech. Dig.*, 2009, 140.
29. P. Packan et al., *IEEE Intl. Elec. Dev. Meet.*, 2009, 659.
30. C.H. Li et al., *IEEE Conf. and WS Adv. Semicon. Manufac.*, 2002, 21.
31. A. Ono et al., *IEEE Intl. Elec. Dev. Meet.*, 1997, 227.
32. R. Williams et al., *Proc. NSTI WS Compact Model.*, 2006, 858.
33. S. Zafar et al., *IEEE Trans. Dev. Mat. Reliab.*, 2005, 45.
34. E. Cartier et al., *IEEE Symp. VLSI Tech. Dig.*, 2009, 42.
35. Y.-W. Chang et al., *IEEE Elec. Dev. Lett.*, 2006, 390.
36. S.J. Han et al., *IEEE Intl. Elec. Dev. Meet.*, 2008, 585.

37. S.E. Thompson et al., *IEEE Trans. Elec. Dev.*, 2006, 1010.

38. H.S. Yang et al., *IEEE Intl. Elec. Dev. Meet.*, 2004, 1075.

39. M. Horstmann et al., *IEEE Intl. Elec. Dev. Meet.*, 2005, 233.

40. S.-D. Kim et al., *IEEE Trans. Elec. Dev.*, 2002, 1748.

41. C.B. Barber et al., *ACM Trans. Math. Soft.*, 22, 1996, 469.

42. *TCAD Sentaurus User Manual*, Ver. 2010.03, Synopsys.

43. V.S. Basker et al., *IEEE Symp. VLSI Tech. Dig.*, 2010, 19.

44. J. Lohstroh, *IEEE J. Solid State Circ.*, SC-14(3), 1979, 591.

45. K. Agarwal and S. Nassif, *Proc. Des. Auto. Conf.*, 2006, 57.

46. A. Bhavnagarwala et al., *IEEE Intl. Electron. Dev. Meet.*, 2005, 659.

47. X. Song et al., *IEEE Intl. Electron. Dev. Meet.*, 2010, 3.5.1.

2 Complete Guide to Multiple Upsets in SRAMs Processed in Decananometric CMOS Technologies

Gilles Gasiot and Philippe Roche

CONTENTS

2.1   INTRODUCTION

Susceptibility to radiation environment of advanced electronic devices is often responsible for the highest failure rate of all reliability concerns (electromigration, gate rupture, NBTI, etc). In modern SRAMs, the two predominant single event effects (SEEs) are the single event upset (SEU) and multiple upsets (MUs). MUs are topological errors in neighboring cells. If the cells belong to the same logical word, they are named multiple bit upsets (MBUs); otherwise they are labeled as multiple cell upsets (MCUs). MUs have received increased scrutiny in recent years [1–8] because MBUs are uncorrectable by simple ECC scheme and therefore threaten the efficiency of EDAC.

As technologies scale down, the amount of transistors per mm² doubles at each generation while the radioactive feature size (ion track diameter) is constant. This is illustrated in Figure 2.1 with 3D TCAD simulation showing an ion impacting a single cell in 130 nm while several are impacted in 45 nm. Moreover, the SRAM

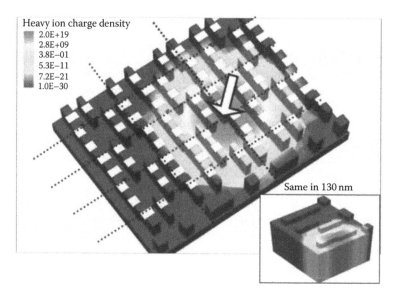

FIGURE 2.1   Three-dimensional TCAD simulation of ion impact (single LET) in a single SRAM bitcell in 130 nm and 12 SRAM bitcells in 45 nm.

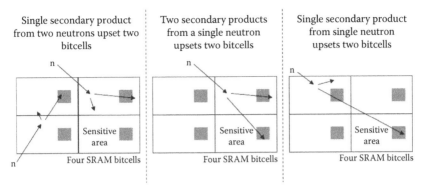

Single secondary product from two neutrons upset two bitcells

Two secondary products from a single neutron upsets two bitcells

Single secondary product from single neutron upsets two bitcells

Sensitive area

Four SRAM bitcells

Sensitive area

Four SRAM bitcells

Sensitive area

Four SRAM bitcells

FIGURE 2.2 Scheme of neutron interaction that can cause multiple cell upset in SRAM array. (Derived from Wrobel, F. et al., *IEEE Trans. Nucl. Sci.*, 48(6), 1946, 2001.)

ability to store electrical data (critical charge) is reduced as technology feature size and power supply are jointly decreased. The probability that a particle upsets more than a single cell is therefore increased [9–11].

Mechanism for MCU occurrence in SRAM arrays is more than "enough energy was deposited to upset two cells" and depends upon the radiation used. Directly ionizing radiation from single particles (alpha particles, ions, etc.) deposits charges diffusing in wells that can be collected by several bitcells. This phenomenon is enhanced by using tilted particles either naturally (alpha particles whose emission angle is random from the radioactive atom) or artificially (heavy ions can be chosen during experimental tests from 0° to 60°). Nonionizing radiation such as neutrons and protons can have different MCU occurrence mechanism (Figure 2.2). A nonionizing particle can produce one or more secondary products. Several cases have to be considered: two secondary ions from two nucleons upset two or more bitcells, two secondary ions from a single nucleon upset two or more bitcells, and a single secondary ion from a single nucleon upsets two or more bitcells (in this case, the phenomenon is close to the previously described direct ionizing mechanism). It has been shown that type 1 mechanism was negligible, but that type 2 and type 3 mechanisms coexist [12]. However, the proportion of MCUs due to these two mechanisms has never been precisely assessed.

One of the first experimental evidence of MBU was reported in 1984 in a 16 × 16 bit bipolar RAM under heavy-ion irradiation [13]. It is noteworthy that as many as 16-bit errors in columns from a single ion strike were detected. This means that 6% of the entire memory array was in error from a single particle strike. Since this first experimental evidence, multiple bit errors were detected in several device types such as DRAM [14], polysilicon load SRAM [15], and antifuse-based FPGA [16], and under various radiation types: protons [17], neutrons [18], laser [19], etc.

The goal of this work is first to experimentally quantify MCU occurrence as a function of several parameters such as radiation type, test conditions (temperature, voltage, etc.), and SRAM architecture. These results will be used to sort the parameters driving the MCU susceptibility by order of importance. Second, 3D TCAD simulations will be used to investigate the mechanisms leading to MCU occurrence and to determine the most sensitive location to trigger a 2-bit MCU as well as the cartography of MCU sensitive areas.

2.2   DETAILS ON THE EXPERIMENTAL SETUP

The design of experiment included different test patterns and supply voltages. The test procedure is compliant with the JEDEC SER test standard JESD89 [20] for alpha and neutrons, and ESA test standard n°22900 for heavy ions and protons [21].

2.2.1   NOTE ON THE IMPORTANCE OF TEST ALGORITHM FOR COUNTING MULTIPLE UPSETS

When experimentally measuring MCUs, it is mandatory to distinguish (1) multiple independent failures from a cluster of nearest neighbor upset from a single multi-cell upset caused by a single energetic particle and (2) signature of errors due to a hit in redundancy latch or sense amplifier that may upset an entire row or column from an MCU signature. Test algorithm allows separating independent events due to multiple particle hits from single events that upset multiple cells. Dynamic testing of memory usually involves writing once and then reading continuously at a specified operating frequency at which events are recorded one at a time. This gives a real insight on MCU shapes and occurrence. However, with static testing of memory, test pattern is written once and stored for an extended period before reading the pattern back out. The result is a failure bitmapping in which events due to multiple particle hits and single events that upset multiple cells cannot be distinguished. However, statistical tools can be applied to quantify the rate of neighboring upsets due to several ions [22,23]. One of these tools is described in detail in Annex 1.

2.2.2   TEST FACILITY

2.2.2.1   Alpha Source

The tests were performed with an alpha source, which is a thin foil of americium 241 that has an active diameter of 1.1 cm. The source activity was 3.7 MBq as measured on February 1, 2002. The alpha particle flux was precisely measured in March 2003 with a Si detector, which was placed at 1 mm from the source surface. Since the atomic half-life of Am241 is 432 years, the activity and flux figures are still very accurate. During SER experiments, the americium source lies above the chip package in the open air.

2.2.2.2   Neutron Facilities

Neutron experiments were carried out with the continuous neutron source available at the Los Alamos Neutron Science Center (LANSCE) and Tri University Meson Facility in Vancouver (TRIUMF). The neutron spectrums closely match the terrestrial environment for energies ranging from 10 up to 500 and 800 MeV for TRIUMF and LANSCE, respectively. The neutron fluence is measured with a uranium fission chamber. The total number of produced neutrons is obtained by counting fissions and applying a proportionality coefficient.

2.2.2.3 Heavy-Ion Facilities

The heavy-ion tests were conducted at the RADiation Effect Facility (RADEF) [24] cyclotrons. The RADEF facility is located in the Accelerator Laboratory at the University of Jyväskylä, Finland (JYFL). The facility includes beam lines dedicated to proton and heavy-ion irradiation studies of semiconductor materials and devices. The heavy-ion line consists of a vacuum chamber with component movement apparatus inside and ion diagnostic equipment for real-time analysis of beam quality and intensity. The cyclotron used at JYFL is a versatile, sector-focused accelerator for producing beams from hydrogen to xenon. The accelerator is equipped with three external ion sources. There are two electron cyclotron resonance ion sources designed for high-charge-state heavy ions. Heavy ions used at the RADEF facility have stopping ranges in silicon much larger than the whole stack of back-end metallization and passivation layers (~10 μm).

2.2.2.4 Proton Facility

Proton irradiations were performed at the Proton Irradiation Facility (PIF) at Paul Scherrer Institute. This institute was constructed for the testing of spacecraft components. The main features of PIF are that irradiation takes place in air, the flux/dosimetry is about 5% of absolute accuracy, and beam uniformity is higher than 90%. The experiments have used the low-energy PIF line whose energy range is 6–71 MeV, and the maximum proton flux is 5E8 p/cm^2/s.

2.2.3 Tested Devices

Most of the data presented in this work were performed using a single testchip (Figure 2.3). This testchip embeds three different bitcell architectures, two single port (SP) and one dual port (DP). It was manufactured in a 65 nm commercial CMOS technology with low-power process option. Main features of tested devices are summarized in Table 2.1. Each bitcell was processed with and without the triple well (TW) process option.

TW layer consists of either an N+ or P+ buried layer in respectively a p- or n-doped substrate. As most devices are processed in a P-substrate, TWs are often referenced to as deep N-well or N+ buried layers (Figure 2.4). For years, TW layers have been used to electrically isolate the P-well and to reduce the electronic noise from the substrate. The TW is biased through the N-well contacts/ties connected to VDD while the P-wells are grounded. The well ties are regularly distributed along the SRAM cell array as depicted in Figure 2.5. The TW process option has two main effects on the radiation susceptibility. First, it allows for decreasing the SEL sensitivity since the PNP base resistance is strongly reduced (Figure 2.1). TW makes accordingly the latchup thyristor more difficult to trigger on. In the literature, full latchup immunity

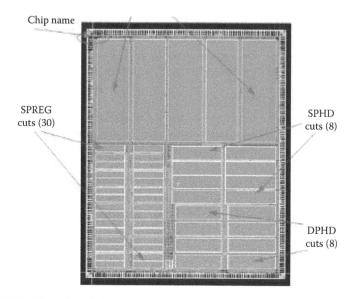

FIGURE 2.3 Floorplan of the test vehicle designed and manufactured in a 65 nm CMOS technology.

TABLE 2.1
Content of the Test Vehicle

Bitcell	Bitcell Area (μm^2)	Capacity (Mb)	DNW
Single-port SRAM high density	0.52	2	No
Single-port SRAM high density	0.52	2	Yes
Single-port SRAM standard density	0.62	2	No
Single-port SRAM standard density	0.62	2	Yes
Dual-port SRAM high density	0.98	1	No
Dual-port SRAM high density	0.98	1	Yes

Note: Three different bitcell architectures were embedded. Every bitcell is processed with and without triple well layer.

is reported even under extreme conditions (high voltage, high temperature, and high LET) [25,26]. Second, this buried layer allows for concurrently decreasing the SEU/SER sensitivity since the electrons generated deep inside the substrate are collected by the TW layer and then evacuated through the N-well ties. The improvement of the SER using TW is reported in several papers [27–29]. However, other research teams have published an increased SER sensitivity due to the TW in a commercial CMOS 0.15 μm technology [30,31].

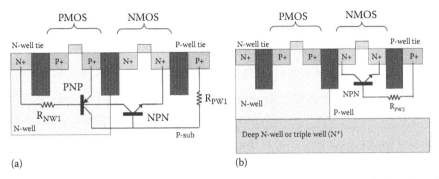

(a)   (b)

FIGURE 2.4 Schematic cross section of a CMOS inverter (a) without triple well and (b) with triple well. The PNP base resistance R_{NW1} is lowered by the TW: the PNP cannot be triggered. Conversely, the TW layer pinches the P-well and increases the NPN base resistance R_{PW2}: the NPN triggering is facilitated.

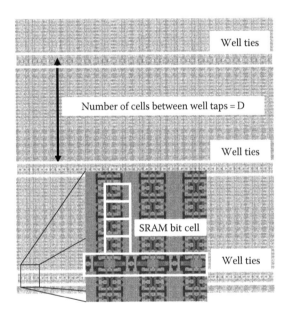

FIGURE 2.5 Layout of an SRAM cell array showing the periodical distribution of the well tie rows every 32 cells.

2.3   EXPERIMENTAL RESULTS

MCUs were recorded during the SER experiments on the 65 nm SRAM, but no MBU was ever detected as the tested memory uses bit interleaving or scrambling. All the MCU percentages reported in this work were computed in dividing the number of upsets from MCU by the total number of upsets (single bit upsets [SBUs], plus MCUs). Note that, in the literature, events are sometimes used instead of upsets [31]; the MCU percentages are in this case significantly underestimated. Otherwise specified, tests were performed at room temperature, in dynamic mode with checkerboard and uniform test patterns. In addition to the usual MCU percentages, we report in this work the failure rates due to MCU (also called MCU rate). MCU rates allow comparing quantitatively MCU occurrence between different technologies and test conditions.

2.3.1 MCU AS A FUNCTION OF RADIATION SOURCE

The four radiation sources have different interaction modes, which are either directly ionizing (alpha and heavy ions) or nonionizing (neutrons and protons). However, it is of interest to compare the MCU percentage from these radiations on the same test vehicle. The test vehicle chosen is SP SRAM of standard density (SD) processed without TW. MCU percentages are synthesized in Table 2.2, which shows that alpha particles lead to the lower MCU occurrence. Moreover, heavy ions lead to the higher MCU percentages while neutrons and protons are similar. Heavy ions are the harshest radiation MCU-wise.

2.3.2 MCU AS A FUNCTION OF WELL ENGINEERING: TRIPLE WELL USAGE

Table 2.3 synthesizes and compares MCU rates and percentage for the SD SP SRAMs processed with and without TW. Table 2.4 indicates first that the usage of TW increases the MCU rate by a decade and the MCU percentage by a factor ×3.6. Usage of MCU rate is mandatory since MCU percentages can lead to incomplete information. As presented in Figure 2.6, devices without TW have lower number of bits involved per MCU event ($\leq$8) compared to those with TW. This figure also indicates that for SRAMs with TW, 3-bit and 4-bit MCU events are more likely than 2-bit events.

TABLE 2.2

Percentage of MCU for the Same Single-Port SRAM under Several Radiation Sources

Radiation source	Single-port SRAM
	Standard density
	CKB pattern
	No triple well
Alpha	0.5%
Neutron	21% at LANSCE
Proton	4% at 10 MeV
	20% at 40 MeV
	25% at 60 MeV
Heavy ions	0% at 5.85 MeV/cm²·mg
	87% at 19.9 MeV/cm²·mg
	99.8% at 48 MeV/cm²·mg

TABLE 2.3
MCU Rates and Percentages of a
Single-Port SRAM Processed with
and without Triple Well

	MCU Rate	%MCU
SP SRAM standard density No triple well	100 (norm)	21
SP SRAM standard density Triple well	1000	76

Note: MCU rate is normalized to its value without triple well.

TABLE 2.4
MCU Percentages and Rates after Neutron
Irradiation at Nominal Voltage and Room
Temperature for Two Different Test Patterns

Technology	Bitcell Area (μm^2)	CKB Pattern	
		MCU%	MCU Rate (au)
Bulk	2.5	16.90	100
SOI	2.5	2.10	10

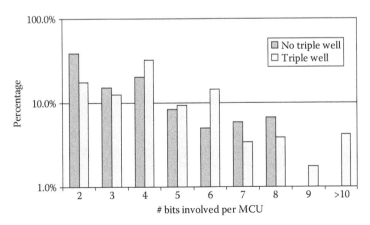

FIGURE 2.6 Number of bits involved in MCU events for high-density SP SRAM under neutron irradiation.

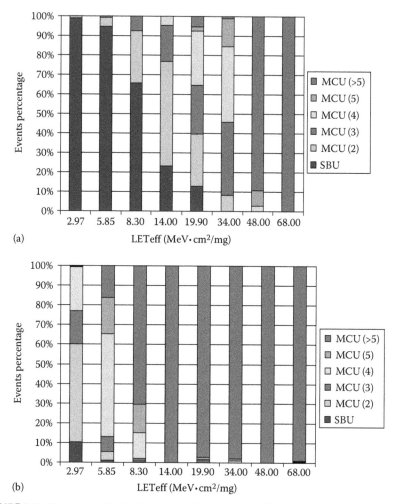

FIGURE 2.7 Proportion for single and multiple events for (a) high-density SP SRAM without triple well option and (b) high-density SP SRAM with triple well option. (From Giot, D. et al., *IEEE Trans. Nucl. Sci.*, 2007.)

The effect of a TW layer on MCU percentages under heavy ions is reported in Figure 2.7. The SRAM under test is a high-density (HD) SP SRAM. For the smallest LET, MCUs represent 90% of the events with TW but less than 1% without TW. For LET_{eff} higher than 5.85 $MeV/cm^2 \cdot mg$, there is no SBU in the SRAM with TW. For LET higher than 14.1, all the MCU events induce more than five errors with TW. With TW, the significant increase in MCU amount and order causes an increase in the error cross section.

Whatever the radiation source, the usage of TW strongly increases the occurrence of MCU. This increase is so high that it can be seen in the total bit error rate for neutrons and error cross section for heavy ions.

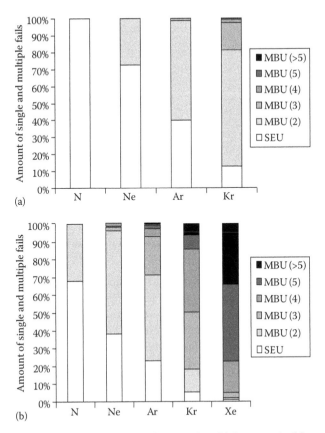

FIGURE 2.8   Amount of bit fails due to single and multiple events in 90 nm SP SRAM: (a) with heavy-ion beam not tilted and (b) with heavy-ion beam tilted at 60°.

2.3.3   MCU AS A FUNCTION OF TILT ANGLE DURING HEAVY ION EXPERIMENTS

Figure 2.6 shows respectively the amount of single and multiple bit fails induced by a given ion species (N, Ne, Ar, Kr) whose tilt angle is either vertical (Figure 2.8a) or tilted by 60° (Figure 2.8b). Tilt angle from 0° to 60° increases the MBU percentages for each ion species. For nitrogen, the MBU% is increased from 0% to 30% with a tilt = 60°. For neon and argon, the amount of MBU fails is doubled at 60° compared to vertical incidence. For krypton, the increase in MBU% with the tilt is less pronounced (+10% from 0° to 60°) because of the progressive substitution of low-order MBUs (order 2, order 3) by higher-order MBUs (order 5, order >5).

On average, the amount of bit fails due to MBU is doubled for 60° tilt compared to normal incidence [41].

2.3.4   MCU AS A FUNCTION OF TECHNOLOGY FEATURE SIZE

Figure 2.9 shows the experimental neutron MCU percentages as a function of technology feature size and compares data from this work with data from the literature. These data show that technologies with TW have MCU percentages higher than

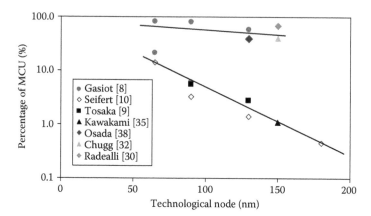

FIGURE 2.9 Neutron-induced MCU percentages as a function of technological node from this work and from the literature. Triple well usage is not indicated in the data from the literature. (From Chugg, A.M., *IEEE Trans. Nucl. Sci.*, 53(6), 3139, 2006.)

50% while technologies without TW have MCU percentages lower than 20%. Data from the literature fit either our set of data with TW or without TW. Consequently, Figure 2.7 suggests that MCU percentages can be sorted with a criterion of TW usage. Moreover, the MCU percentages increase both with and without TW when the technologies scale down. This slope is higher without TW since for old technologies, MCU percentages were very low (~1% in 150 nm).

2.3.5 MCU as a Function of Design: Well Tie Density

TCAD simulations on 3D structures built from the layout of the tested SRAMs have been performed as shown in Section 2.4. Simulation results for the ratio between drain collected charge with and without TW are plotted in Figure 2.10. This figure

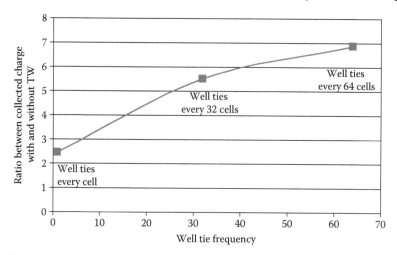

FIGURE 2.10 Simulation results for the ratio between collected charge by the N-off drain with and without triple well. This ratio is plotted as a function of well tie frequency.

indicates first that the collected charge with TW is higher than without whatever the well tie frequency. Second, the charge collection increase ranges from ×2.5 to ×7 for the highest and the lowest well tie frequency respectively. This demonstrates that when TW is used, increasing the well tie frequency mitigates the bipolar effect and therefore the MCU rate and SER.

2.3.6   MCU AS A FUNCTION OF SUPPLY VOLTAGE

The effect of supply voltage on the radiation susceptibility is well known: the higher the voltage, the lower the susceptibility since the charge storing the information is increased proportionally to the supply voltage. However, the effect of the supply voltage on the MCU rate is not documented. Experimental measurements were performed at LANSCE on an HD SRAM processed with and without TW option at different supply voltage ranging from 1 to 1.4 V. Results are synthesized in Figure 2.11. It shows that when the supply voltage is increased, the device with TW MCU rate remains constant within the experimental uncertainty. However,

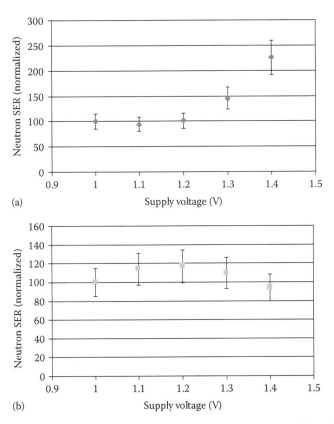

(a)

(b)

FIGURE 2.11   MCU rate as a function of supply voltage for the HD SRAM processed (a) without triple well and (b) with triple well process option. MCU rates are normalized to their value at 1 V.

a different trend is observed for the device without TW layer. When the supply voltage is increased, the MCU rate is constant from 1.0 to 1.2 V and then increases from 1.3 to 1.4 V. The MCU rate increase is 220% for V_{DD} equal to 1.4 V.

2.3.7 MCU AS A FUNCTION OF TEMPERATURE

High-temperature constraint is associated with high-reliability applications such as automotive. Some papers have quantified the temperature effect on SER or heavy-ion susceptibility [34,35]. At the time of this writing, no reference can be found in the literature experimentally measuring the temperature effect on the MCU rate. Experimental measurements were performed at LANSCE on an HD SRAM processed with and without TW option at room temperature and 125°C. Results are synthesized in Figure 2.12. It demonstrates that the MCU rate increases by 65% for the device without TW and by 45% for the device with TW. Note that the usage of MCU percentage would have been misleading since the MCU percentage is constant between room temperature and 125°C for the device with TW.

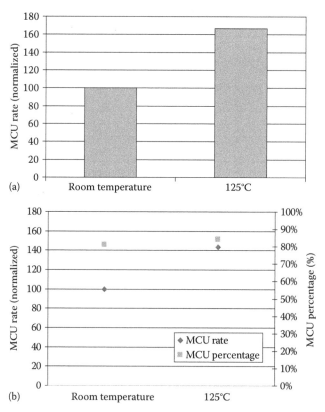

FIGURE 2.12 MCU rate as a function of temperature for the HD SRAM processed (a) without triple well and (b) with triple well process option. MCU rates are normalized to their value at room temperature. Figure xb also displays the MCU percentages.

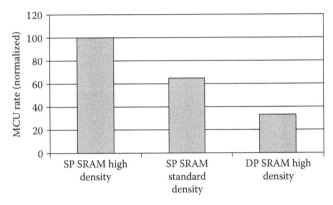

FIGURE 2.13   MCU rate comparison for several bitcell architectures. SP stands for single port, DP for dual port (eight-transistor SRAM). The devices under test were processed without triple well.

2.3.8   MCU as a Function of Bitcell Architecture

Figure 2.13 synthesizes MCU rates for HD and SD SP SRAMs as well as a DP SRAM (eight transistors). These SRAMs were processed without TW. Figure 2.13 indicates that the higher the density, the higher the MCU rate. A decrease in the bitcell area by a factor ×2 (HD SP SRAM compared to DP SRAM) induces a decrease in the MCU rate by a factor ×3.

The effect of bitcell architecture on MCU percentages under heavy ions is reported in Figure 2.14. The devices under test are HD SP SRAMs (Figure 2.14a) and SD SP SRAMs (Figure 2.14b). Figure 2.4a and b show the respective amount of SBU and MCU events for experimental ion LET ranging from 2.97 to 68 MeV/cm^2·mg. For the HD SRAM, the first MCU occurs below 2.97 MeV/cm^2·mg, while for the SD SRAM, it occurs between 5.85 and 8.30. For higher LET, the amount and the order of the MCU events increase while the proportion of single events (SBU) decreases. For every LET, the SBU component is the highest for the lowest density memory (SD SRAM) while the MCU component is the highest for the highest density SRAM (HD SRAM) [32].

2.3.9   MCU as a Function of Test Location LANSCE versus TRIUMF

Several facilities around the world provide white neutron beam for SER characterization. An exhaustive list of these facilities can be found in the JEDEC test standard [20]. The most known facilities are LANSCE and TRIUMF. Experimental measurements on the same testchip embedding an HD SP SRAM processed with TW option were performed at these two facilities. The MCU percentages were perfectly equal to 76% for both facilities. The MCU rates are reported in Figure 2.15. It shows that the MCU rate decreases by 22% at TRIUMF compared to LANSCE. This can be explained by the energy cut-off, which is 800 MeV at LANSCE while it is 500 MeV at TRIUMF.

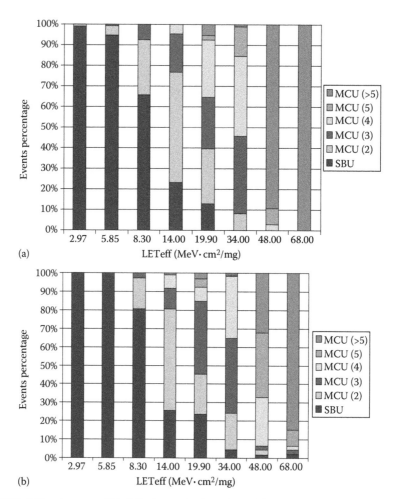

(a)

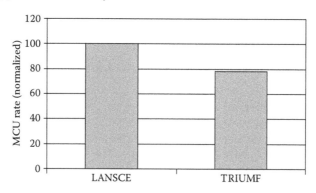

FIGURE 2.14   Amount of bit fails due to single and multiple upsets: (a) for high-density SP SRAM and (b) for standard-density SP SRAM.

FIGURE 2.15   MCU rate comparison between LANSCE and TRIUMF white neutron beam sources. The device under test is a high-density SRAM processed with triple well.

2.3.10 MCU AS A FUNCTION OF SUBSTRATE: BULK VERSUS SOI

SRAMs were manufactured with a CMOS 130 nm commercial technology either bulk or SOI. For comparison purposes, both SRAM designs are strictly identical. The testchip contains in 4 Mb of SP SRAMs in which two different bitcell designs were embedded. In this work, only the SD SRAM will be reported. The bulk technology was processed without TW layer. Table 2.4 therefore synthesizes the failure rates due to MCU (also called MCU rate) and MCU percentage for a single test pattern (CKB). It is noteworthy from Table 2.3 that SOI SRAMs have much lower MCU rate and percentage compared to bulk. More parameters (pattern, bitcell area, and supply voltage) were studied in the following article [36].

2.3.11 MCU AS A FUNCTION OF TEST PATTERN

An HD SRAM was measured at LANSCE with several test patterns using a dynamic test algorithm. Results are synthesized in Figure 2.16, which shows that uniform patterns have higher MCU rate than the CKB. To understand the reason of this discrepancy, it is necessary to plot the topological shape of experimental 2-bit MCU events as a function of pattern filling the memory during the testings (Figure 2.17a and b). The prevailing shape for 2-bit MCU and a checkerboard pattern is "diagonal adjacent" while it is "column adjacent" with uniform pattern (as observed in [37]). Three-dimensional TCAD simulations have shown that 2-bit MCU threshold LET is the lowest for two bitcells in column (see in [41] and in Section 2.4.2). It is therefore consistent that uniform patterns have higher MCU rate since their error clusters are the easiest to trigger.

It is also noteworthy from Figure 2.17a and b that TW usage did not modify the prevailing shape of MCU neither for a checkerboard nor for a uniform pattern.

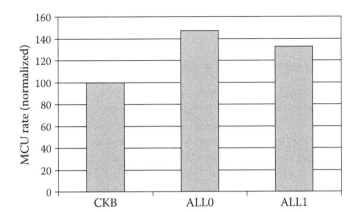

FIGURE 2.16 MCU rate comparison for several test patterns. CKB stands for checkerboard, ALL0 and ALL1 for uniform of 0 and 1 respectively. Note that test patterns are physical. The device under test is a high-density SRAM processed without triple well.

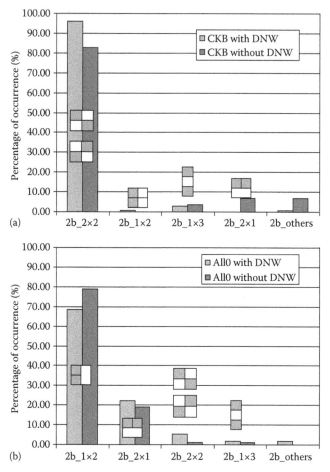

FIGURE 2.17 Two-bit MCU cluster shape on high-density SP SRAM processed with or without triple wells after neutron irradiation when the test pattern is (a) a checkerboard or (b) a uniform pattern.

2.4   3D TCAD MODELING OF MCU OCCURRENCE

Previous part has clearly highlighted the importance of TW in the MCU response. In this part, 3D TCAD simulations are set up to analyze the increased MCU occurrence when TW is used. All 3D SRAM structures in this part were built using a methodology described in [38] and the tool suite v10.0 of Sentaurus Synopsys package [39]. Cell boundaries are defined from the CAD layout and technological process steps. One-dimensional doping profiles are precisely modeled from secondary ion mass spectrometry profiles. Cell boundaries are defined from the CAD layout and technological process steps. One-dimensional doping profiles are included to define N-well, P-well (with a 4 µm epi layer thickness), and active regions of transistors. Mesh refinements are included in regions of interest: channels, LDD, junction boundaries (to tackle short channel effects), and a round ion track (to allow accurate generation of carriers in silicon). Wire connections between the different electrodes of the cell are modeled in the SPICE domain (mixed-mode TCAD simulations) to reduce the CPU burden. The parasitic circuit capacitances due to metallization layers are also taken into account.

Device simulations with ion impacts are performed using the Sentaurus device simulator. For this purpose, several physical models are activated: drift diffusion for carriers' transport, Shockley–Read–Hall and Auger for recombination, electric field and doping-dependent models for mobility, and heavy-ion module for carrier deposition along the particle track. The heavy-ion generation model uses a Gaussian radial distribution of charges with a fixed characteristic radius of 0.1 μm and a Gaussian time distribution centered at 1 ps. An additional assumption consists of taking a constant LET along the track because of the low diffusion depth of transistor active areas (~0.2 μm). Properties of boundaries are defined by the Neumann reflective conditions [38,39].

2.4.1 BIPOLAR EFFECT IN TECHNOLOGIES WITH TRIPLE WELL

For an in-depth analysis of the MCU phenomenon, 3D device simulations were performed on full SRAM bitcells. Ion strikes were located in the most sensitive MCU location (source of the SRAM) for different distances from the well taps, with and without TW. It is noteworthy that Osada et al. [40] already tried to model the effect of the parasitic bipolar amplification on the MCU. A more simple mix of device (2D uniformly extended) and circuit simulations was used but not for the worst sensitive location for MCU occurrence [41].

2.4.1.1 Structures Whose Well Ties Are Located Close to the SRAM

Figure 2.18 presents the 3D SRAM bitcell made up of six transistors (6T), two P-wells, one N-well, and three well ties. The well ties are as close as possible

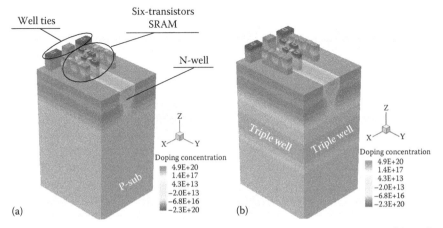

FIGURE 2.18   Full 3D structures of the 65 nm 6T SRAM located as close as possible to the well ties (a) without triple well and (b) with triple well. Two NMOS are embedded per P-well (one is a part of the inverter, and the other is an access transistor).

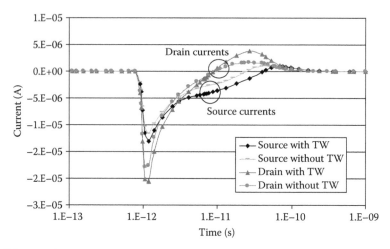

FIGURE 2.19 Full 3D TCAD simulation results on the structure presented in Figure 2.9 (6T SRAM very close to the well taps) show a limited bipolar effect due to the presence of the triple well layer. Heavy-ion LET is 5.5 fC/μm.

to transistors. The simulation results of these structures are presented in Figure 2.19, which compares source and drain currents after an ion impact in the source at 1 ps. The charge collected at the N-off drain is slightly higher with TW when well ties are located close to the SRAM transistors. With TW, a limited bipolar effect (see next part for details on bipolar triggering) is observed for structures close to the ties. These simulation results are consistent with the experimental results presented in [22,30], which have shown that MCU occurrence is less likely close to well ties.

2.4.1.2 Structures Whose Well Ties Are Located Far from the SRAM

A second set of 3D structures were built to model the effect of the spacing between well ties and SRAM cells with and without the TW doping profiles. Figure 2.20a and b illustrates four structures dedicated to well tie frequency modeling. The simulation results are presented in Figure 2.21 for ion features (LET and strike location) identical to those used in Figure 2.19. The charge injected by the source and the charge collected at the N-off drain are much higher with TW when well ties are located away from the SRAM transistors.

Injected carriers by the source are forerunners of the bipolar transistor triggering. Ion-deposited majority carriers flow toward the well ties. The well resistance causes a voltage drop beneath source diffusion. If enough carriers are deposited or if there is enough distance between well ties and ion impact (the higher the distance, the higher the voltage drop), the source–well junction will therefore be turned on, and additional carriers will be injected in the well (Figure 2.22). Most of these additional carriers will be collected at the drain junction and thus increasing the collected charge at the drain. The additional charge collection due to the source injection and due to the parasitic bipolar action is responsible for the bitcell upset. Moreover, voltage drop in the well can turn on several sources along the well,

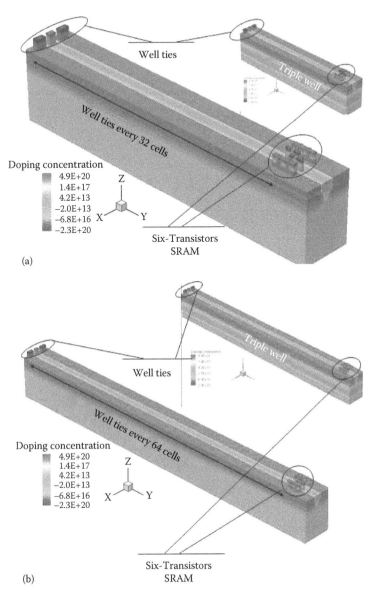

(a)

(b)

FIGURE 2.20 Full 3D structures of the 65 nm 6T SRAM whose well ties are located (a) 32 cells and (b) 64 cells away from the well taps without triple well. Same structures with triple well are shown in the upper right inserts.

which will upset several bitcells and be responsible for MCU pattern experimentally reported in Section 2.3.11.

The simulations have shown that with TW, a strong bipolar effect (electron injection from the sources) is observed for the structure away from the ties. These simulation results are consistent with the experimental results presented in [22,30], which have shown that the MCU occurrence is more likely away from well ties.

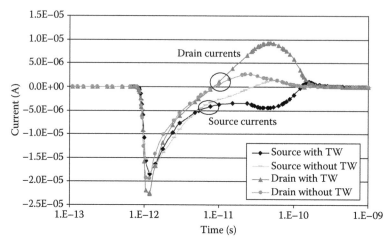

FIGURE 2.21　Full 3D TCAD simulation results on the structure presented in Figure 2.11a. Source current shows a strong bipolar effect due to the presence of the triple well layer. Heavy-ion LET is 5.5 fC/μm.

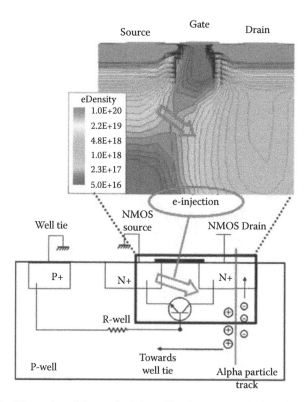

FIGURE 2.22　Illustration of the carrier injected by the source and triggering of the parasitic bipolar transistor after an alpha particle strike in the drain. Insert is from device simulation of the 65 nm 3D structure.

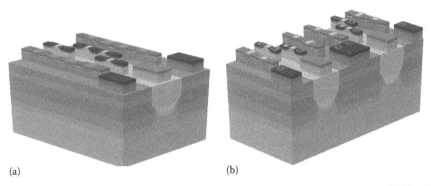

(a)   (b)

FIGURE 2.23   SRAM 3D structures (STI not displayed for clearness): (a) double 6T bitcells in column and (b) double 6T bitcells in row.

2.4.2   REFINED SENSITIVE AREA FOR ADVANCED TECHNOLOGIES

This part aims at showing that by means of 3D TCAD simulations, bitcell SEE sensitive area is not restricted to the area of reverse-biased junctions. Figure 2.23 shows the 3D TCAD final structures of two SP bitcells arranged in "column" (a) and "row" (b). These continuous TCAD domains include respectively 710,000 and 580,000 elements. The double bitcell structures are dedicated to double MBU studies. CPU burden is respectively around 1 week to simulate a double SRAM structure with up-to-date high-performance workstations.

Figure 2.24 shows an area of four SP bitcells. Two bitcells of the same column share the sources of their MOS transistors whereas two bitcells of the same row do not share any P/N junction and are isolated with shallow trench isolation (STI). At first order, an MBU of two adjacent cells is horizontal, vertical, or diagonal (configurations 1, 2, and 3 in Figure 2.24). The third case of diagonal double MBU was not simulated. Indeed, diagonal MBU would provide a higher MBU LETth than one computed for row MBU because of the longer distance between the adjacent SEU sensitive areas (both are separated with STI) (Table 2.5).

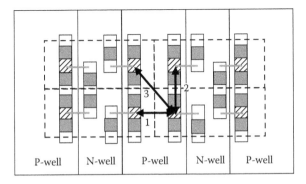

FIGURE 2.24   Four contiguous SRAM bitcells: dashed rectangles are bitcells. Connected striped and white squares are respectively drains of NMOS and PMOS transistors. Single gray and white squares are gates and sources of NMOS and PMOS transistors.

TABLE 2.5
Simulated MCU Threshold LET for Two Single-Port SRAMs Arranged in Row and in Column

TCAD Structure	Ion Location	LETth (MeV·cm²/mg)
Double row MBU	NMOS drain	13.5 ± 0.5
	Mid-distance between NMOS drains	8.5 ± 0.5
Double column MBU	NMOS drain	11.5 ± 0.5
	Mid-distance between NMOS drains	3.75 ± 0.25
	Mid-distance between PMOS drains	5.25 ± 0.25

2.4.2.1 Simulation of Two SRAM Bitcells in Row

The most efficient memory pattern to trigger a double row MBU is to reverse-bias neighboring drains. This is obtained with the logical pattern "01" (Figure 2.25). In row configuration, PMOS cannot trigger MCU since they are separated by two reverse-biased N-well/P-well junctions. MCU threshold LET were computed for two ion locations shown in Figure 2.25. Table 2.6 synthesizes these LETth and shows that an ion crossing an NMOS drain requires at least an LET of 13.5 MeV·cm²/mg to create an MCU, while an ion at mid-distance between two NMOS drains requires a lower LET (8.5 MeV·cm²/mg). Gray area in Figure 2.25 shows the extrapolated spread out of the sensitive area for row MBU until a LET of 13.5 MeV·cm²/mg.

2.4.2.2 Simulation of Two SRAM Bitcells in Column

For the configuration depicted in Figure 2.26, the most efficient memory pattern to induce MBU is "11" or "00" because the transistors of adjacent bitcells (particularly SEU sensitive areas) share the same well region and are separated by the same distance. Note that MCU can be triggered by NMOS as well as PMOS.

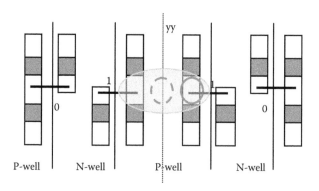

FIGURE 2.25 Scheme of the layout for two SRAM bitcells arranged in row. Plain circle is an ion impact in the NMOS drain (most sensitive single bit upset location) while dashed circles are an impact at mid-distance between two NMOS drains. Gray area is the spread of MCU sensitive area at a LET of 13.5 MeV·cm²/mg.

TABLE 2.6

Relative Neutron MCU Rate Variation as a Function of Several Parameters

Parameter	Details in Section	Relative MCU Rate
SOI substrate[a]	3.10	10
Bitcell architecture	3.8	30
Reference 65 nm single-port SRAM without triple well	—	100
Test location	3.9	125
Test pattern	3.11	145
Temperature	3.7	165
Supply voltage	3.6	230
Triple well usage	3.2	1000

[a]   Experimental results in 130 nm technology.

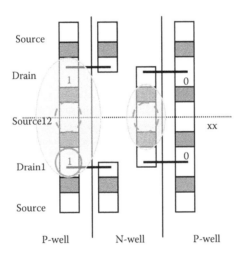

FIGURE 2.26   Scheme of the layout for two SRAM bitcells arranged in column. Plain circle is an ion impact in the NMOS drain (most sensitive single bit upset location) while dashed circles are an impact at mid-distance between two NMOS or PMOS drains. Gray area is the spread of MCU sensitive area at a LET of 11.5 MeV·cm²/mg.

MCU threshold LET were computed for three ion locations schematized in Figure 2.26. MCU LETth values are synthesized in Table 2.6. As already observed for row configuration, the lowest LETth is obtained for an ion impact at mid-distance between NMOS drains (3.75 MeV·cm²/mg). MCU LETth for an impact at mid-distance between PMOS drains is however slightly higher (5.25 MeV·cm²/mg). Gray areas in Figure 2.9b shows the extrapolated spread out of the sensitive area for column MCU until a LET of 11.5 MeV·cm²/mg.

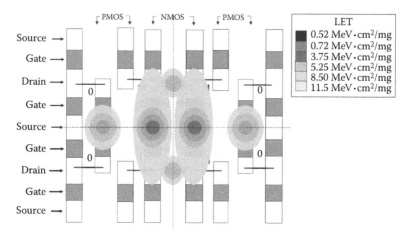

FIGURE 2.27   SEE sensitive area cartography as a function of ion LET.

2.4.2.3   Conclusions and SRAM Sensitive Area Cartography

Despite a lower distance between two adjacent SEU sensitive areas, the row MCU LETth is twice higher compared to column MBU LETth. This is explained by the incidence of the ion that crosses through 0.3 μm of STI in the first case (dashed circle in Figure 2.25), whereas it directly strikes the active area of NMOS transistor in the second case (dashed circle in Figure 2.26). As a consequence, there is less silicon volume for carrier deposition in the case of row MBU. Row and column LETth show that the layout of the memory cells (STI regions and silicon regions) strongly impacts their sensitive area.

SEE sensitive area cartography as a function of ion LET can be drawn from TCAD results shown in Sections 2.4.2.1 and 2.4.2.2. This cartography is shown in Figure 2.27. It is noteworthy that double MBU sensitive area extends beyond a single bitcell area.

2.5   GENERAL CONCLUSION: SORTING OF PARAMETERS DRIVING MCU SENSITIVITY

SEE testings carried out with alpha, neutrons, heavy ions, and protons on several SRAMs are reported in this work. These SRAMs were processed by STMicroelectronics in a CMOS 65 nm technology and embedded in several test vehicles. MCU percentages and MCU rates were given as a function of a dozen of parameters. These parameters are either technological (feature size, process option, etc.) or design (bitcell architecture, well tie density, etc.) or related to experimental test conditions (supply voltage, temperature, test pattern, etc.). Table 2.6 synthesizes the relative neutron MCU rate variations as a function of these parameters. It is noteworthy that the use of SOI substrate is as the solution decreasing the most MCU rate by taking benefit from its fully isolated transistors. Parameter which worsens the more the MCU rate is the usage of TW layer process option. On the other hand, it has to be remembered that the usage of TW allows suppressing the single event latchup occurrence even in harsh environment (high temperature, high voltage, heavy ions).

Full 3D structures were built from a layout of 65 nm SRAM bitcells. The use of TCAD structures whose SRAM bitcells are located away from the well ties was mandatory to confirm that the bipolar effect enhances the collected charge with TW. The simulations have additionally confirmed that bipolar effect is reduced by increasing the well tie frequency and therefore efficiently mitigate MCU and SER.

Other 3D structures embedding two SRAM bitcells were built. Bitcells were arranged either in column or in row to reproduce the actual SRAM array. Simulation of these structures has allowed building a SEE sensitive area cartography as a function of ion LET. This cartography shows that sensitive area extends beyond a single bitcell area.

2.6 ANNEX 1

After radiation testing with a static algorithm, bitmap error can have thousands of SEUs. With such a high density of SEUs, the key question is therefore how many upsets are "true" MCU (i.e., several SEUs simultaneously created by a single ion), and how many are "false" MCU (i.e., sequentially created in the same vicinity by different ion strikes)?

MCU rates and shapes depend on the test pattern filling the memory. It was experimentally verified that checkerboard, All1 and All0 test patterns have similar MCU rates. The following analyses and MCU counting are given for CKB pattern. A cell spacing (CS) criterion (k) is chosen when analyzing a postirradiation error bitmap for MCU detection. This criterion corresponds to the upset-to-upset spacing (maximum number of cells between two SEUs in the X and Y directions to count an MCU). The effect of this criterion on the number of counted MCU is illustrated in Figure 2.28. This figure points out that the MCU number (zero or one bitflip) and type (two or three cells) are a function of the CS criterion value: the larger this value (5, 6...), the higher the MCU number. However, large k value would lead to count two single SEU in neighboring cells created by two different events as an MCU, that is, not simultaneously generated. This would lead to a large overestimation of the MCU rates.

For this reason, formula (2.1) is proposed for quantifying the rates of "false" MCU in order to correct raw experimental data to count only the "true" MCUs. The formula is further detailed in Annex 1. We believe that this result should

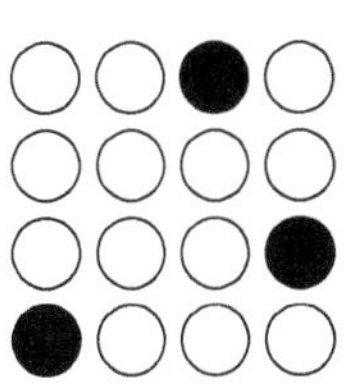

	Cell spacing criterion	MCU detected
	k = 1	No MCU
	k = 2	1 MCU of 2 cells
	k = 3	1 MCU of 3 cells

FIGURE 2.28 Illustration of the impact of cell spacing criterion on the MCU detection efficiency.

be useful in hardness assurance processes: for the total number of fails to target before stopping the irradiation and for the choice of the radiation source intensity (here a radioactive alpha source).

$$\text{false MCU \%} = 1 - e^{-E_{SRP} \times (AdjCell/Nbit)} \tag{2.1}$$

where

E_{SRP} is the number of SEU recorded after irradiation

AdjCell is the number of cells around each SEU, which are inspected to detect an MCU

Nbit is the size of the memory array

The probability to count a "false" MCU is given by

$$P = E_{SRP} \times \frac{AdjCell}{Nbit} \tag{2.2}$$

where

E_{SRP} is the number of SEU recorded after irradiation (from a single readout period)

AdjCell is the number of cells around each SEU, which are inspected to detect an MCU; this number is a function of the CS criterion (Table 2.7)

Nbit is the total number of bits in the memory array

The probability that an MCU occurred is the complementary probability that no MCU occurred (n = 0) is given using the cumulative Poisson probability by

$$MCU_{proba} = 1 - \sum_{i=0}^{n} \frac{e^{-P} \times P^i}{i!} = 1 - e^{-P} \text{ for } n = 0 \tag{2.3}$$

Multiplying this probability by the total number of SEU gives the number of MCUs. This number divided by the total number of SEU is the percentage of MCU. Using formulas (2.1) and (2.2), the percentage of "false" MCU (SEUs from two different events are counted as an MCU) is

---

TABLE 2.7

Number of Adjacent Cells Inspected for MCU around Each SEU as a Function of the Cell Spacing Criterion

Cell Spacing Criterion	k = 1	k = 3	k = 5	k = 8
No. of adjacent cells = AdjCell	8	48	120	288

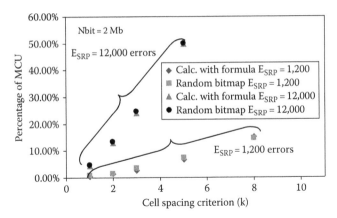

FIGURE 2.29 Comparison of MCU percentages obtained from either a randomly generated bitmap or formula (2.1) for a 2 Mb memory array (Nbit = 2 Mb).

$$\text{false MCU \%} = 1 - e^{-E_{SRP} \times (AdjCell/Nbit)} \tag{2.1}$$

In order to double-check the relevance of this model, MCU percentages obtained from formula (2.1) are compared to MCU percentages counted from randomly generated error bitmaps (Figure 2.29). This figure shows that whatever the CS criterion, the MCU percentages match perfectly.

Formula (2.1) is very convenient as it is easy to use, and it can be used for different devices (SRAM, DRAM, etc.) and many radiation sources (alpha, neutrons, heavy ions, etc).

REFERENCES

1. X. Zhu, X. Deng, R. Baumann, and S. Krishnan, A quantitative assessment of charge collection efficiency of N+ and P+ diffusion areas in terrestrial neutron environment, *IEEE Transactions on Nuclear Science*, 54(6), 2156–2161, Part 1, December 2007.
2. A.D. Tipton et al., Device-orientation effects on multiple-bit upset in 65 nm SRAMs, *IEEE Transactions on Nuclear Science*, 55(6), 2880–2885, Part 1, December 2008.
3. V. Correas et al., Simulations of heavy ion cross-sections in a 130 nm CMOS SRAM, *IEEE Transactions on Nuclear Science*, 54(6), 2413–2418, Part 1, December 2007.
4. D.G. Mavis et al., Multiple bit upsets and error mitigation in ultra-deep submicron SRAMS, *IEEE Transactions on Nuclear Science*, 55(6), 3288–3294, Part 1, December 2008.
5. X. Franz Ruckerbauer and G. Georgakos, Soft error rates in 65 nm SRAMs—Analysis of new phenomena, *13th IEEE International On-Line Testing Symposium (IOLTS 2007)*, Washington, DC, pp. 203–204.
6. D. Heidel et al., Single-event-upset and multiple-bit-upset on a 45 nm SOI SRAM, *IEEE International Conference NSREC*, Québec City, Canada, July 20–24, 2009.
7. S. Uznanski, G. Gasiot, P. Roche, J.-L. Autran, and R. Harboe-Sørensen, Single event upset and multiple cell upset modeling in a commercial CMOS 65 nm SRAMs, *Presented at IEEE International RADECS Conference*, Bruges, Belgium, September 14–18, 2009.

8. G. Gasiot, D. Giot, and P. Roche, Multiple cell upsets as the key contribution to the total SER of 65 nm CMOS SRAMs and its dependence on well engineering, *44th Annual International NSREC 2007*, Honolulu, Hawaii, July 2007.

9. T. Merelle et al., Monte-Carlo simulations to quantify neutron-induced multiple bit upsets in advanced SRAMs, *IEEE Transactions on Nuclear Science*, 52(5), 1538–1544, October 2005.

10. Y. Tosaka et al., Comprehensive study of soft errors in advanced CMOS circuits with 90/130 nm technology, *IEEE International Electron Devices Meeting IEDM Conference, Technical Digest*, 2004.

11. N. Seifert et al., Radiation-induced soft error rates of advanced CMOS bulk devices, *Presented at IRPS Conference*, San Jose, CA, 2005.

12. F. Wrobel et al., Simulation of nucleon-induced nuclear reactions in a simplified SRAM structure: Scaling effects on SEU and MBU cross sections, *IEEE Transactions on Nuclear Science*, 48(6), 1946–1952, December 2001.

13. T.L. Criswell, P.R. Measel, and K.L. Wahlin, Single event upset testing with relativistic heavy ions, *IEEE Transactions on Nuclear Science*, NS-31(6), 1559–1561, December 1984.

14. J.A. Zoutendyk et al., Single-event upset (SEU) in a DRAM with on-chip error correction, *IEEE Transactions on Nuclear Science*, NS-34(6), 1310–1315, December 1987.

15. Y. Song et al., Experimental and analytical investigation of single event multiple bit upsets in polysilicon load 64k NMOS SRAMs, *IEEE Transactions on Nuclear Science*, NS-35(6), 1673, 1988.

16. J.J. Wang et al. Single event upset and hardening in 0.15 μm antifuse-based field programmable gate array, *IEEE Transactions on Nuclear Science*, 50(6), 2158, December 2003.

17. R.A. Reed et al., Heavy ion and proton-induced single event multiple upset, *IEEE Transactions on Nuclear Science*, 44(6), 2224–2229, Part 1, December 1997.

18. N. Seifert, B. Gill, K. Foley, and P. Relangi, Multi-cell upset probabilities of 45 nm high-k + metal gate SRAM devices in terrestrial and space environments, *IEEE International Reliability Physics Symposium IRPS*, 2008, 181–186, April 27–May 1, 2008.

19. O. Musseau et al., Analysis of multiple bit upsets (MBU) in CMOS SRAM, *IEEE Transactions on Nuclear Science*, 43(6), 2879–2888, Part 1, December 1996.

20. JEDEC standard n° JESD 89, Measurement and Reporting of Alpha Particles and Terrestrial Cosmic Ray-Induced Soft Errors in Semiconductor Devices, August 2001.

21. Single Event Effects Test Method and Guidelines, European Space Agency, ESA/SCC Basic Specification No. 22900, 1995.

22. G. Gasiot, D. Giot, and P. Roche, Alpha-induced multiple cell upsets in standard and radiation hardened SRAMs manufactured in 65 nm CMOS technology, *IEEE Transactions on Nuclear Science*, 53(6), 3479–3486, December 2006.

23. E.H. Cannon, M.S. Gordon, D.F. Heidel, A.J. KleinOsowski, P. Oldiges, K.P. Rodbell, and H.H.K. Tang, Multi-bit upsets in 65 nm SOI SRAMs, *Presented at the IRPS Conference*, Phoenix, AZ, May 2008.

24. A. Virtanen, R. Harboe-Sorensen, H. Koivisto, S. Pirojenko, and K. Rantilla, High penetration heavy ions at the RADEF test site, *Presented at the RADECS*, 2003.

25. H. Puchner, R. Kapre, S. Sharifzadeh, J. Majjiga, R. Chao, D. Radaelli, and S. Wong, Elimination of single event latchup in 90 nm SRAM technologies, *IEEE International Reliability Physics Symposium Proceedings*, pp. 721–722, March 2006.

26. P. Roche and R. Harboe-Sorensen, Radiation evaluation of ST test structures in commercial 130 nm CMOS bulk and SOI, in commercial 90 nm CMOS bulk and in commercial 65 nm CMOS bulk and SOI, *European Space Agency QCA Workshop*, January 2007.

27. T. Kishimoto et al., Suppression of ion-induced charge collection against soft-error, *Proceedings of 11th International Conference Ion Implantation Technology*, Austin, TX, pp. 9–12, June 16–21, 1996.

28. D. Burnett et al., Soft-error-rate improvement in advanced BiCMOS SRAMs, *Proceedings of 31st Annual International Reliability Physics Symposium*, Atlanta, GA, pp. 156–160, March 23–25, 1993.

29. P. Roche and G. Gasiot, invited review paper, Impacts of front-end and middle-end process modifications on terrestrial soft error rate, *IEEE Transactions on Device and Materials Reliability*, 5(3), 382–396, September 2005.

30. H. Puchner et al., Alpha-particle SEU performance of SRAM with triple well, *IEEE Transactions on Nuclear Science*, 51(6), 3525–3528, December 2004.

31. D. Radaelli et al., Investigation of multi-bit upsets in a 150 nm technology SRAM device, *IEEE Transactions on Nuclear Science*, 52(6), 2433–2437, December 2005.

32. D. Giot, P. Roche, G. Gasiot, J.-L. Autran, and R. Harboe-Sørensen, Heavy ion testing and 3D simulations of multiple cell upset in 65 nm standard SRAMs, *IEEE Transactions on Nuclear Science*, 2007.

33. A.M. Chugg, A statistical technique to measure the proportion of MBU's in SEE testing, *IEEE Transactions on Nuclear Science*, 53(6), 3139–3144, December 2006.

34. D. Tryen, J. Boch, B. Sagnes, N. Renaud, E. Leduc, S. Arnal, and F. Saigne, Temperature effect on heavy-ion induced parasitic current on SRAM by device simulation: Effect on SEU sensitivity, *IEEE Transactions on Nuclear Science*, 54(4), 1025–1029, 2007.

35. M. Bagatin, S. Gerardin, A. Pacagnella, C. Andreani, G. Gorini, A. Pietropaolo, S.P. Platt, and C.D. Frost, Factors impacting the temperature dependence of soft errors in commercial SRAMs, *IEEE Transactions on Nuclear Science*, 47(6), 100–106, 2008.

36. G. Gasiot, P. Roche, and P. Flatresse, Comparison of multiple cell upset response of BULK and SOI 130 NM technologies in the terrestrial environment, *IRPS Conference*, Phoenix, AZ, May 2008.

37. Y. Kawakami et al., Investigation of soft error rate including multi-bit upsets in advanced SRAM using neutron irradiation test and 3D mixed-mode device simulation, *IEDM 2004*, 945–948, 2004.

38. P. Roche et al., SEU response of an entire SRAM cell simulated as one contiguous three dimensional device domain, *IEEE Transactions on Nuclear Science*, 45(6), 2534–2543, December 1998.

39. Synopsys Sentaurus TCAD tools, World Wide Web: http://www.synopsys.com/products/tcad/tcad.html.

40. K. Osada et al., Cosmic-ray multi-error immunity for SRAM, based on analysis of the parasitic bipolar effect, *2003 Symposium on VLSI Circuit Digest of Technical Papers*.

41. D. Giot, G. Gasiot, and P. Roche, Multiple bit upset analysis in 90 nm SRAMs: Heavy ions testing and 3D simulations, *RADECS Conference*, Athens, Greece, September 2006.

3 Radiation Hardened by Design SRAM Strategies for TID and SEE Mitigation

Lawrence T. Clark

CONTENTS

3.1 CHAPTER OVERVIEW

3.1.1 EMBEDDED SRAMs IN IC DESIGN

Static random access memory (SRAM) is ubiquitous in modern system-on-a-chip integrated circuits (ICs). Due to its value in programmable systems, providing fast scratchpad memory in embedded and real-time applications, and providing space for large working sets in microprocessor designs, IC SRAM content continues to grow. As ICs surpass 1 billion transistors, and given the high relative design and power efficiency of memory arrays compared to random logic, SRAM was projected to comprise as much as 90% of the total die area in 2013 [1]. For instance, the Itanium processor has progressed from 6 and 9 MB L3 caches on 130 nm fabrication processes, to 24 MB caches on the 65 nm technology generation [2–4]. The Xeon processors include 16 MB caches [5]. Consequently, ICs designed for space and other radiation environments require robust SRAM designs if they are to track the size and performance of commercial ICs.

3.1.2 RADIATION SPACE ENVIRONMENT AND EFFECTS

The earth's radiation environment consists of electrons, protons, and heavy ions. The former two are trapped by the earth's magnetic field where they follow the field lines, where these particle fluxes are the highest. Eighty-five percent of galactic cosmic ray particles are protons with the rest composed of heavy ions [6]. Cosmic ray flux is essentially omnidirectional, so microelectronics may be affected by particles impinging at any angle. Importantly, this means that ions can transit an IC parallel to the device surface, since there is no practical level of shielding that can stop all protons and heavy ions. Solar cycles also strongly affect the radiation environment. Ordinarily the helium ions in the solar emitted particle fluxes comprise 5%–10%, and heavier ion fluxes are very small, well below the galactic background. During major solar events, some heavy ion fluxes may increase by up to four orders of magnitude above the galactic background for as long as days at a time.

The dominant radiation effects on microcircuits in space are due to deposited charge from ionization tracks produced by single particles. These produce two primary effects: First, collected charge from a single particle can upset circuit state, referred to as a single event effect (SEE). Second, changes in the charge state of dielectrics due to total accumulated ionization can alter device characteristics, referred to as total ionizing dose (TID) effects [7].

Both protons and heavy ions can deposit charge that can upset the circuit state. Upsetting a feedback (state storage) node such as a memory bit is defined as a single event upset (SEU). Heavy ions affect the circuit state through direct ionization due to columbic interaction with the substrate material, producing about 10 fC of charge per µm of track length per LET. Memory cells are often characterized for SEU by the total charge Q_{crit} that is required to upset their state. Charge that temporarily disrupts a logic node results in an incorrect voltage transient of a magnitude and duration determined by the node capacitance and the driving circuit's ability to remove the charge. These are referred to as a single event transient (SET). An SET can only affect the IC architectural state (the state that is visible to the surrounding system) if sampled by a latch whose output is subsequently used.

Protons interact with the silicon through multiple mechanisms, predominantly by direct ionization, but also through secondary nuclear particle emission due to Si recoil. The former generates relatively small amounts of charge, but the latter can upset circuits hardened to high LET. Approximately 1 in 100,000 protons impinging will produce a nuclear reaction. Moreover, the multiple secondary particles may interact with the circuit after moving in multiple directions. A single particle produces charge in linear tracks. Charge is collected by diffusion and by drift, with the latter due to the device depletion regions. Charge collection is enhanced by "funneling," which is a third field-driven collection mechanism that extends the field-driven collection by the redistribution of the deposited carriers. Parasitic bipolar action can also increase the current collected at a specific node, greatly increasing the upset rate and extent.

Impinging particles can also permanently disable the microcircuit by excessive displacement damage or by rupturing the gates. Such permanent effects are not pertinent to the discussions in this chapter.

3.1.3 CHAPTER OUTLINE

This chapter focuses on SRAM design using RHBD techniques. Both TID and SEE hardening are covered. The latter approaches described assume that error detection and correction (EDAC) is used to mitigate individual SEU, as RHBD hardened cell approaches have diminishing value in modern highly scaled fabrication processes. Small, dense geometries make simultaneous upset of multiple circuit nodes from a single particle strike increasingly likely. A primary focus, therefore, is on mitigating SETs that can cause upsets that confound the EDAC, or otherwise cause incorrect SRAM operation. All of the approaches examined in this chapter have been fabricated and tested—measurements quantifying their effectiveness are also described and discussed.

The last section briefly outlined the space radiation environment. Subsequent sections include a discussion of basic SRAM cell design, which is tutorial in nature. Test structures to characterize SRAM cells are then described. This is important, particularly for RHBD SRAM cells, which do not undergo the same rigorous testing and validation during the fabrication process development that the foundry-provided cells do. The TID response of SRAM cells hardened by various techniques and that of an unhardened version are examined, as are the tradeoffs in cell size and hardness for various TID hardening approaches. Heavy ion beam testing results show the importance of multiple bit upset (MBU) and SET response. The design of an SET hardened SRAM is then described, as well as its response in ion beam testing, which is compared to that of an unhardened device. We then briefly summarize the results to conclude the chapter.

3.2 RADIATION HARDENING

All hardened designs should mitigate four issues: first, single event latch-up (SEL) due to ion-strike-induced substrate currents; second, single event logic upset due to the capture of SETs in sequential circuits, for example, latches and flip-flops; third, SEUs of storage nodes, which include storage latches in registers and SRAM memories; the fourth issue is the TID, which can affect the individual device and isolation characteristics. These device changes, in turn, may deleteriously affect the circuit behavior.

There are two basic approaches to fabricate radiation-tolerant ICs—hardening by process [8] and hardening by design [9]. Hardening by process uses a specialized fabrication process that has features specifically added to mitigate radiation effects, such as SOI substrates, special body-ties, and dense high-value resistors [10–12]. Radiation hardening by design (RHBD) allows radiation-tolerant circuits to be fabricated on commercially available state-of-the-art CMOS manufacturing processes [9,13] to reduce cost and improve circuit performance. It relies exclusively on special circuit topologies and layouts rather than specialized process features and devices to provide hardening. For example, P-type guard rings around NMOS diffusions, similar to those used for I/O ESD protection, provide increased SEL immunity. Of course, actual designs may utilize a combination of approaches. For instance, SOI substrates are available on commercial unhardened processes. Furthermore, specific rad-hard circuits and layouts are still required when using rad-hard fabrication processes.

3.2.1 Total Ionizing Dose Effects

In modern processes with sub-3 nm thick gate oxides, TID primarily increases leakage under isolation oxides and at the gate edges, that is, at the thin gate oxide to isolation oxide interfaces. This slowly increases leakage from a parasitic transistor at the transistor edge as its threshold voltage (V_{th}) decreases with TID. Since the trapped charge is positive, only NMOS transistors suffer from increased leakage due to these parasitic devices along the gate edges. Similarly, leakage between N-type diffusions, for example, between the N-well and NMOS drains, can be increased by reduction of the field oxide V_{th}. [14,15]. These increases in leakage are manifest in a given IC as increased I_{DD} measured in the quiescent state, commonly referred to as standby current (I_{SB}). TID has been shown to cause functionality loss in SRAMs [16]. Increased leakage currents can interfere with proper precharging or small swing bit-line signal development. Leakage within the cell can also affect the read stability by changing the cell static noise margin (SNM) [17].

TID is mitigated by higher doping at the oxide interfaces or, when using RHBD approaches, by using annular or edgeless NMOS transistor gates. The standard

RHBD technique for mitigating TID increased leakage in the parasitic edge transistors is to use "edgeless" or annular transistor geometries. The annular transistor fully encloses the drain or source, so the same potential is at both sides of the transistor edge to isolation oxide interface.

3.2.2   Single Event Effects in SRAMs

SRAMs are prone to both SEU and SET generated errors. The former has been addressed by the use of the dual interlocked storage cell (DICE) or other approaches that add transistors to the storage-cells [18]. The DICE circuit adds redundant storage nodes and a self-correction mechanism, which allows three correct storage nodes to correct one incorrect node. Other approaches generally attempt to limit the ability of collected charge to affect the latch feedback state. It is important to realize that at any critical node separation errors may occur in space, where particles can be incident at any angle. In any of these approaches, the cell storage (critical) nodes must be spaced sufficiently far apart to minimize the probability, by increasing the incident ionizing radiation particle track angle required to upset multiple latch nodes with a single ionizing particle strike [19,20]. This becomes more difficult as process dimensions are scaled down, as this naturally places critical nodes closer together. Temporal latches mitigate both types of upset [21] but with so much added circuit area and path delay that they are impractical for high-speed, high-density memory arrays.

Easily the most common approach in hardened processes is the addition of resistors in the SRAM cell latch feedback path [10,22]. Since charge can only be collected at a PN junction, the series RC produces a delay that allows the collected charge to be removed by the latch feedback transistors before the feedback node can transition, thus restoring the original cell state. Since undoped polysilicon conductivity is constant, it becomes increasingly difficult to produce resistors providing sufficiently long time constants as fabrication processes scale. Thus, more modern versions use special high-resistivity vias or via layers [12]. Hardened processes often use SOI substrates or special implants to limit the track length of the collected charge from impinging ionizing radiation particles [23]. Limiting the track length reduces the required RC time constant by attenuating the collected charge and thus the duration of an upset. Similarly, it mitigates SET durations [12].

SRAMs are not only susceptible to cell state upset (SEU). It has also been long known that an SET on word line (WL) signals [24,25] can cause improper operation, by asserting the wrong, or more than one of the normally one-hot WLs high. Similarly, any SET in the control, clocking, or decode paths may cause the wrong operation or the wrong address to be accessed. In the worst case, for example, when the wrong memory address is read, the parity or ECC may be correct. Referring to Figure 3.1a, WLs act as selects that allow a row of SRAM cells to discharge the appropriate precharged bit lines (BL and BLN) in each column. In the event of a WL SET, a number of circuit-level behaviors may occur. If two WLs are asserted high simultaneously during a read operation, for example, WLn and WL0 in Figure 3.1a, then the BLs will logically OR the values, as a BL is a dynamic NOR multiplexer. If the SET occurs late in the read or write phase, and one of the BL pairs in each column has discharged to V_{SS}, a subsequent WL misassertion may write the BL

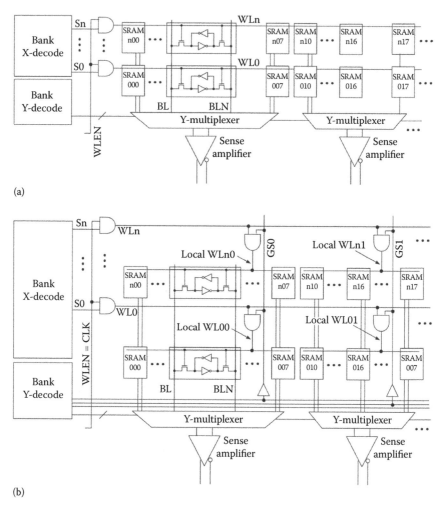

FIGURE 3.1 (a) Conventional SRAM bank word line architecture and (b) divided word line architecture. A SET may affect all or a number of local WLxx by incorrectly asserting the global WLx attached to each of them.

state into the row controlled by that misasserted WL. When this value inadvertently enters the IC architectural state and is subsequently read, this undetected error is termed a silent data corruption (SDC).

The "column group," which is the basic design unit that can read or write 1 bit, generally contains many SRAM cell columns. There are eight SRAM cell column, labeled SRAM000–SRAM007 in the leftmost column group in the examples in Figure 3.1, sharing one sense amplifier and associated write circuitry through the column or "Y" multiplexer. By convention WLs are the X multiplexer selects. Multiple SRAM cell columns per sense/write circuit are required primarily by the fact that the former are large. Thus, the layout is eased by not trying to fit large sense circuits into the tight SRAM cell pitch. It also forces spacing between individual cells containing data from the same word, assuming a given word is read in one cycle. This separation due to Y multiplexing makes it less likely that an MBU will upset multiple bits in a single protected codeword, as has been common knowledge in the SRAM design community since the 1980s. Commercial designs have tended to use at least four SRAM columns per group, but that may need to increase in the future as SRAM cells scale to smaller dimensions [26].

Figure 3.1b shows a technique that has been employed to mitigate such control and WL SET-induced errors. Each column group again contains multiple SRAM cell columns, but each WL (and control line, not shown) is individually buffered. Thus, an SET on the local WL, for example, LocalWLn0 in Figure 3.1b, will affect only that local column group. This scheme was applied, and errors due to local WL ion strikes were recorded [27]. Since only 1 bit can be read at a time from a column group, EDAC can correct such an error, whether on a read or write operation. However, the global WLs (WL0–WLn in Figure 3.1b) are not so protected. Sufficient WL capacitance and drive will provide some SET immunity, but in general, large array sizes are required to raise the threshold LET sufficiently [28].

One approach that has been put forward to mitigate this issue is the so-called bit per array architecture, where each bit in an EDAC protected word is stored in a separate SRAM bank. Conventionally, a "bank" is a stand-alone unit containing clocking, control, decode, and array circuits. However, this in itself is insufficient to protect against such errors in all cases. An example of this case is shown in Figure 3.2a. Here, all bits in an EDAC protected codeword reside in separate SRAM banks, providing excellent critical node separation and greatly limiting the probability of a single ionizing particle strike upsetting multiple nodes in a single EDAC word. Referring to Figure 3.2, note that all addresses and control signals fan out from single registers, which we may assume are protected against errors on the inputs or clocks, for example, by the use of temporal or other techniques [18,21]. However, the output node, or one of the inverters that provides fan up to drive the heavily loaded address bus, is not so protected. Consequently, an SET on one of these nodes propagates to all arrays, which can manifest as an SDC by having all the arrays perform the wrong operation or access the wrong memory location. The way to protect against this, albeit small cross section, failure scenario is to place the SET mitigating latches

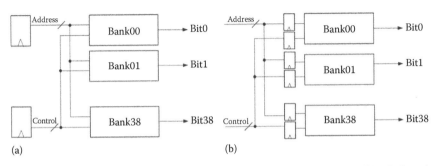

(a)   (b)

FIGURE 3.2 (a) SRAM bit per bank architecture. A SET may still affect all banks by misasserting the address or (b) control signals unless mitigating latches are placed at each bank input as in (b).

in each memory array or subarray, as shown in Figure 3.2b. However, the address setup time, measured at the array input, increases commensurately, as it must be greater than 2 t_{SET}, where t_{SET} is the SET duration [29]. Consequently, such retiming must be comprehended by the overall micro-architecture. Power dissipation obviously increases with the bit per array architecture, as many SRAM banks must be activated in a single clock cycle, where otherwise only one might be. The SRAM caches in embedded commercial microprocessors account for as much as 43% of the total power dissipation [30]. This can be driven down to 15% by fine-grained clock gating, that is, activating only the necessary smaller banks [31]. Thus, the designer is faced with a number of circuit and micro-architectural challenges that may profoundly affect the speed and power dissipation, when implementing radiation-hard SRAMs.

3.3 RADIATION HARDENING BY DESIGN IN SRAMs

As mentioned earlier, edgeless NMOS transistors effectively mitigate a major TID-induced leakage current increase component. The drawback of using annular NMOS transistor gates is that the topology imposes a relatively large minimum transistor width, since a contact must be placed within the edgeless gate. The necessary gate to contact spacing sets the inside perimeter, while the gate length, which must usually include extra margin on the 45° angle gate edges, sets the outer perimeter. For the 130 and 90 nm processes used in most examples in this chapter, the minimum widths are increased by 11.1 times and 9.1 times, respectively. For high drive gates, for example, inverters driving clock or large control nodes, the RHBD penalty is negligible [32].

In SRAM cells, RHBD using conventional techniques imposes a significant increase in size. To avoid a significant area impact for RHBD SRAMs, more clever TID mitigation techniques must be found. Furthermore, commercial foundries offer smaller SRAM cells that violate the standard process layout design rules. These tighter SRAM cell layouts are optimized by fabricating large numbers of cells and optimizing the required "array rules" during the process technology development. Essentially, which rules can be "cheated" for these highly regular SRAM layouts is determined experimentally. This makes the RHBD impact even greater, since no RHBD design will be able to similarly validate the use of such aggressive design rules.

3.3.1 SRAM CELL READ AND WRITE MARGINS

The commonly used six-transistor SRAM cell is shown in Figure 3.3a. The figure also illustrates the layout of the SRAM cells, designed on a 130 nm foundry technology. The key SRAM cell design requirements are to ensure write-ability and read margin. To this end, the SRAM cell device sizes are a compromise between those that result in the smallest cell but still provide adequate read and write margins. When sizing the transistors, all process, voltage, and temperature corners must be considered. The SRAM transistors are small enough that random dopant fluctuations have a considerable effect on the actual cell margins [33]. The large on-die memory and cache sizes noted in Section 3.1.1 mandate that 10–14 sigma manufacturing

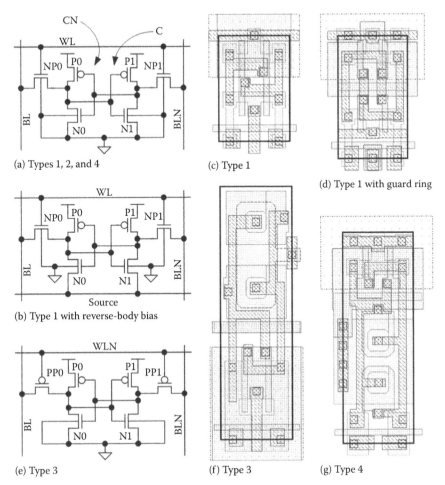

(a) Types 1, 2, and 4

(b) Type 1 with reverse-body bias

(e) Type 3

(c) Type 1

(d) Type 1 with guard ring

(f) Type 3

(g) Type 4

FIGURE 3.3 Hardened and unhardened 130 nm SRAM cell designs. (a) 4-NMOS and 2-PMOS standard SRAM designs and (b) with body bias capability. (c) Layout type 1 and type 1 with guard ring and (d) reach near foundry densities, but using annular NMOS pull-downs (g) is much larger. (e) Using 4-PMOS and 2-NMOS transistors (f) does not save area since the PMOS devices must be very large to overpower the edgeless NMOS transistors to write the cell. (After Clark, L. et al., *IEEE Trans. Nucl. Sci.*, 54(6), 2028, 2007.)

variability must be considered. To comprehend the increasing variability and diminishing cell margins, a number of statistical methodologies for SRAM cell and array design have been proposed [34,35]. The difficulty in designing commercial SRAM cells with adequate margins points to the importance of advanced techniques for RHBD memory cells, since there are less validation resources available for the small rad-hard market, and the impact of more difficult to pattern and fabricate annular gate geometries must be comprehended.

The cell is written differentially, where one BL is at the high precharge potential and the other (BLN) is driven low (or vice versa). During a write, the NMOS access transistor (NP1) must overpower the PMOS pull-up transistor—the cell is a ratioed circuit during writes. Adequate write margin requires that the access transistor NP1 in Figure 3.1 be stronger than the pull-up device P1. The write margin is typically defined in one of two ways. The DC approach is to measure the BL voltage required to flip the SRAM cell state, by keeping BLN high and lowering BL from V_{DD} toward V_{SS} until the cell state is flipped. Alternatively, the delay to write the cell when the BL is driven to V_{SS} may be measured [36].

When the SRAM is read, the low storage node rises due to the voltage divider composed of the two series NMOS transistors in the read current path (N0 and NP0 in Figure 3.3a). The storage node CN is between them, rising above V_{SS} during a read. This reduces the SRAM cell SNM as measured by the smallest side of the square with largest diagonal that can fit in the small side of the static voltage curves [37]. The worst-case SNM, shown in Figure 3.4, is usually determined by Monte Carlo simulation or response surface models [38] based on the measured process variation parameters. For the unhardened cell simulated here, the SNM in Figure 3.4 is quite small at 58 mV, even at the nominal V_{DD} of 1.2 V. The large transistor mismatch due to both systematic (die-to-die) and random (within-die) variation causes asymmetry in the SNM plot. Other noise margin definitions have been developed, based on imbalance created across the cell by disturb voltages [36]. Read margin is also ensured by transistor sizing. Typically, the pull-down NMOS transistors are drawn wider than the access transistors. The access transistor NP0 is also frequently drawn with a longer channel than that of the pull-down N0 and pull-up P0. Consequently, there is limited design latitude—the PMOS pull-up must be weaker than the NMOS access device, which, in turn, must be weaker than the NMOS pull-down transistor.

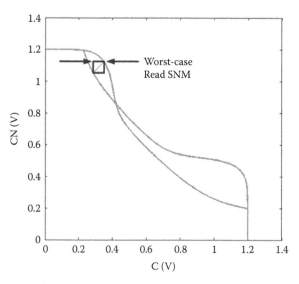

FIGURE 3.4 SRAM SNM plot. The line depicts the worst case found in a Monte Carlo simulation to five sigma variations. Note the severe asymmetry. (After Yao, X. et al., *IEEE Trans. Nucl. Sci.*, 55(6), 3280, 2008.)

3.3.2 Reverse-Body Bias

Reverse-body bias (RBB) with and without simultaneous power supply collapse (SC) has been used in commercial integrated circuits for full-chip [39,40] and memory [41,42] standby power reduction. RBB is presently used on commercial ICs for low standby power (LSP) state retention on processes varying from 250 through 65 nm process generations [39,41,43,44]. These modes often combine RBB and SC, here termed RBB+SC, since the latter helps to mitigate emerging transistor leakage paths such as direct band-to-band tunneling through the gate oxide [44]. RBB electrically increases the transistor threshold voltage by the body effect, which is also applied to the parasitic edge and field oxide FET. Its use for TID mitigation has been demonstrated at the transistor level on low V_{th} 0.35 μm bulk CMOS transistors [45]. RBB mitigation of TID on advanced fabrication process technologies allows the use of the smallest foundry-optimized cells and thus eliminates RHBD SRAM size penalties.

3.3.3 RHBD SRAM Cell Design

A number of RHBD SRAM cell designs have been investigated on 130 and 90 nm technologies [46,47]. As mentioned, any SRAM cell must have adequate write margin and read stability. For a fair analysis, these are constrained to be as good or better than that of the baseline two-edge foundry cells when evaluating potential circuit topologies and layouts. A number of potential SRAM cell schematics and layouts are shown in Figure 3.3, which were used in a 130 nm study [47].

The type 1 cell (see Figure 3.3a and c) is essentially a conventional (commercial non rad-hard) design with two-edge NMOS pull-down transistors and two-edge NMOS access devices. Figure 3.3b shows a variation that separates the NMOS source from the substrate taps (i.e., allowing RBB). Since in SRAM arrays the well and substrate taps are placed in special tap rows (or columns for vertical well designs), RBB support does not add size to the array, as shown in Figure 3.3c. Note, however, that in an RHBD IC these tap spacings should be considerably smaller to avoid SEL.

The type 2 SRAM cell employs annular NMOS pull-down transistors and annular NMOS pass gates. The type 3 SRAM cell uses edgeless NMOS pull-downs and two-edge PMOS access transistors (Figure 3.3e and f). The gate bias dependence of NMOS transistor TID degradation, where the leakage current increase is suppressed when the gate is biased at 0 V so the electric fields do not repel the positive trapped charge toward the oxide/Si interface, suggests the type 4 SRAM cell (see Figure 3.3a and g). It has annular NMOS pull-down transistors and two-edge NMOS access transistors. This variation relies on the fact that most of the time all but one of the SRAM WLs are de-asserted at 0 V.

The type 2 through type 4 cells include PMOS guard rings between the NMOS transistor drains and the N-well to limit TID-induced leakage between the two. No guard rings are used to limit leakage between the NMOS drains at different potentials in any of the cell designs investigated. For instance, in the type 4 cell shown in Figure 3.3g, the NMOS pull-down sources, at V_{SS}, are near the access transistor drains and not separated by a P-type guard ring. If these guard rings are necessary, the cells must grow to accommodate them.

For each of the designs in Figure 3.3, simulations were used to determine write margins and read stability. Since write margin can be increased at the expense of read stability, the designs are optimized by minimizing total transistor width at similar write margin, but forcing read stability to meet the baseline set by the foundry SRAM cell. This analysis assumes that the total cell size is proportional to the required transistor widths.

3.3.3.1 Conventional Two-Edged Transistor Cell (Type 1)

In a conventional SRAM design where all transistors are two-edged, all devices can be drawn at or near minimum width as in the commercial foundry cell. The large NMOS to PMOS mobility ratio and use of minimum width PMOS pull-ups provide adequate write margin. Adding a guard ring to mitigate SEL and TID-induced NMOS drain-to-N-well leakage increases the cell size by about 20%, as evident in Figure 3.3d. If guard rings are unnecessary, standard production SRAMs employing even tighter design rules and smaller cell size can be used—the foundry-supplied 130 nm cell is 27% smaller than the cell in Figure 3.1c, which is drawn to the logic layout rules.

3.3.3.2 Annular NMOS-Based SRAM Cell (Type 2)

The simple analysis shows that using a conventional cell with four annular NMOS and two PMOS transistors results in a total transistor width approximately seven times that of the conventional two-edge cell at the same write margin and read stability. While annular NMOS layout eliminates the source-to-drain leakage path formed at the shallow trench isolation (STI) to channel interface, their greater minimum size increases the preirradiation cell leakage commensurately. One potential design is shown in Figure 3.5. This cell, implemented on a 90 nm foundry bulk CMOS technology, is 5.1 times the size of the foundry cell, which uses tighter SRAM design rules, and 3.6 times the size of a cell drawn to the same (90 nm) logic design rules. Of this, about 20% of the size is attributable to providing portability between process versions, which have different gate lengths. Thus, the cell could have been 20% smaller if portability were not a requirement. Another 20% is attributable to the guard rings, similar to the impact on the 130 nm cells described earlier. The aspect ratio helps this, since the P-type guard rings are oriented vertically. The wide cell increases the critical node spacing in the key horizontal dimension, making a column group wider, with the same n:1 Y multiplexing.

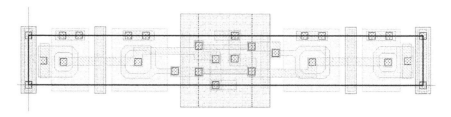

FIGURE 3.5 Full edgeless 90 nm NMOS SRAM cell layout. This cell is 5.1 times larger than the smallest available foundry cell on this 90 nm process, due to compatibility with the LSP version with longer gate length, P-type guard rings, and annular NMOS gates.

The particular RHBD NMOS layout choice also affects the transistor leakage. For the "ring" topology used in the NMOS transistors in Figures 3.3f,g and 3.5, placing the transistor drain inside the ring and source outside minimizes the area used as the sources can be shared between cells or transistors within the cell. Extra guard rings are then unnecessary. Transistor I_{DS} vs. V_{DS} measurements of ring gate transistors show three times greater I_{OFF} and reduced I_{DSAT} with the source inside the ring [47]. The former is due to much higher drain-induced barrier lowering, which is a function of the drain/gate vs. source/gate interface widths. It is therefore important when determining stability to account for the specific gate geometries used and their particular I_{DS} vs. bias characteristics. Standard foundry transistor models do not provide this level of modeling detail, particularly for the edgeless gate geometries, so appropriate test transistor arrays must be used for modeling and validation.

3.3.3.3 PMOS Access Transistor SRAM Cell (Type 3)

The TID immunity of two-edge PMOS transistors on modern processes suggests their use as the SRAM access transistors. However, to write the cell, the PMOS pass transistors must overpower the NMOS pull-downs, which is made difficult by the large NMOS/PMOS mobility difference. Consequently, using annular NMOS pull-downs implies wide PMOS access transistors as shown in Figure 3.3f. In this cell, an NMOS long-channel "ring" gate pull-down transistor is used to make it as weak as possible, as the channel comprises only one edge of the NMOS transistor and has a long channel length L and narrow width W for a low W/L ratio. The SRAM cell area is still large. One advantage of the large NMOS gate area is increased capacitance and thus increased Q_{crit}. For the conventional bulk CMOS processes used here, this cell is very large and has sufficient drawbacks that further analysis is unnecessary. Future processes may have similar PMOS and NMOS drive currents [48]. For such a fabrication process, this cell topology may be very good.

3.3.3.4 Two-Edged NMOS Access Transistor SRAM Cell with Annular Pull-Down Transistors (Type 4)

Using two-edge NMOS access transistors (NP0 and NP1 in Figure 3.3a) and annular pull-down transistors N0 and N1 is an interesting alternative since the WL is high for only the row being accessed. A low gate voltage minimizes TID-induced leakage at the transistor edges, and this is the bias condition on access transistors NP0 and NP1 over 99% of the time. This cell has good stability and write margin even for narrow access transistor widths. Cell size is reduced as shown in Figure 3.3g but is still significantly larger than the conventional SRAM cell. A drawback to this configuration is that depending on the layout, the access transistor source and drain nodes may be difficult to shield from the other cell transistors using guard rings. Fortunately, on sub-150 nm processes, this TID component may be tolerable or mostly mitigated by avoiding polysilicon crossing from N+ to N-well diffusions [49].

3.3.4 SNM Test Structure

Since read stability is so important, and it is unlikely that an RHBD IC designer will be able to include RHBD SRAM cells on the process development test chips as the

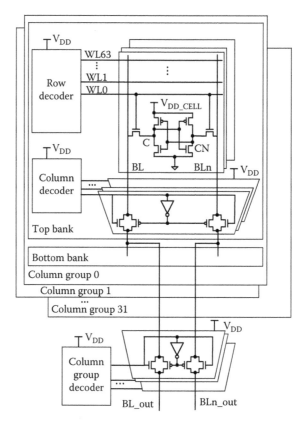

FIGURE 3.6 Ninety nanometer test array allowing the measurement of individual cell margins. (After Yao, X. et al., *IEEE Trans. Nucl. Sci.*, 55(6), 3280, 2008.)

standard foundry cells are, appropriate test structures to rapidly determine the cell quality are essential. A test structure that allows direct measurement of the SNM of individual SRAM cells is shown schematically in Figure 3.6. It is based on the standard SRAM array and can be integrated with a production design. To allow accurate analog signal propagation, two supply voltages, V_{DD} and V_{DD_CELL}, are used. V_{DD} is independent of V_{DD_CELL} allowing the gate overdrive of the cell and access devices to be controlled independently. By applying $V_{DD} > V_{DD\_CELL}$, the resistance of the analog signal path multiplexers is reduced, limiting their effect on the measurements. During the test, nodes WL and the access multiplexer enables are asserted high. The high WL voltage allows single-ended writes of the SRAM cell, unlike the normal operating condition, where writes must be differential. The test is DC, so there is no time dependence in the measurement. Consequently, the BL and BLN voltages, when used as outputs, accurately represent that of the SRAM cell storage nodes C and CN, respectively. Thus the circuit allows direct measurement of the as-fabricated SRAM cell P/N ratios through observation of the switching points when driving the BL (or BLN) high or low.

An analog multiplexer under software control is used to connect the test structure to a digital to analog converter. The multiplexer allows switching either the BL or BLN attached to the FPGA driver, with its complement BLN or BL attached to the analog to digital converter to measure the cell state. The measurements can be made with the device under test (DUT) inside the Co-60 irradiator so measurements vs. TID can be made in situ, allowing the determination of the TID impact on the individual SRAM cell read and write characteristics. Measuring the DUT in situ, that is, while being irradiated, avoids relaxation of the TID effects that would occur when removing the device from the irradiator to make a measurement. Measured TID results from unhardened and hardened SRAM cells are presented as follows.

3.3.5 Experimental TID Testing Results

Test die were fabricated on both 130 and 90 nm CMOS bulk processes at the same foundry. The test die included both SRAM arrays and transistor test structures. The latter are laid out in the SRAM cell layouts, so that the results are representative of the device responses that would occur in the actual SRAM arrays. SRAM test structures with annular pull-down transistors (type 4) were also fabricated.

A 90 nm 1.2 Mb SRAM design fabricated on a low leakage (LSP) variation of the process can provide a baseline for the discussion [17]. This SRAM exhibited a 131 times increase in I_{SB} after irradiation to 1 Mrad(Si) as shown in Figure 3.7. Note, however, that significant leakage increase does not occur until after 300 krad(Si), which may be sufficient for many spaceborne IC applications. The design is fully

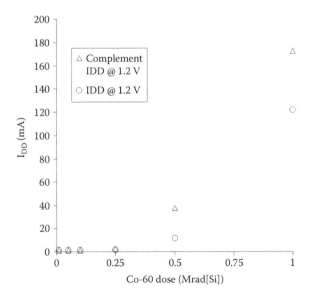

FIGURE 3.7 Ninety nanometer 1.2 Mb conventional (nonhardened) SRAM TID results. The increase at 1 Mrad(Si) for the array in the state opposite that when irradiated shows a 131 times increase in standby I_{DD} (I_{SB}) for this device fabricated on the LSP process version. (After Yao, X. et al., *IEEE Trans. Nucl. Sci.*, 55(6), 3280, 2008.)

functional despite this large leakage increase after 1 Mrad(Si) dose. While prior SRAM designs have failed at relatively low TID levels [16], careful circuit design can avoid this. In general, since size scaling requires increasing doping levels, and V_{DD} scaling requires lower V_{th} to maintain gate overdrive, smaller geometry processes exhibit less TID-induced I_{SB} increases. The higher I_{OFF} and I_{gate} leakage currents in more highly scaled processes tend to mask what TID-induced increase there is until higher doses.

It is important to know the exact SRAM cell organization in order to apply the worst-case TID conditions. In particular, horizontally adjacent cells may have adjacent NMOS diffusions biased the same or differently with a solid or checkerboard pattern. This is not just a matter of geographic cell location, but also a function of whether or not the BL and BLN are stepped or folded in the layout. For example, the 90 nm SRAM uses a pattern BL0 BLN0, BL1 BLN1, ... BL7 BLN7. However, the 130 nm design uses BL0 BLN0, BLN1 BL1, ... BLN7 BL7. In the former case, a solid array pattern of all 1s or 0s is worst case for TID leakage increase, while in the latter, a *physical* checkerboard is. Finally, the physical and logical organization can be quite different, so knowledge of the physical layout is critical here, as it is in choosing appropriate production SRAM test patterns.

3.3.5.1 Impact of V_{DD} Bias on TID Response

The 1.2 Mb 90 nm SRAM, fabricated on the foundry LSP process version, irradiation results were described earlier. Additionally, 5 kB SRAM test arrays were fabricated on the standard process version that supports the shorter gate length. The test SRAMs include an array without RBB and an array with RBB capability, with the latter configured with node SOURCE (see Figure 3.3b) biased at V_{SS} during these initial irradiations. Two bias conditions, $V_{DD} = 1.0$ V and 1.3 V, were used (see Figure 3.8).

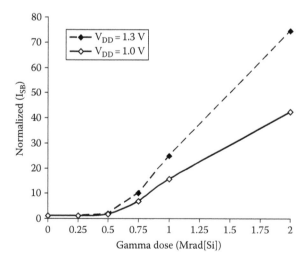

FIGURE 3.8 Measured effect of V_{DD} on the TID-induced standby I_{DD} (I_{SB}) increase in 90 nm 5kB SRAM array fabricated on the standard process. The 2 Mrad(Si) value is 75 times the initial I_{SB}. (After Clark, L. et al., *IEEE Trans. Nucl. Sci.*, 54(6), 2028, 2007.)

The I_{SB} is normalized to the initial values for each die, which exhibit substantial (and expected) die-to-die variations. The $V_{DD} = 1.3$ V TID-induced I_{SB} increase of 75 times is 1.8 times the TID-induced increase at $V_{DD} = 1.0$ V. This indicates that if sufficient SEU tolerance can be provided, there is substantial TID response benefit to low V_{DD} operation. The TID leakage increase at 1 Mrad(Si) is larger on the LSP process is in whole or part due to the substantially lower original leakage.

3.3.5.2 Impact of TID on Cell Margins

Using the test structure described earlier in Section 3.3.4 (see Figure 3.6), a 90 nm LSP 4 kB test array of unhardened SRAM cells was exposed to Co-60 radiation at an approximate rate of 20 rad/s [17]. During this exposure, the BL switch points were measured continuously, and during irradiation (after each measurement switched the cells), the cells were rewritten to the state where node C is 0 V and node CN is 1 V. The BL switch point response vs. the applied dose is plotted in Figure 3.9, where the SRAM node is being pulled high by the BL. It is unaffected until about 300 krad(Si), where TID-induced leakage becomes significant compared to the inherent leakage components, consistent with the full SRAM results in Figure 3.7. The TID impact on the measured switching voltage and hence cell write margins saturates near 1.5 Mrad(Si).

At doses of 1.5 Mrad(Si) and 3.0 Mrad(Si), both the BL and BLN switch points were measured, indicating a strong downward shift in the BLN switch point across the cells, again indicative of a shift in the cell effective P/N ratios due to the irradiation. The cells were irradiated with the gate of transistor N0 high, and as expected, this device exhibits the most degradation, that is, increased leakage due to TID. The SRAM access transistors, NP0 and NP1, are assumed to be largely unaffected, since they have 0 V gate bias most of the time. One key observation is a lower BLN voltage

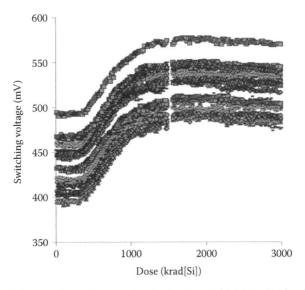

FIGURE 3.9 Behavior of sample two-edge (unhardened) SRAM cell trip points vs. TID. (After Clark, L. et al., *IEEE Trans. Nucl. Sci.*, 54(6), 2028, 2007.)

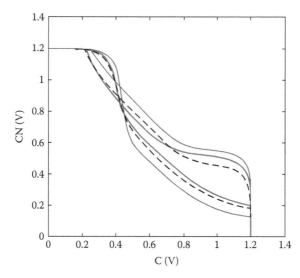

FIGURE 3.10 Simulated worst-case Monte Carlo derived read SNM pre- and postirradiation. The thin solid and thin dashed lines show the postirradiation SNM, while the thick gray lines show the preirradiation response. (After Clark, L. et al., *IEEE Trans. Nucl. Sci.*, 54(6), 2028, 2007.)

required to write the cell state, indicating diminishing write margin—that greater drive is required to write the cell in this direction over time. This result, consistent with an increase in the drive of transistor N0, presents a possible failure mechanism due to TID.

The impact on the SRAM cell read margins are shown in Figure 3.10, which compares the pre- and postirradiation SNM as simulated by changing the leakage of transistor N0 to match the TID measurement results. The test structure does not allow direct measurement of the SRAM read SNM, which must be inferred from the write margin measurement results. To determine the impact of TID on the read margin response, the NMOS response was modeled from transistor TID measurements on the same process. Two responses with the degradation on each NMOS pull-down transistor were simulated independently. Immediately evident is the closing of the larger "eye" post-TID. In one case (the dashed lines), the initially weaker NMOS pull-down transistor is made slightly stronger by the TID-induced leakage, and the worst-case read SNM is slightly improved (from 58 to 59 mV). The read SNM on the other node is diminished to 53.7 mV (note the smaller "eye" at the top outlined by the thin lines) when the TID-induced increased leakage is on the initially stronger NMOS pull-down transistor. Whether TID mitigation is necessary to maintain SRAM cell read margins is thus determined by the initial as-fabricated margins—a larger cell with large margins may still be smaller than that required by annular transistor layout and guard rings, as well as the TID environment expected.

The TID switching point response vs. irradiation dose of the RHBD cell of Figure 3.5 is shown in Figure 3.11. Since the leakage currents are completely mitigated, as indicated by I_{SB} measurements vs. TID, the switch points are stable up to 2 Mrad(Si). Clearly, allowing sufficient cell size, the RHBD techniques are effective.

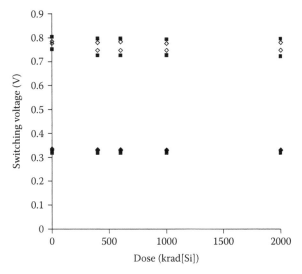

FIGURE 3.11 Sample SRAM cell trip points vs. TID for the all annular 90 nm NMOS SRAM cell. No variation due to irradiation is measured for this cell.

3.3.5.3 Type 4 Cell

In general, a worst-case experiment uses a high NMOS gate voltage to maximize TID effects. However, this condition simply cannot occur on the NMOS access transistors in an SRAM, as the WL decoder ensures only one can be active, and then only during one clock phase. To validate that WL bias at 0 V for all unaccessed cells is sufficient to mitigate TID-induced leakage in the NMOS pass devices NP0 and NP1, experiments were made using test transistors fabricated on both the 130 nm and 90 nm processes. On the former, four 0.28 μm width, minimum-length two-edge NMOS transistors connected in parallel with $V_{DS} = 1.2$ V and $V_{GS} = 0$ V, the relevant access transistor bias, exhibit TID-induced increase in I_{OFF} less than three times after exposure to 500 krad(Si). A 1.5 μm effective width annular transistor (the minimum for this geometry) has five times greater I_{OFF} current preirradiation than this narrower two-edge device. These experiments indicate that 130 nm SRAM cells using annular pull-down NMOS and two-edge pass transistors exhibit less total leakage current after exposure to 500 krad(Si), with less area than a cell using PMOS access transistors. The same experiments were carried out on transistors fabricated on a 90 nm foundry process to 1 Mrad(Si) and indicate that for the off-state bias condition, the TID-induced I_{OFF} increase is less than two times [46].

3.3.5.4 Type 1 Cell with RBB: Array Design Considerations

The low WL activity factor makes annular gates less important for the SRAM cell access transistors, as shown earlier experimentally, but TID-induced leakage remains an issue if two-edge NMOS pull-down transistors are used. This can be mitigated by applying RBB or RBB+SC. By setting V_{DD} to be 1.2 V and the external V_{SOURCE} to be 0.7 V, the SRAM cells have 0.5 V V_{DS} storing the SRAM cell state. The SRAM cell supply voltage can be varied by raising the NMOS sources (cell node SOURCE in Figure 3.3b and row by row SOURCE0 to SOURCE63 in Figure 3.12a) while

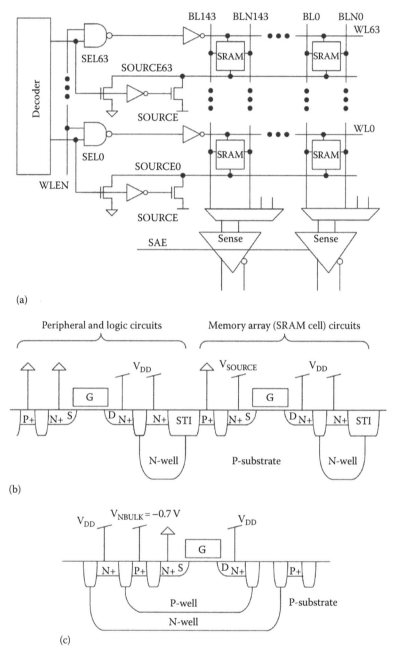

(a)

(b)

(c)

FIGURE 3.12 (a) Circuits providing dynamic RBB in the 130 and 90 nm SRAMs for NMOS transistor TID mitigation. (b) The connections of the well and substrates for the periphery are shown. (c) A triple-well process allows continuous negative NMOS transistor bulk bias. These well bias conditions were simulated in the arrays by applying high V_{DD} to the peripheral circuits and raising VSS above the bulk voltage in some of the TID experiments. (After Clark, L. et al., *IEEE Trans. Nucl. Sci.*, 54(6), 2028, 2007.)

maintaining the NMOS transistor bulk connection at 0 V. This allows RBB+SC to reduce the NMOS transistor leakage in the pull-down transistors N0 and N1 as well as significantly reducing I_{OFF} in transistors NP0 and NP1 through negative V_{GS}. Alternatively, since the channel surface potential is pinned, that is, the gate fields are unaffected by the bulk potential, V_{DD} and V_{SOURCE} can be raised to provide reduction in TID-induced leakage without affecting the V_{DS} and hence cell Q_{crit}.

Low cell voltages during operation reduce TID effects as shown earlier, but modern SRAM cells are not read stable at low voltages. Consequently, to employ reduced biases for TID mitigation, the cell bias must be changed dynamically to full V_{DD} during reads. The circuits providing this ability are shown in Figure 3.12. By driving the row SOURCE node to 0 V dynamically before the WL is selected, the SRAM cells in that row can be read without upset that might otherwise occur since SNM diminishes rapidly with decreasing V_{DD}. Sufficient address setup time ensures that the row SOURCE node is driven to 0 V before the WL is asserted. The raised source structure was chosen since it can apply RBB with power SC. This is applied dynamically to allow full read stability in the selected row. This configuration can also simulate a triple-well SRAM by using the appropriate bias conditions as shown in Figure 3.12b and c by holding the bulk (P substrate) at 0 V and making V_{SS} and V_{DD}, 0.7 and 1.7 V respectively.

3.3.5.5 Type 1 Cell with RBB: Transistor-Level Measurements

Four 0.28 µm wide 130 nm minimum-length NMOS transistors connected in parallel were irradiated with $V_{DS} = V_{GS} = 1.2$ V ($V_{DD} = 1.9$ V), V_S (the transistor source) = 0.7 V, and $V_{bulk} = 0$ V, the worst-case condition for the raised V_{SS} TID mitigation scheme. The I_{OFF} was measured preirradiation and after a total dose of 750 krad(Si). An operational SRAM will have data that change, so this was applied as 250 krad(Si) biased on ($V_G = V_{DD}$), 250 krad(Si) biased off ($V_G = V_S$), and then 250 krad(Si) biased on ($V_G = V_{DD}$). The 130 nm NMOS transistors exhibit an 8 times increase in TID-induced leakage, as opposed to over 200 times increase for $V_{DD} = 1.2$ V and no RBB applied as shown in Figure 3.13. The pull-down transistors account for about one-third of the total leakage, so when this result is combined with the access transistor TID response, the SRAM cell will exhibit about three times I_{OFF} increase after exposure to 750 krad(Si). The postirradiation I_{OFF} with $V_{SB} = 0.7$ V and $V_{GS} = V_{DS} = 0.5$ V is nearly one order of magnitude less than the preirradiation I_{OFF} at $V_{SB} = 0$ V and $V_{GS} = V_{DS} = 1.2$ V—by using dynamic source biasing, a conventional SRAM cell will exhibit less post-TID leakage than cells employing annular gates preirradiation. The I_{SB} for this condition is lower at 1 Mrad(Si) than the preirradiation annular (type 4) SRAM. The voltage-collapsed SRAM with RBB applied has lower I_{SB} at 1 Mrad(Si) than the conventional two-edge transistor SRAM (type 1) cell has preirradiation. We attribute the annular (type 4) SRAM cell I_{SB} increase to leakage under the field oxide at the two-edge access transistors, since as mentioned, no cells had guard rings between adjacent N-type source and drain diffusions.

3.3.5.6 Test SRAM Designs and Experiments

The 5 kB RHBD test SRAMs implemented in 0.13 µm bulk CMOS contain 4 kB for data and 1 kB for EDAC parity bits. A single 40-bit read and write port is organized

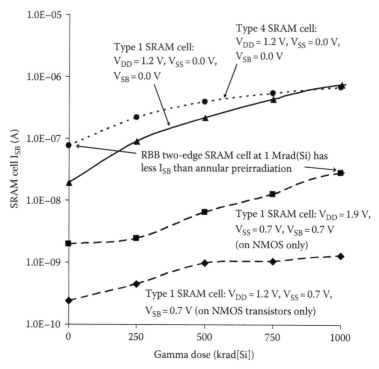

FIGURE 3.13 Co-60 irradiation results of different SRAM cell I_{SB} responses. Note that the type 4 cell, which has annular NMOS pull-down transistors, has higher I_{SB} preirradiation than the type 1 with RBB and full V_{DS} postirradiation. The higher preirradiation leakage of the type 4 eliminates much of its advantage over the type 1 (unhardened) cell. The type 4 cell has leakage increase attributable to the lack of guard rings blocking STI leakage paths. The type 1 with RBB+SC has little leakage increase and reduced I_{SB} at all dose levels. (After Clark, L. et al., *IEEE Trans. Nucl. Sci.*, 54(6), 2028, 2007.)

as 32 data bits plus 8 EDAC bits. EDAC protects against SEU in the array while dual-redundant control lines are used to detect and prevent SET data corruption due to WL SET, as described later. The 90 nm SRAM design utilizes similar circuits to apply RBB and supports the same (40, 32) single error correct and double error detect EDAC. Both interleave the storage bits by eight cells to avoid multiple bit error upsets in the same EDAC codeword. The 90 nm design does not mitigate SET-induced errors.

In commercial designs, the most important leakage-induced failure is the case where during a read operation, a single bit is driving the BL high, but leakage on the other cells is driving the same BL low, significantly slowing the BLN–BL voltage signal development. If the total BL leakage approaches the cell read current, small signal differential sensing may fail. This is even more important in memory designs that use single-ended sensing, since the output high BL node may register as a logic 0 after being discharged by leakage within the timing window. This failure mechanism is avoided in the designs here by using full swing differential sensing and relatively short BLs with 64 cells attached, as well as the high I_{ON}/I_{OFF} ratio of the foundry process.

The 90 nm SRAM design uses cross-coupled NAND gate set–reset latches to sense. This allows a robust timer-free sense, eliminating one timing signal that could otherwise be upset by a SET. The SRAM cell high node provides current through the on pass transistor (NP0 in Figure 3.3) clamping the BL reading a logic "1" at approximately $V_{DD} - V_{tN}$. Meanwhile, the BLN reading a logic "0" discharges completely to V_{SS}. The 130 nm design uses BL keepers to hold the nondischarging BL to V_{DD}.

3.3.5.7   Type 1 Cell with RBB: SRAM Measurements

Triple-well processes, which are common, allow RBB to be applied statically without collapsing V_{DS} (see Figure 3.12c). Full cell V_{DS} maintains the cell Q_{crit}, allowing better SEU response. The effect of this bias condition on TID response was investigated in an experiment comparing the 130 nm SRAM TID performance on a nontriple-well process with $V_{DD} = 1.9$ V, $V_{DS} = 1.2$ V, and $V_{SB} = 0.7$ V (referring to Figure 3.12, $V_{DD} = 1.2$ V and $V_{SOURCE} = -0.7$ V) to the case without RBB. The experimentally measured SRAM array I_{SB} vs. irradiation level on the 130 nm test chip (Figure 3.14) also clearly shows that RBB is a viable approach to mitigate TID-induced increase in I_{SB} up to 1 Mrad(Si). Since measuring I_{SB} encompasses all leakage components, both I_{OFF} and conduction under the STI field oxide is mitigated by RBB. This allows the use of the foundry cells. Both RBB and RBB+SC improve I_{SB} compared to the standard bias and RBB+SC reduces leakage sufficiently such that the total I_{SB} of the array drops below that of the SRAM control logic, indicated by the dashed line at the bottom of Figure 3.14. The control logic was not reverse-body biased. Consequently,

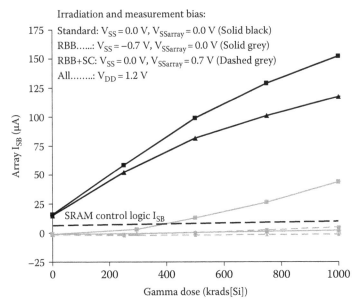

FIGURE 3.14   One hundred and thirty nanometer SRAM array I_{SB} vs. TID and irradiation bias for chips irradiated with and without RBB, and with RBB+SC. I_{SB} was measured at the irradiation bias. Irradiation was performed with the array programmed to a checkerboard pattern, and I_{SB} measured with the same (triangles) and opposite (squares) patterns. (After Mohr, K. et al., *IEEE Trans. Nucl. Sci.*, 54(6), 2092, 2007.)

further decreases in SRAM array I_{SB} will not provide significant improvements in the overall SRAM static power unless all circuits have RBB applied.

Note that using RBB+SC allows the postirradiation leakage to be less than the preirradiation leakage without it. On this 130 nm SRAM, some single bit failures were observed for the standard bias condition starting at 725 krads(Si) indicating that increased leakage currents had destabilized some of the SRAM cells, presumably those that were least stable to begin with. No failures were observed for the RBB+SC bias up to 1 Mrad(Si).

This TID mitigation approach was also investigated on a 90 nm bulk CMOS 5 kB SRAM using two-edge transistor cells. Test SRAMs were irradiated with $V_{DD}=0.9$ V (the non-RBB case, i.e., with $V_{SB}=0$ V), and with $V_{DD}=1.6$ V and node SOURCE—the source of the SRAM NMOS pull-down transistors (N0 and N1 in Figure 3.3b)—at 0.7 V and the P-type bulk at 0 V (the RBB case). This latter bias condition applies a 0.7 V NMOS RBB with the cell $V_{DS}=0.9$ V, the same as the non-RBB case. The SRAM array SOURCE node (see Figure 3.12a) current was measured with all 1s and with all 0s stored in the array to determine the NMOS pull-down transistor I_{OFF} pre- and postirradiation for both cases. The SRAM was written with all 1s during irradiation. The measured results are shown in Figure 3.15 for TID up to 1 Mrad(Si) [46]. While the non-RBB SRAM I_{SOURCE} increases by 10 times for array data opposite to that during irradiation, no increase is observed in the RBB I_{SOURCE} at that irradiation level. Between 0 and 500 krad(Si), the high intrinsic leakage delays the onset of noticeable TID impact to higher irradiation.

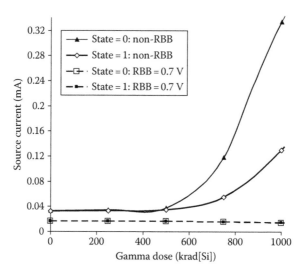

FIGURE 3.15 Measured current on the RBB two-edge 90 nm SRAM SOURCE node (I_{OFF} through the pull-down transistors) after Co-60 irradiation to 1 Mrad(Si) with $V_{GS} = 1.0$ V and RBB using $V_{SB} = 0.7$ V in the State = 1 condition. Measurement biases are given in the legend. A large increase in I_{OFF} is evident, exacerbated when the measurement is in the opposite state. Application of RBB fully mitigates the I_{OFF} increase. (After Clark, L. et al., *IEEE Trans. Nucl. Sci.*, 54(6), 2028, 2007.)

The same measurements with irradiation at $V_{DD} = 1.6$ V and node SOURCE = 0.6 V are shown in Figure 3.16 [46]. I_{SOURCE} with RBB of 0.5 V and 0.6 V shows the sensitivity to the amount of RBB bias applied. The cell $V_{DS} = 1$ V for both cases. Two key points are evident. First, the applied RBB strongly affects the measured I_{OFF}. Thus, RBB does not mitigate the trapped charge or TID-induced interface traps; it merely raises the parasitic edge transistor V_{th} sufficiently to alleviate the increased leakage. Second, the current required by node SOURCE can decrease with TID. This suggests that current can be delivered by another path separate from the V_{SOURCE} node. This current path is clearly dependent on the SRAM cell state during irradiation, despite its symmetrical nature.

The reduction in I_{SOURCE} at high TID suggests that the cell bias current is being fed from another path, other than the test chip pin (see Figure 3.16). This is surmised to be due to leakage under the STI from a V_{SS} node to the SOURCE node. The small magnitude of the current difference between the two cell states suggests that this will not limit chip level I_{SB}. Previous work has reported that the SRAM source can be completely floated without the SRAM losing state [50]. This was proven on our 90 nm SRAM as well—gate leakage is sufficient to maintain the bias in a standby mode. I_{SB} measurements of irradiated NMOS transistors suggests that much of the large leakage increase in the 5 kB arrays is under the STI between N-type diffusions, for example, between NMOS drains and the N-well. This further validates the conclusion that the RBB is effective in mitigating this leakage component in the 90 nm process and that this under STI component is responsible for the slight I_{SOURCE} at high TID. Figure 3.16 shows higher I_{SB} for arrays measured without RBB

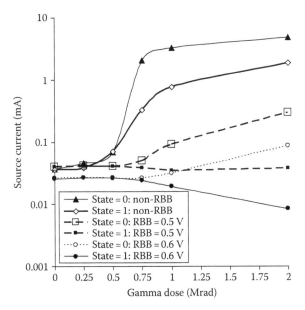

FIGURE 3.16 Measured SRAM I_{SOURCE} on the 90 nm 5 kB SRAM. The part was irradiated with $V_{SOURCE} - V_{SS} = 0.6$ V and $V_{DD} = 1.6$ V. (SRAM cell $V_{DS} = 1.0$ V). (After Clark, L. et al., *IEEE Trans. Nucl. Sci.*, 54(6), 2028, 2007.)

but irradiated with it. This response, similar to that of the 130 nm SRAM, indicates that the RBB does not mitigate STI interface charging or trap formation, but that the net effect is again mitigated by the RBB application.

3.3.5.8   90 nm Transistor-Level Response

Two-edge transistor arrays were also measured pre- and postirradiation to help determine details of the TID leakage increase in the SRAMs. These arrays have the same narrow-width NMOS transistors as the SRAM cells. The NMOS transistor arrays were irradiated with 0.7 V RBB applied and $V_{DS} = 0$ V, and with no RBB and $V_{DS} = 0$ V, with high transistor gate voltage applied, that is, $V_{GS} = 1.0$ V to determine the worst-case response. The results of measurements with RBB applied during irradiation but with $V_{SB} = 0$ V during the measurement sweeps showed that application of RBB during irradiation subtly enhances the increase in transistor I_{OFF} due to TID. Since only one SRAM row is accessed at a time, the impact of a higher I_{OFF} in the non-RBB condition is negligible, and RBB application reduces the leakage by much more than the actual STI oxide degradation.

When irradiated, the measured SRAM cell leakage increase is greater than the NMOS transistor I_{OFF} increase experimentally measured on transistors. If the primary SRAM TID effect was increased parasitic NMOS drain-to-source leakage increase, the transistor increase would be higher than that of the SRAM. Since no P-type guard rings were used (all SRAM cells use the layout in Figure 3.3c), this suggests that leakage under the STI field oxides is a significant contributor. Subsequent work has shown that the field oxide FET formed by the polysilicon bridging from the NMOS transistors to the N-well is a dominant TID-induced leakage path [49].

When used to reduce circuit standby leakage, the actual leakage improvement can be limited by both gate leakage and drain-to-source leakage at the drain edge, either I_{GIDL} or I_{ZENER} [51]. All are direct band-to-band tunneling effects—the former through the thin oxide and the latter two are due to sharp band bending caused by the steep doping profile at the drain-to-bulk transition region. The steep doping profiles are from halo implants used to control short channel effects [44,52]. Since the RBB bias creates higher drain-to-bulk bias conditions, it is important to determine if this leakage component will limit the available improvement that can be provided by using RBB. For example, if the I_{ZENER} increase with RBB is larger than I_{OFF}, RBB application will actually only mask TID-induced leakage by increasing the baseline value. Experiments on the 90 nm foundry process showed that this was not the case. Since doping increases exponentially as processes are scaled, and the precise fields are process dependent, RBB TID mitigation on future processes may be limited by this mechanism.

3.3.6   SINGLE EVENT EFFECTS IN UNHARDENED SRAM

SEE was investigated on a 90 nm 1.2 Mb unhardened SRAM using the ion beam at the SEE facility at Texas A&M University. Since no SET mitigation techniques are used, this design provides a baseline for comparison with a SET mitigated design. This design uses small signal sensing and conventional circuits commonly used in commercial SRAMs, with two exceptions. First, tighter well and substrate tap spacing to avoid SEL in the beam testing were used. Second, RHBD I/O was used to avoid test failures due to the TID-induced I/O failure before the core TID effects could be seen.

Figure 3.17 shows some measured SEU patterns for the LET = 59 MeV-cm²/mg beam normal to the surface of the die ($0°$ angle) with different stored patterns. Each strike is noted by one of eight different shades. The MBU detection algorithm assumes that bits more than eight cells apart are from different strikes, as the cycle count may not be indicative, since multiple particles may strike between read sweeps through the array. Note the prevalence of MBUs. Figure 3.17a shows that for checkerboard patterns, diagonal upsets predominate. This is due to the cell nodes storing the same logic one or zero being on the diagonal—for a hit between them, both can collect charge and be upset by one particle strike. Note that this can occur for both PMOS and NMOS collection, but referring to Figure 3.3c for strikes at opposite ends of the cells. The figure also clearly shows one row completely upset. This is due to an SET-induced WL assertion, which wrote the BL values into the cells. Luckily, the test pattern had alternating rows of 010101…. and 101010… (since it is a checkerboard) so this error could be caught. Figure 3.17b shows the pattern measured with vertical stripes programmed into the array, while Figure 3.17c shows the resulting patterns from a solid 0s pattern. In the latter, since the BLs alternate as BL, BLN, BLN, BL, up to four adjacent diffusions may collect charge simultaneously, as evident. Stripes predominate in the stripes case. The likelihood that a strike generates an MBU is thus dependent on the stored data pattern, with the likelihood of a 4-bit upset rising from under 5% for the checkerboard and stripe patterns, to over 16% for the solid pattern. No 5-bit upsets were detected for the two former cases, but they composed over 15% of the strikes in the solid pattern $0°$ angle tests at the same LET. Again, knowing the exact physical organization of the array down to whether cells are tiled or folded in the horizontal direction is critical to accurate SEU analysis.

How many bits are upsets by a given strike vs. beam incident angle with the checkerboard pattern, again at LET = 59 MeV-cm²/mg, is shown in Figure 3.18. Here, 1- and 2-bit upsets predominate at normal incidence, while at $42°$, the majority of hits upset two or more cells. At the higher angles (see Figure 3.18d), most hits are 2-bit MBU, with up to 5-bit upset by one particle strike.

Figure 3.19 shows the measured SRAM cell cross section vs. effective LET (LET_{eff}) at different V_{DD} voltages. As expected, since increasing V_{DD} raises the cell Q_{crit}, the cross section diminishes with increasing V_{DD} at low LET. However, at high V_{DD}, as indicated by the points connected by lines, the cross section rises considerably at LET_{eff} above 70 MeV-cm²/mg. This is due to enhanced charge generation due to amplification by parasitic bipolar transistor action, which can cause very large MBU extents, particularly down SRAM wells [53].

3.3.7 SINGLE EVENT EFFECT MITIGATION

In this section, the SEE mitigation circuits implemented in the 130 nm design and their operation are described.

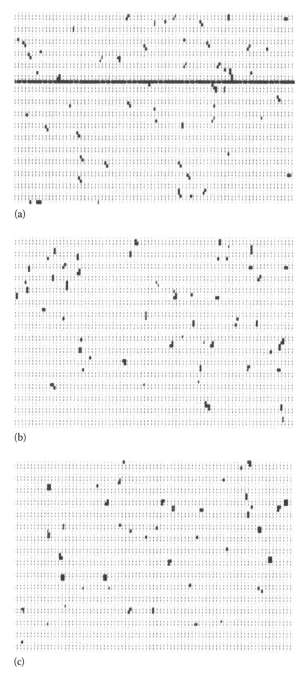

(a)

(b)

(c)

FIGURE 3.17 MBU extent vs. stored memory patterns observed in testing the unhardened 1.2 Mb 90 nm SRAM at normal beam incidence for (a) checkerboard, (b) stripes, and (c) all zeros patterns, respectively. Note the entire row disrupted, presumably by a SET that in turn asserted a WL, which overwrote the contents.

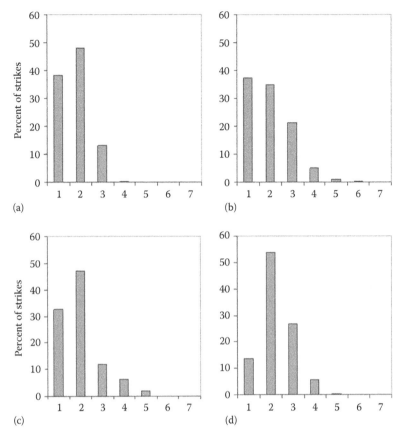

FIGURE 3.18 The effect of particle angle on MBUs in the unhardened 90 nm 1.2 Mb SRAM: (a) At 0°, 1-, and 2-bit upsets predominate, with few 3- and 4-bit upsets. (b) As the angle increases to 42°, a larger number of 3-bit upsets occurs, (c) until at 53° (d) and 65°, MBUs predominate.

3.3.7.1   130 nm SRAM Design with RBB+SC Support and SEE Mitigation

As mentioned earlier, the 130 nm 5 kB RHBD SRAM used dual-redundant control lines to detect and prevent SET data corruption due to control or WL SET. The dual-redundant control logic SET mitigation used here imposes no absolute maximum clock frequency limits and allows operating frequencies above 500 MHz in this implementation.

Each row in the SRAM array is controlled by 1 of 128 dual-redundant WL drivers labeled L for left and R for right (see Figure 3.20). Separate decoders provide SelLx and SelRx with timing set by WL enable signals RowENLx and RowENRx, respectively. Each left dual-redundant driver for row x controls the WLLNx and, in part, the SOURCEx signals for one row. The redundant drivers are spatially separated to reduce the probability of a single ionizing particle strike affecting both of them. The test array uses RBB+SC (optionally RBB) to determine its value for TID mitigation on this 130 nm process.

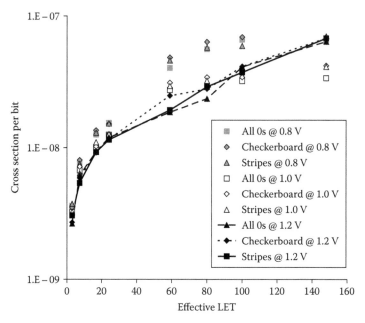

FIGURE 3.19 Ninety nanometer 1.2 Mb SRAM cross section vs. LET_{eff}. Note that at low LET, the cross section is higher at lower V_{DD}. At very high LET, higher V_{DD} increases the cross section.

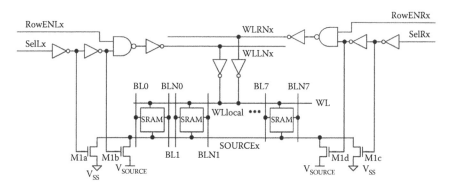

FIGURE 3.20 SRAM row design. Each subgroup of 8 bits is driven by two local WL drivers. The left side row driver (not shown) is an exact mirror image of this one, but drives WLbarL.

As described earlier, when a row is inactive, it is biased to the higher V_{SOURCE} voltage that applies the RBB as the NMOS bulks are all at $V_{SS} = 0$ V. During a read or write cycle, SOURCEx is driven to V_{SS} through transistor M1a in order to ensure read stability and fast writes. The row drivers must be immune to an SET that could cause the SOURCEx node voltage to rise above V_{SOURCE}, which would collapse the SRAM cell supply voltage, potentially upsetting the stored state in the entire row. SETs that drive the SOURCE node voltage low are not a concern as they momentarily increase that row's SRAM cell supply voltage magnitude. To avoid a strike raising a row's SOURCE bias node SOURCEx, it is connected to only N-type diffusions. These can only collect ionizing radiation-deposited electrons and thus only drive SOURCEx to a lower potential. Strikes on upstream logic that controls the bias are also a concern. For example, referring to Figure 3.20, enabling transistor M1a in one of the two redundant row drivers and M1d in the other creates contention between them. In this case, SOURCEx takes on an intermediate voltage between V_{SS} and V_{SOURCE} reducing read speed and margin. Two approaches are taken to ensure that this does not affect functionality. First, the M1a and M1c transistors are sized wider than M1b and M1d, keeping the contention voltage low. Second, this SOURCEx contention condition bias value is used as the worst case when determining read margin and speed.

To avoid an ionizing particle strike asserting an inactive WL signal during a read operation as discussed in Section 3.2.2, the design uses two redundant active low WL signals per row, labeled as WLLNx and WLRNx that in turn control 40 local WL (labeled WLlocal) signals in each row (see Figure 3.20). WLRNx is driven by the right row driver and signal WLLNx by the left row driver. These two redundant signals are combined locally every 8 bits, driving the WLlocal signals, which control eight SRAM cells. An SET assertion of a single WLlocal signal corrupts at most 8 bits. Since each of the 8 bits resides in a different EDAC codeword, all 8-bit errors are correctable. Unlike the design in Figure 3.1b, which could still activate all local WLs by incorrectly asserting the global WL due to an SET, if either the WLbarL or WLbarR signal is corrupted due to an SET, a contention condition is created in the local WL driver inverters. The transistor sizing in the local WL drivers ensures that under contention, the WLlocal signals will not rise to a high enough voltage to write the SRAM cells they control. Consequently, an SET-enabled WL signal in a row that is not active cannot cause a false write. This local WL driver circuit was chosen over an AND gate because it is smaller. An erroneously disabled WL signal is detected, as described in the following text, allowing the write or read operation to be repeated. Of course, the controlling circuitry micro-architecture must comprehend this condition, by buffering write data for a retry and appropriately rerunning the operation as needed.

An SET-induced WL assertion is detected by two additional columns connected only to WLLN or WLRN signals, rather than the local WLs. These cells always discharge the BLs in their column and are read on both read and write operations. One of WLbarL or WLbarR incorrectly asserted is indicated by either being incorrect. In this case, the cycle is repeated. Of course, a "false-positive" error can be induced by an ionizing particle strike on the BL itself, in which case the data are correct, but rewritten nonetheless. BL development is much faster during a write, owing to the stronger write drivers, than during a read. This ensures that a write has successfully completed if both WLchk outputs are valid.

3.3.7.2 SRAM Column Circuits

Similar to the WL protection, dual-redundant column decoders generate the Y multiplexer control signals ColEnLWy and ColEnRWy for each of 16 words y (8 above and 8 below the column sense and write circuits). Two decoders—one on the left and one on the right form a dual-redundant pair, with one pair for each of the top

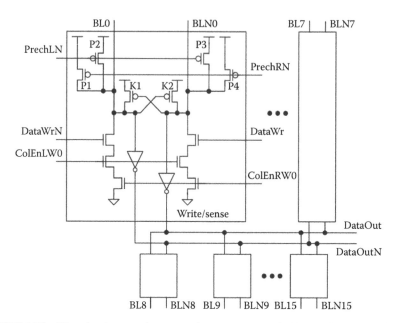

FIGURE 3.21 BL redundant precharge transistors, cross-coupled keeper PMOS transistors, and sixteen-to-one 1-bit column multiplexer with redundant select and write control. If a precharge fails to turn off due to an SET, this condition is detected by the read/write detection columns.

and bottom subarrays. There are sense and write circuits in each column of this design. The ColEnL and ColEnR signals are combined as shown in Figure 3.21. If either control signal is corrupted, an incorrect write cannot be generated, but a valid write may be aborted. Such a write abort is detected since the write data are monitored by the read sense circuits as it is being driven on to the BLs. These data are sampled at the end of the operation clock phase and compared to the data that were to be written. Any difference triggers an error, allowing the write to be repeated.

The read sensing is single ended using full-swing BLs. Both BL and BLN signals are sensed by high skew tristate inverters as shown in Figure 3.21. The outputs of these tristate inverters form the column (Y) output multiplexer. A SET that asserts one inadvertently will upset at most one output bit, which the EDAC will correct. Dual-redundant signals ColEnLW0 and ColEnRW0 enable DataOutN and DataOut, respectively, if word 0 is being read. If either of the ColEnLW0 or ColEnRW0 signals is corrupted due to a SET, the DataOut and DataOutN signals in a column will match, that is, one not reflecting the discharged BL as appropriate, signaling that the read must be repeated. EDAC may also detect this but cannot be guaranteed to do so. Other dynamic BL read errors caused by SETs, for example, on the precharge signals PrechLN and PrechRN are detected similarly.

Cross-coupled keeper transistors K1 and K2 shown in Figure 3.21 ensure that once the bit cell discharges either of the two BLs, the opposite line maintains a full rail logic "1." Dual-redundant precharge circuits preclude SET events from deasserting PrechLN or PrechRN during BL precharge cycles.

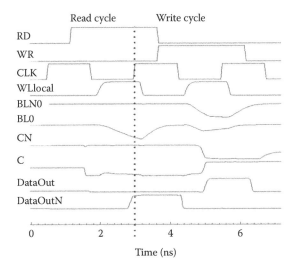

FIGURE 3.22 Simulation results showing the SRAM read and write cycles. Note the large BL read swing in the read cycle and that in the write cycle, the BL discharge begins to discharge until the cell is written, whereupon it is restored by the BL keeper transistors.

3.3.7.3 SRAM Operation with RBB+SC

Figure 3.22 shows a simulated read cycle followed by a write cycle. The SRAM reads and writes in the low clock phase, with BLs precharged in the high clock phase. At about 1.5 ns, node C voltage of the storage cell node holding a logic 0 transitions from the elevated SOURCE voltage to V_{SS} in preparation for the read cycle. Note that this is controlled by the address input, and at lower clock frequencies, this may occur much earlier. At 2 ns, node C rises—this is due to the read current, which reduces the cell stability during a read as discussed in Section 3.3.1. The first falling clock signal CLK edge initiates a read of a stored logic 0 resulting in each of the 40 WLlocal nodes being asserted high, followed by the BL discharging and a logic 1 driven out on DataOutN. The second falling CLK edge at about 4.2 ns initiates a write cycle. Here a logic 1 is written into the SRAM cell, inverting the C and CN storage node signals.

3.3.8 EXPERIMENTAL SEE MEASUREMENTS

While both RBB and RBB+SC improve standby leakage, they also reduce the SRAM cell drive strength, reducing Q_{crit} and making the cell more susceptible to SEU. The SEU impact of varying the V_{SS} and V_{SOURCE} potentials was quantified by cyclotron measurements.

Figure 3.23 compares the standard bias, $V_{SOURCE} = 0.0$ V, SRAM cell cross sections to those with RBB with V_{SS} voltages of −0.4, −0.6, and −0.8 V. No increase in cross section is observed for $V_{SS} = -0.4$ V. Measurements with $V_{SS} = -0.6$ and −0.8 V exhibit up to a 15% SRAM cell cross-section increase at high effective LET. This is expected due to higher NMOS V_{th} as well as extended depletion regions, which improve funneling efficiency [6]. However, most of the change should occur

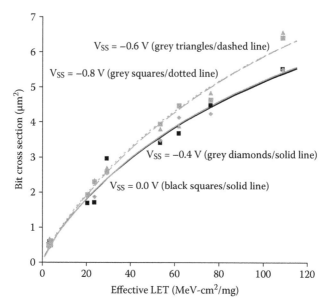

FIGURE 3.23 Measured RBB effect on SRAM cross section vs. effective LET and bias: $V_{DD} = 1.2$ V, $V_{SOURCE} = 0.0$ V, $V_{SS} = 0.0, -0.4, -0.6,$ and -0.8 V. (After Mohr, K. et al., *IEEE Trans. Nucl. Sci.*, 54(6), 2092, December 2007.)

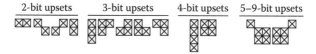

FIGURE 3.24 MBU patterns at the standard bias $V_{DD} = 1.2$ V, $V_{SOURCE} = 0.0$ V, $V_{SS} = 0.0$ V, that is, no RBB applied. (After Mohr, K. et al., *IEEE Trans. Nucl. Sci.*, 54(6), 2092, 2007.)

as RBB is applied, that is, from 0.0 to -0.4 V, where the V_{th} and depletion depth are most affected by the applied back bias, since it increases with the square root of V_{BS}. The MBU patterns produced by these tests, where no RBB is applied, are shown in Figure 3.24. At these angles, limited by shielding of the package to less than 60°, the MBU extent is limited.

Figure 3.25 shows the effect of RBB+SC on the measured SRAM cell cross sections at V_{SOURCE} biases of 0.4, 0.6, and 0.8 V relative to the standard bias of $V_{SOURCE} = 0.0$ V. RBB+SC has a significant effect on bit cell cross section. Cross section increases less than 60% for a V_{SOURCE} potential of 0.4 V, while at V_{SOURCE} of 0.8 V, the SEU cross section triples. This is easily attributable to the lower V_{GS} and V_{DS} of the transistors maintaining the SRAM cell state in these bias conditions, which significantly reduce Q_{crit}. Additionally, due to multi-bit errors (MBEs) at these biases, the cell cross section is considerably larger than the physical extent of one SRAM cell.

The bias dependence of MBUs was examined to ensure that increases in the bit cell cross section due to changes in V_{SOURCE} bias can be effectively mitigated by increasing the EDAC scrub frequency. Required scrub frequencies are quite low [54],

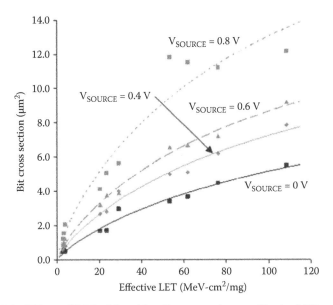

FIGURE 3.25 Effects of RBB+SC on bit cell cross section vs. effective LET and biases of $V_{DD} = 1.2$ V, $V_{SS} = 0.0$ V, with $V_{SOURCE} = 0.0$, 0.4, 0.6, and 0.8 V. (After Mohr, K. et al., *IEEE Trans. Nucl. Sci.*, 54(6), 2092, 2007.)

so small cross-sectional increases can be easily dealt with. MBUs whose extent spans across EDAC codewords at these low incident ion angles would result in an increase in the uncorrected error rate to unacceptable levels—recall that in space any angle is possible, so limiting the angles that can produce large MBU extent is the goal. The error cross section with error pattern size as a parameter for the standard bias ($V_{DD} = 1.2$ V, $V_{SS} = V_{SOURCE} = 0.0$ V) is shown in Figure 3.26. Single bit errors dominate at low LET. Above LET = 30 MeV-cm²/mg, strikes are more likely to result in MBUs causing the frequency of single bit upsets to drop and 2-, 3-, and 4-bit upsets to increase as shown. The 8-SRAM bit cell codeword interleaving used in this SRAM appears sufficient at the incident ion angles below 60°, which was limited by shielding of the IC package.

Using RBB+SC produces new MBU phenomena. Figure 3.27 shows that this significantly increases the frequency of MBUs with more than 5-bit upsets. Raising the V_{SOURCE} voltage to only 0.4 V increases the SRAM cell cross section MBU extent and frequency, allowing very large MBUs not observed without RBB+SC. The largest MBU observed at this bias, with 0.4 V RBB applied and 0.8 V across the SRAM transistors, was 11 bits as shown in Figure 3.28. The large MBUs tend to be long and slender, and oriented in the BL direction. Since words on different rows of this SRAM are always in different EDAC codewords, all codewords are still correctable as only one upset bit resides in each. Additionally these upsets cross many N-well boundaries. The N-wells should provide favorably biased charge collection nodes that collect deposited charge and thus mitigate upsets. We believe that the source of such long slender errors is ion strikes on the BL or write driver diffusions, causing the BL to glitch below the WL voltage. This is a classical signaling noise

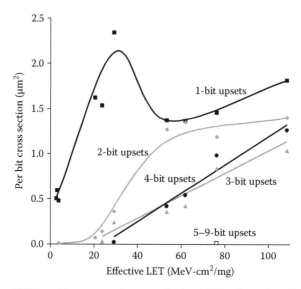

FIGURE 3.26 MBU per bit cross section vs. effective LET and number of upsets per particle strike at the $V_{DD} = 1.2$ V, no RBB applied. (After Mohr, K. et al., *IEEE Trans. Nucl. Sci.*, 54(6), 2092, 2007.)

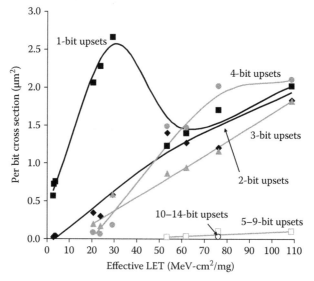

FIGURE 3.27 Effects of RBB+SC on MBU per bit cross section vs. effective LET and number of cells upset per particle strike for $V_{DD} = 1.2$ V, $V_{SOURCE} = 0.4$ V, $V_{SS} = 0.0$ V. The cross section of 5 or more bits upset becomes noticeable at $LET_{eff} > 50$ MeV-cm²/mg. (After Mohr, K. et al., *IEEE Trans. Nucl. Sci.*, 54(6), 2092, 2007.)

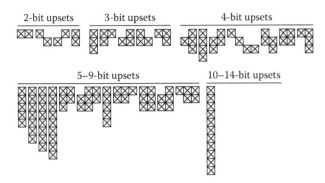

FIGURE 3.28 MBU patterns observed for V_{DD} = 1.2 V, V_{SOURCE} = 0.4 V, V_{SS} = 0.0 V, that is, 0.4 V RBB bias. The periphery circuits can drive the BLs to 0 V, below the SRAM access transistor V_{th}, which may account for the large vertical MBUs. (After Mohr, K. et al., *IEEE Trans. Nucl. Sci.*, 54(6), 2092, 2007.)

scenario that causes the access transistors of multiple cells to be asserted on as their gates are at V_{SS} = 0.4 V and sources glitch to 0 V or less, writing multiple cells in the same column.

Further reductions in cell storage voltage by raising V_{SOURCE} increase the SEU cross section of the SRAM cell and affect the size and shape of single strike MBEs. The trend continues, and when V_{SOURCE} is raised to 0.8 V, very long thin errors are observed extending over all 64 bits in a column, that is, the entire BL lengths were observed. Consequently, while very effective at mitigating TID, RBB+SC must be used with caution as it can allow large MBU increases. RBB alone, however, appears to be effective at mitigating TID and does not appreciably affect the SEU rate.

3.4 CONCLUSIONS

The SRAM cell designs presented here were designed with similar read and write margins to ensure a fair comparison of the size impact of RHBD. Test layouts of the more promising designs show that RHBD SRAM cells using annular NMOS or PMOS access transistors are at least three times and potentially greater than five times larger in area than the foundry-optimized unhardened cells. It is worth noting that many tradeoffs are related. For instance, the use of two-edge access transistors may make the use of P-type guard rings superfluous, as the guard ring cannot mitigate flow between the two-edge transistor source and drain. In Co-60 accelerated TID experiments, 130 nm SRAM arrays showed 75 times I_{SB} increases at 2 Mrad(Si), and a 90 nm SRAM fabricated on an LSP process, with a fully commercial design style, exhibited 131 times I_{SB} increase after 1 Mrad(Si).

These clearly indicate that some form of mitigation is necessary to limit I_{SB} increase at high, for example, Mrad level doses. Of course, the narrow SRAM transistors provide a worst case, and spaceborne ICs with low memory content may find such large I_{SB} increases acceptable. Additionally, most satellite requirements are met with lower specifications, for example, 300 krad(Si), which modern sub-100 nm processes may provide intrinsically. However, such a choice must be made cautiously.

Recall from Section 3.3.5.7 that experiments showed that the cell stability can be affected by TID, where the 130 nm unhardened SRAM arrays had TID-induced bit failures starting at 750 krad(Si). Using a novel test structure, the switching points of SRAM cells using two-edge transistors were shown to change considerably as they were dosed to 1.6 Mrad(Si). Conversely, no cell switch point changes or I_{SB} increase was observed for a fully annular NMOS P-type guard ringed 90 nm SRAM cell design. However, such larger hardened cells do have naturally higher SRAM leakage as a consequence of their wider transistors.

Measurements of fabricated 130 nm and 90 nm transistors and SRAM cells before and after TID irradiation indicate that two-edge NMOS access transistor cells are superior to PMOS access transistor designs at low radiation levels, that is, those below 500 krad(Si) and are probably adequate to higher doses. Experiments show that increasing SRAM V_{SS} to apply RBB reduces postirradiation leakage at 1 Mrad(Si) in conventional cells below the preirradiation leakage for the annular pull-down cell. At full $V_{DD} = 1.2$ V, SEE is not adversely affected, and at 1 Mrad(Si) at the same bias conditions, the postirradiation two-edge 130 nm SRAM cell I_{SB} is below the preirradiation level for the annular design. The measurements show that relatively low, that is, on the order of 500 mV of RBB is sufficient to mitigate TID-induced I_{SB} increase up to 1 Mrad(Si) on a 130 nm process. RBB+SC was shown to introduce a novel SEE failure where all cells on a single BL are upset. While the bits are all in separate EDAC words in any rational array organization, this may still be problematic, as large MBUs affect the required scrub rates. However, the RBB scheme appears effective at essentially no SEE penalty.

TID measurements of an operating 90 nm 5 kB SRAM show that RBB is effective in limiting I_{OFF} increase at least to 1 Mrad(Si) and probably at higher doses, where the current increases were still small. High intrinsic I_{OFF} and gate leakage on the 90 nm process limit the overall current savings. Concurrently, the high leakage floor masks TID-induced I_{OFF} increase until higher, that is, greater than 500 krad(Si) irradiation levels, suggesting that two-edge cells without RBB may be acceptable for many hardened systems. If RBB is used, the experiments presented here show that there is latitude in the choice of V_{DS} magnitude when RBB is applied. This will be a function of the required SEE hardness and the impact of V_{DS} on the cell Q_{crit}. The experiments presented here have also shown that band-to-band tunneling at the junction edge does not limit the use of RBB.

Fabricated line widths have moved considerably beyond the lithographic generation. For example, 193 nm lithography is used to fabricate 35 nm polysilicon gates [55] in production by using resolution enhancement techniques and phase shift masks. Consequently, support for the polysilicon shapes required for RHBD enclosed geometry gates on sub-90 nm processes has ceased. RBB applied to the NMOS transistors promises a potential RHBD approach that is compatible with such highly scaled fabrication processes, not just for SRAM, but for logic as well.

Simulation studies have shown that the logic delay and active power increase over an unhardened design when using NMOS RBB is less than that when using enclosed geometry transistors [32]. The former causes less than 5% increase in logic delay, at reduced leakage, and similar active power, compared to a commercial two-edge-only design. Nearly all modern ICs use higher I/O voltages for compatibility and lower scaled V_{DD} in the core logic transistors. For example, $V_{DDIO} = 1.8–2.5$ V, $V_{DD} = 1.2$ V (the core V_{DD}), and $V_{SS} = 0$ V (gnd) are common. RBB can easily be applied to an entire IC to provide NMOS transistor TID hardness by placing the core circuits in a domain between $V_{DDIO} = 1.8$ V and $V_{SS(CORE)} = 0.6$ V. The resulting IC still as three power supplies as before. It is straightforward to convert the I/O circuit level shifters, which presently convert the upper rail between V_{DD} and V_{DDIO} to convert the lower rail between $V_{SS(CORE)}$ and V_{SS}.

REFERENCES

1. International Technology Roadmap for Semiconductors, 2003. http://www.itrs.org [online].
2. S. Rusu, J. Stinson, S. Tam, J. Leung, H. Muljono, and B. Cherkauer, A 1.5-GHz 130-nm itanium-2 processor with 6-MB on-die L3 cache, *IEEE J. Solid-State Circ.*, 38(11), November 2003, 1887–1895.
3. J. Chang et al., A 130-nm triple-Vt 9-MB third-level on-die cache for the 1.7-GHz Itanium-2 processor, *IEEE J. Solid-State Circ.*, 40(1), January 2005, 195–203.
4. J. Wuu et al., The asynchronous 24MB on-chip level-3 cache for a dual-core itanium family processor, *Proc. Int. Solid-State Circ. Conf.*, 48, 2005, 488–489.
5. S. Rusu et al., A dual core multi threaded xeon processor with 16MB L3 cache, *IEEE Int. Solid-State Circ. Conf.*, 2006, 315–324.
6. J. Srour and J. McGarrity, Radiation effects on microelectronics in space, *Proc. IEEE*, 76(11), November 1988, 1443–1469.
7. E. Stassinopoulos and J. Raymond, The space radiation environment for electronics, *Proc. IEEE*, 76(11), November 1988, 1423–1442.
8. S. Kerns, B. Shafer, L. Rockett, J. Pridmore, D. Berndt, N. van Vonno, and F. Barber, The design of radiation-hardened ICs for space: A compendium of approaches, *Proc. IEEE*, 76(11), November 1988, 1470–1509.
9. G. Anelli et al., Radiation tolerant VLSI circuits in standard deep submicron CMOS technologies for the LHC experiments: Practical design aspects, *IEEE Trans. Nucl. Sci.*, 46(6), December 1999, 1690–1696.
10. H. Weaver, C. Axness, J. McBrayer, J. Browning, J. Fu, A. Ochoa, and R. Koga, An SEU tolerant memory cell derived from fundamental studies of SEU mechanisms in SRAM, *IEEE Trans. Nucl. Sci.*, 34(6), December 1987, 1281–1286.
11. J. Schwank, M. Shaneyfelt, B. Draper, and P. Dodd, BUSFET—A radiation-hardened SOI transistor, *IEEE Trans. Nucl. Sci.*, 46(6), December 1999, 1809–1817.
12. S. Liu et al., The effect of active delay element resistance on limiting heavy ion SEU upset cross-sections of SOI ADE/SRAMs, *IEEE Trans. Nucl. Sci.*, 54(6), December 2007, 2480–2487.
13. R. Lacoe, J. Osborne, R. Koga, and D. Mayer, Application of hardness-by-design methodology to radiation-tolerant ASIC technologies, *IEEE Trans. Nucl. Sci.*, 47(6), December 2000, 2334–2341.
14. T. Oldham and F. McLean, Total ionizing dose effects in MOS oxides and devices, *IEEE Trans. Nucl. Sci.*, 50(3), June 2003, 483–499.
15. H. Barnaby, Total-ionizing-dose effects in modern CMOS technologies, *IEEE Trans. Nucl. Sci.*, 53(6), December 2006, 3103–3121.
16. J. Felix, P. Dodd, M. Shaneyfelt, J. Schwank, and G. Hash, Radiation response and variability of advanced commercial foundry technologies, *IEEE Trans. Nucl. Sci.*, 53(6), December 2006, 3187–3194.

17. X. Yao, N. Hindman, L. Clark, K. Holbert, D. Alexander, and W. Shedd, The impact of total ionizing dose on unhardened SRAM cell margins, *IEEE Trans. Nucl. Sci.*, 55(6), December 2008, 3280–3287.

18. T. Calin, M. Nicolaidis, and R. Velazco, Upset hardened memory design for submicron CMOS technology, *IEEE Trans. Nucl. Sci.*, 43(6), December 1996, 2874–2878.

19. R. Koga, K. Crawford, P. Grant, W. Kolasinski, D. Leung, T. Lie, D. Mayer, S. Pinkerton, and T. Tsubota, Single ion induced multiple-bit upset in IDT 256K SRAMs, *Proc. RADECS*, September 1993, 485–489.

20. C. Underwood, R. Ecoffet, S. Duzeffier, and D. Faguere, Observations of single-event upset and multiple-bit upset in non-hardened high-density SRAMs in the TOPEX/ Poseidon orbit, *IEEE Rad. Effects Data Workshop*, July 1993, 85–92.

21. D. Mavis and P. Eaton, Soft error rate mitigation techniques for modern microcircuits, *Proc. IRPS*, 40, 2002, 216–225.

22. W. Jenkins, R. Martin, and H. Hughes, Characterization of an ultra-hard CMOS 64K static RAM, *IEEE Trans. Nucl. Sci.*, 34(6), Part I, December 1987, 1455–1459.

23. Z. Kai, L. Zhongli, Y. Fang, X. Zhiqiang, and H. Genshen, Radiation Hardened 128K PDSOI CMOS Static RAM, *Proc. ICSICT*, 2006, 1922–1924.

24. P. McDonald, W. Stapor, A. Campbell, and L. Massengill, Non-random single event upset trends, *IEEE Trans. Nucl. Sci.*, 36(6), December 1989, 2324–2329.

25. L Jacunski et al., SEU immunity: The effects of scaling on the peripheral circuits of SRAMs, *IEEE Trans. Nucl. Sci.*, 41(6), December 1989, 2324–2329.

26. J. Maiz, S. Hareland, K. Zhang, and P. Armstrong, Characterization of multi-bit soft error events in advanced SRAMs, *IEDM Tech. Dig.*, December 2003, 21.4.1–21.4.4.

27. D. Mavis et al., Multiple bit upsets and error mitigation in ultra-deep submicron SRAMs, *IEEE Trans. Nucl. Sci.,* 55(6), December 2008, 3288–3294.

28. K. Mohr and L. Clark, Experimental characterization and application of circuit architecture level single event transient mitigation, *Proc. IRPS*, April 2007, 312–317.

29. J. Knudsen and L. Clark, An area and power efficient radiation hardened by design flip-flop, *IEEE Trans. Nucl. Sci.*, 53(6), December 2006, 3392–3399.

30. J. Montonarro et al., A 160MHz, 32b 0.5W CMOS RISC microprocessor, *IEEE J. Solid-State Circ.*, 31(11), November 1996, 1703–1714.

31. L. Clark, E. Hoffman, J. Miller, M. Biyani, Y. Liao, S. Strazdus, M. Morrow, K. Velarde, and M. Yarch, An embedded microprocessor core for high performance and low power applications, *IEEE J. Solid-State Circ.*, 36(11), November 2001, 1599–1608.

32. L. Clark, K. Mohr, and K. Holbert, Reverse-body biasing for radiation-hard by design logic gates, *Proc. IRPS*, April 2007, 582–583.

33. A. Bhavnagarwala, T. Xinghai, and J. Meindl, The impact of intrinsic device fluctuations on CMOS SRAM cell stability, *IEEE J. Solid-State Circ.*, 36(4), April 2001, 658–665.

34. Grossar, M. Stucchi, K. Maex, and W. Dehaene, Statistically aware SRAM memory array design, *Proceedings of the ISQED*, March 2006, Washington, DC.

35. R. Heald and P. Wang, Variability in sub-100nm SRAM designs, *Proc. ICCAD-2004*, November 2004, 347–352.

36. N. Gierczynski, B. Borot, N. Planes, and H. Brut, A new combined methodology for write-margin extraction of advanced SRAM, *IEEE International Conference on Microelectronic Test Structure*, March 2007, pp. 97–100, New York.

37. E. Seevinck, F. List, and J. Lohstroh, Static-noise margin analysis of MOS SRAM cells, *IEEE J. Solid-State Circ.*, SC-22(5), October 1987, 748–754.

38. R. Myers and D. Montgomery, *Response Surface Methodology: Process and Product Optimization Using Designed Experiments*, 2nd edn., Wiley, New York, 2002.

39. L. Clark, N. Deutscher, F. Ricci, and S. Demmons, Standby power management for a 0.18 μm microprocessor, *Proceedings of the ISLPED*, 2002, pp. 7–12, New York.

40. H. Mizuno and T. Nagano, Driving source-line cell architecture for sub-1V high-speed low-power applications, *IEEE J. Solid-State Circ.*, 31(4), April 1996, 552–557.

41. H. Mizuno et al., An 18-μA standby current 1.8-V, 200-MHz microprocessor with self-substrate-biased data-retention mode, *IEEE J. Solid-State Circ.*, 34(11), November 1999, 1492–1500.

42. N. Kim, K. Flautner, D. Blaauw, and T. Mudge, Circuit and microarchitectural techniques for reducing cache leakage power, *IEEE Trans. VLSI Syst.*, 12(2), February 2004, 167–184.

43. PXA27x Processor Family Power Requirements Application Note, http://www.marvel.com [online].

44. S. Zhao et al., Transistor optimization for leakage power management in a 65 nm CMOS technology for wireless and mobile applications, *IEEE Symp. VLSI Tech. Dig. Tech. Papers*, June 2004, pp. 14–15.

45. M. Xapsos, G. Summers, and E. Jackson, Enhanced total ionizing dose tolerance of bulk CMOS transistors fabricated for ultra-low power applications, *IEEE Trans. Nucl. Sci.*, 46(6), December 1999, 1697–1701.

46. L. Clark, K. Mohr, K. Holbert, X. Yao, J. Knudsen, and H. Shah, Optimizing radiation hard by design SRAM cells, *IEEE Trans. Nucl. Sci.*, 54(6), December 2007, 2028–2036.

47. K. Mohr, L. Clark, and K. Holbert, A 130-nm RHBD SRAM with high speed SET and area efficient TID mitigation, *IEEE Trans. Nucl. Sci.*, 54(6), December 2007, 2092–2099.

48. C. Yang et al., Hybrid-orientation technology (HOT): Opportunities and challenges, *IEEE Trans. Electron Dev.*, 53(5), May 2006, 965–978.

49. H. Barnaby, M. Mclain, I. Esqueda, and X. Chen, Modeling ionizing radiation effects in solid state materials and CMOS devices, *Proc. IEEE CICC*, September 2008, 273–280.

50. A. Agarwal, L. Hai, and K. Roy, A single-V_t low-leakage gated-ground cache for deep submicron, *IEEE J. Solid-State Circ.*, 38(2), February 2003, 319–328.

51. S. Wolf, *Silicon Processing for the VLSI Era, Volume 3—The Submicron MOSFET*, Lattice Press, Long Beach, CA, 1995.

52. B. Haugerud et al., The impact of substrate bias on proton damage in 130 nm CMOS technology, *Rad. Effects Data Workshop Rec.*, July 2005, pp. 117–121.

53. J. Black et al., Characterizing SRAM single event upset in terms of single and multiple node charge collection, *IEEE Trans. Nucl. Sci.*, 55(6), December 2008, 2943–2947.

54. K. Mohr and L. Clark, Delay and area efficient first-level cache soft error detection and correction, *ICCD Proc.*, October 2006, 88–92.

55. C. Bencher, H. Dai, and Y. Chen, Gridded design rule scaling: Taking the CPU towards the 16 nm node, *Proc. SPIE*, 7274, 2009, 0G-1–0G-10.

Part II

Dynamic Random
Access Memory

4 DRAM Technology

Myoung Jin Lee

CONTENTS

4.1 INTRODUCTION TO DYNAMIC RANDOM ACCESS MEMORY

Since its invention in the early 1960s, the metal oxide silicon field effect transistor (MOSFET) [1] has served as the building block of the world's biggest industry, the semiconductor industry. The semiconductor industry has become the most important engine driving the world economy and has distinguished itself by the rapid pace of improvement in its products over the last four decades. The improvement trend at the integration level is usually expressed as "Moore's law," as the number of components per chip has been doubling every 2 years since about 1980 [2–4]. This achievement is attributed to the progress in device scaling that has followed an exponential curve. The minimum feature size in recent complementary metal oxide silicon (CMOS) technology has dropped to the sub-50 nm range. The most recent International Technology Roadmap for Semiconductors (ITRS) forecasts a device gate length of as short as about 25 nm by 2015 [2]. Without a doubt, the technology's leading devices in minimum feature size are memory products, such as the dynamic random access memory (DRAM).

However, as the device size, especially the device used in the memory cell, is scaled-down to the sub-100 nm range, however, numerous challenges have emerged from practical and theoretical points of view [5–10].

The device reliability issue is one of the frequently faced challenges as the scaling progresses [11,12]. If we look at the memory devices in terms of the reliability concerns, the two major issues to be taken into consideration are the data retention time in the DRAM [13,14] and the device degradation related to the gate dielectrics [15]. The former issue becomes more severe as the cell size scales down because the data retention time is proportional to the size of the cell capacitor where the data are stored. Thus, more advanced technologies are required to make capacitors with greater height for stacked type and with greater depth for trench type (compared to the former generation of a DRAM chip) to sustain the cell capacitance. One way to maintain or improve the data retention time is by reducing leakage currents, since the data retention time is inversely proportional to the leakage currents. Therefore, it is very important to understand the leakage current mechanism in a DRAM cell. The tunneling current through the gate dielectric is another important issue, because the electric field between the gate conducting material and source/drain overlap or channel region increases as the thickness of the gate dielectric scales down. Moreover, the gate tunneling current mechanism is rather complicated in the gate structure of a three-dimensional (3D) device, such as a recessed channel structure.

From the viewpoint of DRAM circuit technology, the sense amplifier has become the most important issue for high-density DRAM chips. The electronics industry has continuously demanded lower voltages and higher densities in DRAM chips. In order to satisfy this need, it is desirable to use a low V_{CORE} in the DRAM core, even though with such low voltage it is difficult to sense the cell signal due to an insufficient sensing margin in a high-density DRAM. Thus, it is also necessary to develop a high-performance sense amplifier for improving the sensing margin.

4.1.1   DRAM Cell

In order to meet the requirement of the charge retention time as the storage capacitance tends to decrease in the gigabit DRAM era, the characteristics for a highly scalable cell used in the DRAM should have the following conditions.

First, the off current (i.e., source/drain current) and the junction current should be maintained at a lower current level than that specified to satisfy the DRAM retention operation. Second, other sources of leakage in the current path, such as the tunneling current in the gate oxide and the capacitor cell, should also be lower than the specified level. If we continue to use the planar transistor, it will be difficult to satisfy the first condition mainly because the effort for reducing the drain-induced barrier lowering (DIBL) effect leads to a higher channel doping concentration, which increases the gate-induced drain leakage (GIDL) current.

Figure 4.1 shows a schematic illustration of a plane and cross-sectional view of recent stacked-capacitor structural DRAM cells to explain various leakage current paths from a cell capacitor. The first leakage current path is for the junction leakage, which can become worse with the increasing doping concentration. In addition, the second and third paths are for the cell-to-cell leakage current and the subthreshold leakage current, respectively. The fourth leakage current path (GIDL) is the most important path that leads to bad data retention operation. Finally, the fifth, sixth, and seventh paths are for the capacitor dielectric leakage, interlayer oxide leakage, and insulator leakage current, respectively.

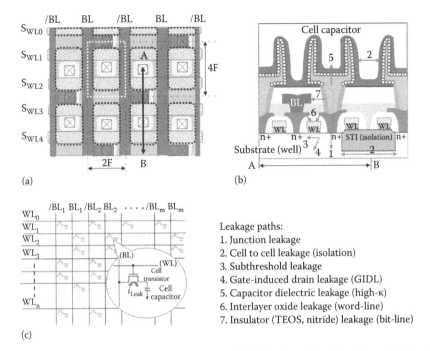

FIGURE 4.1 Schematic illustration of DRAM cell. (a) A plane view of cell array (unit cell: 4F × 2F = 8F2), (b) cross-sectional view of stacked-capacitor structural DRAM cell across line A–B depicted in (a), and (c) symbolic illustration of DRAM cell (1T1C) array. The arrows in (b) represent various leakage current paths causing data losses in a cell capacitor during refresh interval.

In order to overcome these kinds of limitations, many new structures based on the nonplanar structure have been proposed [16–20]. But each structure still has a limitation when the DRAM cell device is further scaled. Thus, it is important to analyze the limitations of the established cell structure and propose a new cell structure that may guarantee superior electrical characteristics.

4.1.2 Sense Operation

Next, we examine sense operations. We begin by assuming that the cells connected to BL1, in Figure 4.2, have logic "1" levels ($+V_{CORE}/2$) stored on them and that the cells connected to BL0 have logic "0" levels ($-V_{CORE}/2$) stored on them. Next, we form a BL (bit-line) pair by considering two BLs from adjacent arrays. The bit-line pair, labeled BL0,/BL0 and BL1,/BL1, are initially equilibrated from the $V_{CORE}/2$ [V].

All word-lines are initially at 0 V, ensuring that the cell transistors are off. Prior to a word-line firing, the bit-lines are electrically disconnected from the $V_{CORE}/2$ bias voltage and allowed to float. They remain at the $V_{CORE}/2$ precharge voltage due to their capacitance.

To read cell data, word-line WL0 changes to a voltage that is at least on transistor V_{TH} above V_{CORE}. This voltage level is referred to as V_{PP}. To ensure that a full logic "1" value can be written back into the cell capacitor, V_{PP} must remain greater than one V_{TH} above V_{CORE}. The cell capacitor begins to discharge onto the bit-line at two

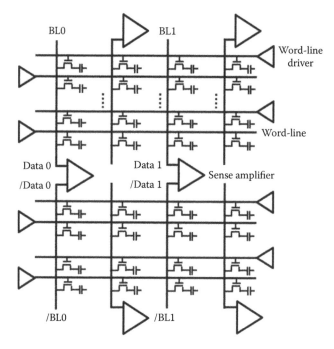

FIGURE 4.2   Open bit line array structure in DRAM.

different voltage levels depending on the logic level stored in the cell. For a logic "1," the capacitor begins to discharge when the word-line voltage exceeds the bit-line precharge voltage by V_{TH}. For a logic "0," the capacitor begins to discharge when the word-line voltage exceeds V_{TH}. Because of the finite rise time of the word-line voltage, this difference in turn-on voltage translates into a significant delay when reading ones, as seen in Figure 4.3.

Accessing a DRAM cell results in charge sharing between the cell capacitor and the bit-line capacitance. This charge sharing causes the bit-line voltage either to increase for a stored logic "1" or to decrease for a stored logic "0." Ideally, only the bit-line connected to the accessed cell will change. In reality, the other bit-line voltage also changes slightly, due to parasitic coupling between bit-lines and between the firing word-line and the other bit-line. Nevertheless, a differential voltage develops between the two bit-lines. The magnitude of this voltage difference is a function of the cell capacitance (C_{CELL}), bit-line capacitance (C_{BIT}), and voltage stored on the cell prior to access (V_{CORE}) (see Figure 4.4). Accordingly,

$$V_{CHARGE\_SHARED} = \text{half} \frac{V_{CORE} \times C_{CELL}}{(C_{CELL} + C_{BIT})}$$

After the cell has been accessed, sensing occurs. Sensing is essentially the amplification of the bit-line signal or the differential voltage between the bit-lines.

Sensing is necessary to properly read the cell data and refresh the cells. Figure 4.5 presents a schematic diagram for a simplified sense amplifier circuit: a cross-coupled

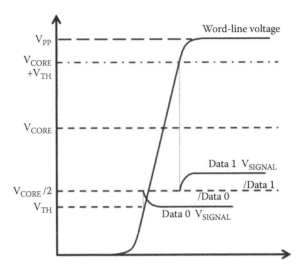

FIGURE 4.3   Cell access waveform in accordance with data polarity.

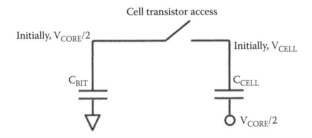

FIGURE 4.4   Charge sharing operation in DRAM cell.

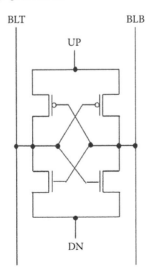

FIGURE 4.5   Bit line sense amplifier schematic.

NMOS pair and a cross-coupled PMOS pair, whose UP and DN provide power and ground. The NMOS latch has a common node labeled DN.

Similarly, the PMOS latch has a common node labeled UP. Initially, DN and UP are biased to $V_{CORE}/2$. When the cell is accessed, a signal develops across the bit-line pair. While the "1" bit-line contains charge from the cell access, the other bit-line does not but serves as a reference for the sensing operation. The sense amplifiers are generally fired and lead to the development of the charge-shared voltage from cell data into the difference of V_{CORE} between the bit-line pair.

4.2 SENSING MARGIN IN DRAM

We treated the sensing operation and cell transistor in the DRAM chip. In the limited operating time for the high-speed DRAM, the insufficient charge-shared voltage should be developed into the V_{CORE} level by the sense amplifier circuit. For the large charge-shared voltage, the cell transistor should show excellent performance in the driving current and the leakage current. These electrical characteristics in the cell transistor guarantee sufficient charge-shared voltage, resulting in the success of the sensing operation. Beyond the cell transistor operation, it is necessary to obtain a sensing circuit immune to the several sensing noises to guarantee the successive operation of the bit-line sense amplifier (BLSA). Several factors guarantee sensing success. These factors are the elements for the sensing margin, which need to be clearly defined for the low-power and high-density DRAM chip.

4.2.1 DEFINITION OF SENSING MARGIN

When the BLSA operates to detect the data stored in the activated cell, this circuit amplifies the charge-shared voltage determined by both the core voltage (V_{CORE}) and the ratio of the cell capacitance to the bit-line capacitance. Because V_{CORE} and cell capacitance need to be small for a low-power and high-density DRAM operation, the charge-shared voltage becomes insufficient for guaranteeing success in the sensing operation. Moreover, the offset voltage of latch transistors in the BLSA becomes an important factor when considering a small area for these transistors. The dopant fluctuation phenomenon is related with the noise immunity in the BLSA including latch transistors with the threshold voltage (V_{TH}) mismatch. However, the sensing noise also induces a serious problem in the BLSA when starting the sensing operation. Therefore, we should take into account all the components affecting the sensing operation.

After all, the sensing margin can be defined as the following expression considering the charge-shared voltage, the BLSA offset, and the sensing noise:

$$\text{Sensing margin} = \frac{V_{CORE} \times C_{CELL}}{(C_{CELL} + C_{BIT})}$$

- BLSA offset (latch transistor V_{TH} mismatch)
- Sensing noise (coupling noise)
- Cell leakage
- Weak write-performance

This sensing margin voltage level means a minimum voltage enough to guarantee success in detecting the data stored in the cell capacitance, even though there are several noise sources affecting the ideal stored charge.

4.2.2 NOISE EFFECT ON THE SENSING MARGIN

In Section 4.2.1, we briefly discussed several factors affecting the sensing margin. In circumstances of low-voltage and high-density DRAM, the charge-shared voltage should be smaller due to a low V_{CORE} and a small cell capacitance. But the noise effects of the BLSA and the cell transistor become more serious as the DRAM technology develops further. In this section, the detailed noise effect on the sensing margin will be taken into account. From the viewpoint of the noise factors in the cell, the most important factor is the DRAM cell leakage, which is caused by a high electric field. Moreover, it is also important to improve the write-performance, which is determined by the current drivability of the cell transistor. On the other hand, the threshold voltage (V_{TH}) mismatch of the latch transistors and the sensing noise becomes worse and more important from the viewpoint of a sensing operation using the BLSA.

4.2.2.1 DRAM Cell Performance (Leakage and Current Drivability)

The leakage in the DRAM cell transistor has been the most important noise factor for high-performance DRAM. As the device feature size shrinks, the channel doping concentration should be higher for guaranteeing a better short channel effect (SCE) of the cell transistor. It leads to a better off-state leakage current. But the high channel doping process induces a high electric field in the drain junction region of the cell transistor. It induces gate-induced drain leakage (GIDL) current in the active area nearby the storage node. As the gate oxide thickness becomes thinner for a better gate controllability, the GIDL effect may be a much more critical issue. On the other hand, the current drivability of the cell transistor should be better because of the narrow width of a high-density DRAM cell. The poor current drivability of the cell transistor leads to a failure in the write operation. This poor write operation induces a weak charge-shared voltage, which is an important factor in the read operation. Overall, both the leakage current and the current drivability are important requirements for the high performance of a cell transistor. In the DRAM industry, several DRAM cell transistors have been developed for high performance in the previously mentioned cell characteristics. In the next section, recently developed 3D cell transistor structures will be introduced with reference to cell leakage and current drivability.

4.2.2.2 High-Performance DRAM Cell Structures

Recently, nonplanar device structures [16,19–24], such as FinFET, recess channel-array transistor (RCAT), and S-Fin, have been applied to the DRAM cell to suppress the junction leakage and the SCE due to the high electric field at channel edge regions as the feature size shrinks [25–27]. Especially, S-Fin, a FinFET device with a recessed channel structure [21], shows improved characteristics on the SCE, the driving current, the subthreshold slope (SS), and the drain-induced-barrier-lowering

(DIBL), compared with the conventional recessed channel structures. However, even though the S-Fin structure has such excellent characteristics owing to the tri-gate effects, it still has some critical problems that need to be resolved from the viewpoint of the drain leakage and the threshold voltage control.

In this section, the representative recessed channel devices, such as the RCAT and the S-Fin, are experimentally analyzed. In these analyses, we considered the following factors: on current, leakage, SCE, and reliabilities as they are the most important determinants of the DRAM cell performance. Based on the measurements, the mechanism and source of the leakage current will be discussed. An optimal recessed channel structure is proposed, and a simulation is conducted by a 3D device simulator, which is well tuned to predict the DRAM leakage distribution, to compare it with conventional structures [28–30].

As the feature size of the DRAM cell shrinks, the RCAT suppresses the SCE by increasing the effective channel length [19,20]. However, since it has poor current drivability, the S-Fin has been developed to enhance current drivability by using the tri-gate technology [21]. While the recessed channel of the RCAT is controlled by a single gate filling the recessed region, the recessed channel of the S-Fin is surrounded by a tri-gate. The 3D views of the RCAT (a) and the S-Fin (b) are illustrated in Figure 4.6. This figure is based on the TEM profile of the S-Fin device (Figure 4.7).

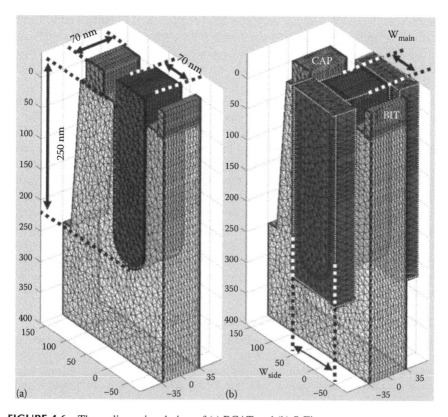

FIGURE 4.6   Three-dimensional view of (a) RCAT and (b) S-Fin.

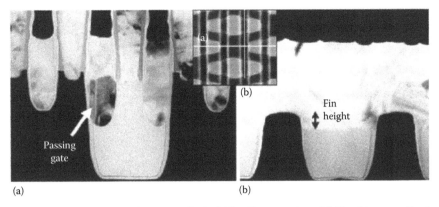

(a) (b)

FIGURE 4.7 Illustration of TEM profile in S-Fin. Cross section of S-Fin: (a) perpendicular to word line; (b) parallel to word line.

The entire recessed channel can be divided into two parts, a bottom channel and a vertical channel in the bit-line side. In this study, all the devices adopt the asymmetric channel doping profile where the channel doping concentrations are much higher in the bit-line side than in the bottom and storage node sides [31–33]. It makes it possible for the DRAM cell to maintain a high threshold voltage as well as suppress the leakage current. The tri-gate structure in the S-Fin structure brings about remarkable improvements in the electric performance compared with the RCAT structure.

First of all, the leakage characteristics, which are probably the most important factor that determines the performance of DRAM cells, have been analyzed through measurements of cell arrays containing about 1 K cells.

The measurement results in Figure 4.8 clearly show the mechanisms of the leakage current in the RCAT and the S-Fin devices. The test bias conditions shown in Figure 4.8a can be grouped into two groups: case 1 ($V_D - V_B = 0.8$ V) and case 2 ($V_D - V_B = 1.6$ V) according to the bias between the storage node and the bulk; in other words, the results of drain to bulk junction leakage current. In addition, both cases (cases 1 and 2) contain two figures that illustrate bias conditions to clearly analyze leakage current mechanism for low and high drain biases.

These bias conditions exclude the junction leakage current from the comparison of drain current by maintaining the same junction voltage ($V_D - V_B$) for each drain bias, respectively. Figure 4.8b shows the drain leakage currents according to $V_D - V_B$ in the 1 K array cells for the RCAT and the S-Fin. Since the junction leakage is dependent only on $V_D - V_B$ value, the differences of the drain leakage between the high V_D and low V_D are ascribed not to the junction but to other regions for each case, that is, the gate-to-channel and the gate-to-drain. Because keeping $V_D - V_B$ constant leads to the same junction leakage current in both cases, that is, high V_D and low V_D, in the case of using high V_D bias, V_B has to be lower in order that junction leakage would be no reason for drain leakage current difference between high V_D and low V_D. Therefore, the leakage current is contributed mainly by the gate-to-drain region. On the other hand, for using low V_D bias, the leakage current mainly originates from the gate-to-channel region. By comparing the differences in current as described at the bottom part of Figure 4.8, we can see which regions are the main causes of the leakages.

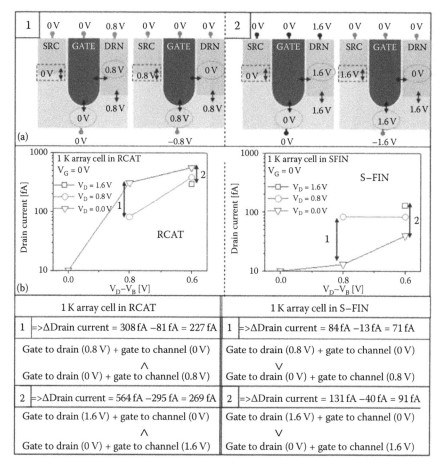

FIGURE 4.8 Mechanism of leakage current for $V_D - V_B = 0.8$ V (case 1) and $V_D - V_B = 1.6$ V (case 2) in RCAT and S-Fin, respectively. (a) Bias conditions of low- and high-drain voltages for each case (cases 1 and 2), respectively, and (b) drain leakage currents according to $V_D - V_B$. In the case of fixed $V_D - V_B$, we can expect the same junction leakage current for high V_D and low V_D. Therefore, the difference of leakage current for high and low V_D is due to two kinds of regions, that is, gate-to-channel and gate-to-drain. Finally, we can distinguish which region is dominant leakage source; gate-to-drain or gate-to-channel region in RCAT and S-Fin. At the same $V_D - V_B$, 1.6 V (0.8 V), for RCAT, the leakage current for V_B of 1.6 V (0.8 V) is larger than that for V_D of 1.6 V (0.8 V). It is because RCAT is dominated by the gate-to-channel leakage. On the other hand, for S-Fin, the leakage current for V_B of 1.6 V (0.8 V) is lower than that for V_D of 1.6 V (0.8 V). It is because S-Fin is dominated by the gate-to-drain leakage. Therefore, the leakage current in RCAT is dominated by the gate-to-channel region, while that in S-Fin is mainly controlled by the gate-to-drain region.

In view of the results so far investigated, we can conclude that the RCAT leakage mainly originates from the bottom channel region. Because the gate oxide of the recessed channel is usually thinner in the hollow bottom region ($t_{ox} = 41$ Å) than in the vertical channel region ($t_{ox} = 57$ Å) [21], there is more leakage generation in the bottom region where strong field takes place. Therefore, it is necessary to lower the doping level of the bottom channel so as to relax the electric field for all kinds of recessed channel devices.

On the other hand, the S-Fin is a slightly different case from the RCAT. The electric field in the bottom channel is mitigated due to the depletion charge sharing by the side gates in the S-Fin channel, so it is essential to lower the doping level of the bottom channel to fully deplete the bottom channel for the S-Fin structure. Instead, the widened gate–drain overlap area and the strong field in that region enhance the leakage generation in the gate–drain overlapped region. Therefore, we can conclude that the main leakage source in the S-Fin structure is the gate–drain overlapped region that is surrounded by the tri-gate. This fact, despite the fact that shows the merit of lower leakage in the bottom channel region, makes the S-Fin less effective in terms of suppressing the off-state leakage. Moreover, considering the statistical retention time distribution, which is the most important factor in the DRAM cells, the strong field distribution in the gate–drain overlapped region formed by the side gate and the main gate would constitute a serious problem for the retention fail cells. Such an insight from both the analyses of leakage characteristics and the expectation of statistical retention problems will be a basis for the proposal of an optimized recessed channel type structure.

On the other hand, the S-Fin drives more on current than the RCAT thanks to the tri-gated channel. The measurements of on current have been done with discrete test patterns. Figure 4.9a and b shows the measured RCAT and S-Fin I–V characteristics with similarly reproduced results by the 3D device simulator.

Comparative analyses between the RCAT and the S-Fin provide valuable insights on the improved device structure for the DRAM cell transistor. Based on these results, we were able to develop an optimized design of the DRAM cell transistor with the recessed channel (RFinFET), which adopts only the positive aspects of the RCAT and the S-Fin. Figure 4.10 shows the 3D view of the proposed RFinFET, which has a tri-gate only in the bottom region so that the whole transistor may have planar gate structures effectively in the S/D region and a tri-gated FinFET structure in the bottom channel region.

Note that the RFinFET does not have the side gates in the S/D overlapping region. Figure 4.11 includes the layout and the cross-sectional views of the RFinFET, revealing the side-gates formation and a possible manufacturing sequence, respectively. After the recessed channel is formed by silicon etch processes [19,20], the oxide surface in the STI region is exposed by second silicon etch (isotropic) for the round channel shape [20]. Then, the isotropic oxide etch is done, followed by the gate oxidation and the gate-material deposition, building the proposed structure [24].

We expect a smaller gate capacitance and lower leakage from the shape of the proposed device that the side-gate and the S/D regions do not overlap. The threshold voltage of the RFinFET can be maintained high enough to reduce off-state current, owing to the existence of the planar-like vertical channel in the bit-line side. 3D device simulations using the frames in Figures 4.6 and 4.10 have been done to compare the RFinFET with the other recessed devices, especially S-Fin. The RFinFET

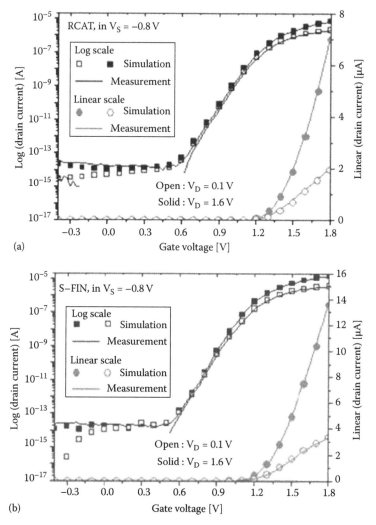

FIGURE 4.9 Comparison of current characteristics for (a) RCAT and (b) S-Fin. To analyze and investigate the feasibility of RFinFET, measurement results of RCAT and S-Fin were fitted by NANOCAD 3D device simulation [28–30], and these are very close to simulation results.

is designed to have the same channel shape as the S-Fin, whose channel is recessed to a depth of 250 nm and a width of 70 nm. For a fair and thorough comparison, many design splits of the S-Fin were simulated with various side-gate widths, that is, S-Fin-57, 00, 60, 90, and 190 whose width difference between main gate and side gate are 0, 114, 234, 294, and 494 Å, respectively. Therefore, the gate–drain overlap region in S-Fin-57 is similar to that of RCAT and RFinFET. But it also has a difference from the viewpoint of the existence of side gate. The S-Fin and the RFinFET were also split by the source junction depth, such as 160 and 200 nm. In the case of the 200 nm source junction, the source diffusion region and the side gate were overlapped so that the whole recessed channel was tri-gated. The channel was doped

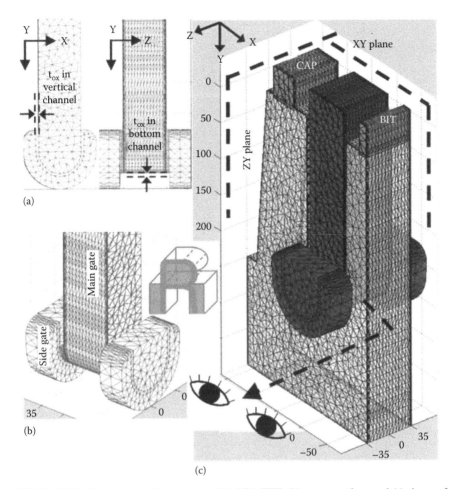

FIGURE 4.10 Three-dimensional view of (a) RFinFET, (b) cross section, and (c) shape of gate in RFinFET shown in detail.

asymmetrically in all the devices so that the junction would always be formed in the tri-gated region in the storage node side.

Simulations were performed on the drift-diffusion models with Lucent mobility [34], Caughey–Thomas expression [35], and Phillips unified mobility [36,37]. The simulations of leakage current levels were done accurately using the TAT (trap-assisted-tunneling) model [38,39].

Figure 4.12 shows the simulation results for the RCAT, the S-Fin, and the RFinFET devices with 160 nm junction depth. The results show that the RFinFET has a much higher threshold voltage than the S-Fin with the same asymmetric channel doping profile ($3.5 \times 10^{18}/cm^3$) (Figure 4.12d). This means that adequate threshold voltage can be achieved with even lower channel doping concentrations. In the case of the 200 nm source junction depth, the threshold voltage of the RFinFET is not higher than that of the S-Fin as shown in Figure 4.13d since the whole channel is tri-gated. The RFinFET and the S-Fin 190 have the same on-current level because they have

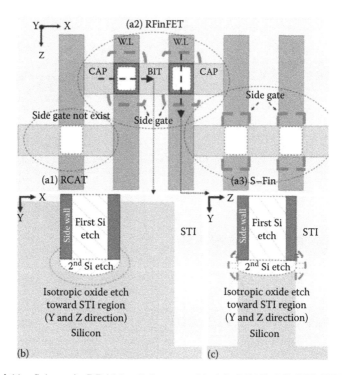

FIGURE 4.11 Schematic DRAM cell layouts with (a1) RCAT, (a2) RFinFET, and (a3) S-Fin. A possible way to achieve RFinFET through schematic cross section: (b) XY plane and (c) ZY plane.

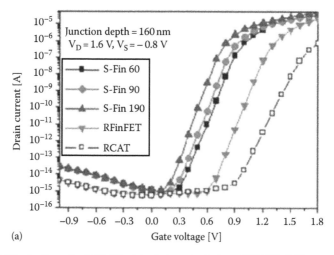

FIGURE 4.12 (a through d) 12 Electrical characteristics of RCAT, S-Fin, and RFinFET structures for the case of 160 nm source junction depth are compared from simulation results, respectively.

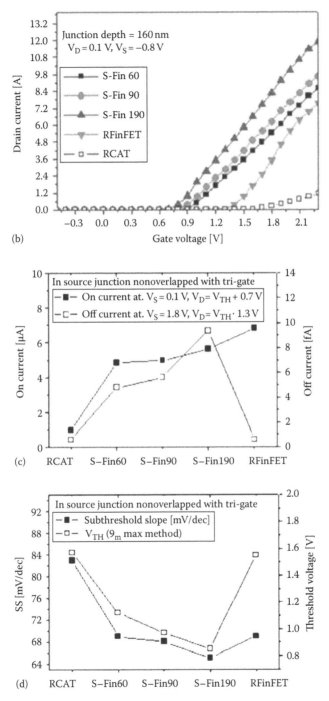

FIGURE 4.12 (continued) (a through d) 12 Electrical characteristics of RCAT, S-Fin, and RFinFET structures for the case of 160 nm source junction depth are compared from simulation results, respectively.

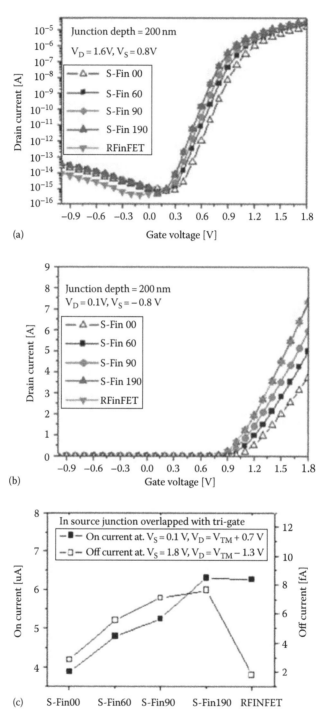

FIGURE 4.13 (a through d) Electrical characteristics of S-Fin and RFinFET for the case of 200 nm source junction depth are compared from simulation results, respectively.

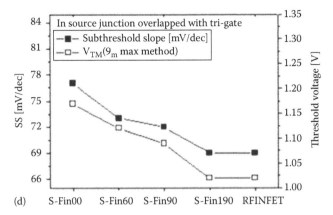

(d) S-Fin00 S-Fin60 S-Fin90 S-Fin190 RFINFET

FIGURE 4.13 (continued) (a through d) Electrical characteristics of S-Fin and RFinFET for the case of 200 nm source junction depth are compared from simulation results, respectively.

the same side-gate width. However, the RFinFET has a lower leakage current than any other device structure as shown in Figure 4.13c.

Assuming that each device has an optimized doping profile, the best on/off current ratio can be obtained by the RFinFET regardless of the source junction depth as shown in Figure 4.14. The optimized doping profiles correspond to very low doping in the bottom region and $2 \times 10^{18}/cm^3$ and $3 \times 10^{18}/cm^3$ in the source side channel region for the case of structures with source junction depths of 160 and 200 nm, respectively. In terms of the on-current defined as the drain current (I_{on}) at $V_G = V_{TH} + 0.7$ [V], where V_{TH} is obtained from g_m max method, the S-Fin 190 is the best choice except for the RFinFET owing to the increased effective channel width. The RFinFET with

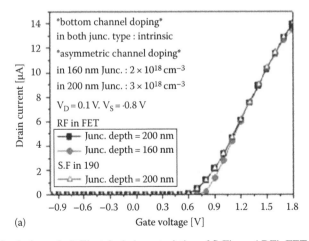

(a)

FIGURE 4.14 (a through d) Electrical characteristics of S-Fin and RFinFET for optimized doping profile case are compared from simulation results, respectively.

(*continued*)

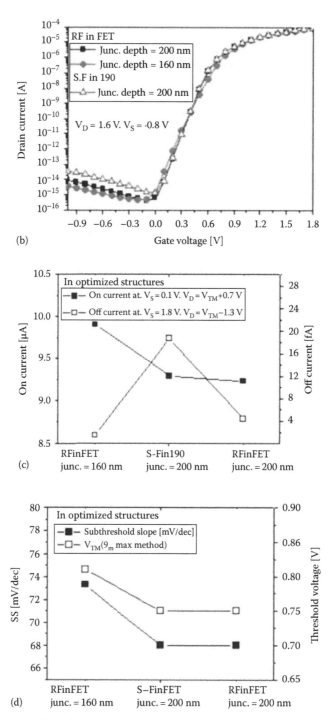

FIGURE 4.14 (continued)   (a through d) Electrical characteristics of S-Fin and RFinFET for optimized doping profile case are compared from simulation results, respectively.

a source junction depth of 160 nm assures the highest threshold voltage with the same doping, implying good current drivability due to low channel doping concentration when the same threshold voltage is assumed, as shown in Figure 4.12c and d.

In addition, in the simulation on the RFinFET, it was found to show less leakage than the S-Fin under the off-state condition. The simulation of the off-state leakage was done with the trap-assisted-tunneling model, and off-state leakage was defined as the drain current (I_{off}) at $V_G = V_{TH} - 1.3$ [V], where V_{TH} is V_G value at $I_D = 10^{-8}$A. The S-Fin trades off the on-current and the leakage level as shown in Figures 4.12c and 4.13c because the extended side gate increases the area of the gate–drain overlapped region and enhances electric field intensity in the corner region while contributing to increase in the on current. However, the RFinFET has less leakage than any type of S-Fin, even with the highest on-current level. It is because the side gate does not overlap with the drain region in the RFinFET.

Figure 4.15 shows the retention time distribution of the RCAT, the S-Fin, and the RFinFET through an in-house 3D device simulation tool especially fit for DRAM simulations including the statistical leakage current distribution, that is, NANOCAD [28–30]. From the viewpoint of the retention time distribution, the RFinFET and

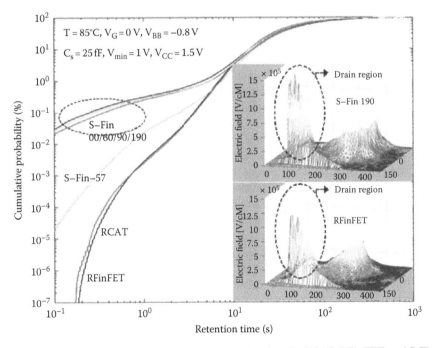

FIGURE 4.15 Cumulative probability versus retention time for RCAT, RFinFET, and S-Fin groups. 3D electric field distribution also shows the cause of retention tail difference between S-Fin groups and RFinFET. For a fair and thorough comparison, many design splits of S-Fin were simulated with various side-gate widths, that is, S-Fin-57, 00, 60, 90, and 190, whose width differences between main gate and side gate are 0, 114, 234, 294, and 494 Å, respectively. Therefore, the gate-drain overlap region in S-Fin-57 is similar to that of RCAT and RFinFET; however, it also has a difference in the viewpoint of existence of side gate.

the RCAT show the best performances as the simulation results. The region with the greatest leakage in the recessed channel structure is the gate–drain overlapped region due to the high electric field profile there [40]. In the S-Fin device, the gate–drain overlapped region is tri-gated so that the electric field intensity is much higher than the RFinFET, especially in the edged region (see the inset in Figure 4.15).

Since the edged part in the gate–drain overlapped region is limited, there is very little probability that a trap exists in the edged region. Therefore, the cell leakage currents are mainly generated in the vicinity of the junction and the gate–drain/channel overlapped regions for most cell transistors, resulting in greater leakage with the wider side gates as shown in Figures 4.12c and 4.13c. However, cell transistors with a trap in the edged region give rise to the tail distribution of the retention time tests, and the leakage currents from the edged region become the dominant factor restricting the chip data-retention time in the real DRAM with giga-level cells.

4.2.2.3 V_{TH} Mismatch in BLSA

After the true bit-line voltage is developed into the charge-shared voltage of ΔV, it can be detected as a logic level of "0" (or "1") by the bit-line sense amplifier (BLSA) circuit. Therefore, the sensing circuit should be able to detect a small voltage difference between true-bit line (BL) and bar-bit line (/BL). As explained in Section 4.1.2, the conventional BLSA type is based on the latch operation. The paired N/PMOS transistors are triggered by the voltage differences between gate (BL, or/BL) and common source (sensing enabled signal), respectively. When the MOS transistors are allowed to operate by the sensing enabled signal, the drain nodes would follow the common source node (sensing enabled signal). Overall, the latch circuits develop the differential voltage between BL and/BL into the V_{CORE} level, based on the relative amplitude of the gate node (BL,/BL) voltages. Therefore, the charge-shared voltage, the difference between BL and/BL, was the key factor for the successful operation of the sensing circuit, when not taking into account the threshold voltage mismatch of latch transistors. However, the small area of the latch transistor induces dopant fluctuation, resulting in the threshold voltage mismatch. The threshold voltage mismatch has the same meaning as the charge-shared voltage, the gap between BL and/BL, with reference to the sensing margin. The dopant fluctuation is a natural phenomenon based on the probability theory. In order to overcome the negative effect of the dopant fluctuation, it is inevitable to adapt low-doped channel engineering to the fabrication of the latch transistors. But it leads to a bad SCE, resulting in a large off-state leakage. There are several solutions for a better threshold voltage mismatch. The 3D transistor such as SOI and a multi-gate structure can be a good candidate for a better latch circuit performance. Thanks to the research on this phenomenon, it is possible to expect a threshold voltage mismatch, analytically and experimentally. It depends on the channel doping concentration, the channel length, and the gate width of the latch transistor. And the gate oxide thickness and temperature also affect the V_{TH} mismatch. The well-known numerical formula for the threshold voltage mismatch is expressed as follows:

$$\sigma V_{TH} \propto \frac{\sqrt{N_A}}{\sqrt{L \cdot W}}$$

4.2.2.4 Sensing Noise in Accordance with Data Pattern

In DRAM, the key solution to high-density chips has been to obtain a large enough cell capacitance to store weak voltage data [41–43]. However, as the DRAM technology develops further it is more difficult to sustain a high cell capacitance. It leads to a small sensing margin due to the small ratio of the cell capacitance to the bit-line capacitance. Therefore, the sensing margin becomes the most important factor in these circumstances. Also what is worse, the margin problem becomes more serious as the core voltage (V_{CORE}) decreases. Therefore, improvement of the sensing margin is inevitable, which consists of a natural threshold voltage mismatch in the latch transistor and the sensing noise in the cell array in accordance with the data pattern. Because the threshold voltage mismatch becomes worse due to the dopant fluctuation, it will be more important to improve the sensing noise in accordance with the data pattern.

The sensing noise is directly related to the DRAM refresh time, which is a key factor for determining the performance of the DRAM. The DRAM refresh time also shows a similar dependence on the type of data pattern to the sensing noise, which changes in accordance with the data pattern [44–46]. Therefore, the focus in this section is on the sensing noise in BLSA. As a result, it is also possible to investigate the DRAM refresh time, dependent on the type of data pattern.

A DRAM cell array consists of a repeated unit cell structure nearby a bit-line, word-line, and storage node. Therefore, it has a very complicated coupling capacitance. Figure 4.16 shows an illustration of the representative sensing noise mechanism in the cell array, which depends on the data pattern. During the early stage of the sensing operation, the transition of the majority of BLs affects the potential of the WL in the cell array, which leads to noise in the target BL. The coupling effect is sufficiently large to deteriorate the target bit-line, when the target BL data are weak. Especially, an open bit-line structure shows a very strong sensing noise due to the plate noise and well noise, and so on [47–49]. Figure 4.17 shows that the sensing operation of the majority of the bit-line affects the transition of the target bit-line. When the polarity of the majority of BL data is opposite to the target BL data,

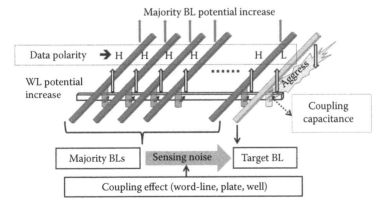

FIGURE 4.16 Sensing noise mechanism in cell array. Majority bit lines affect sensing of weak target bit line through coupling effect in cell array.

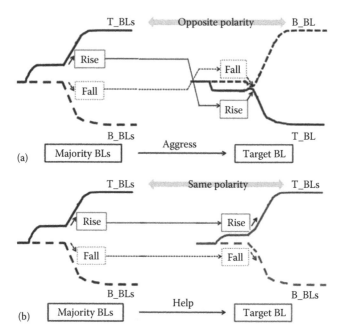

FIGURE 4.17 Polarity of majority bit lines affects sensing of target data. (a) Majority BLs opposite to target BL data interfere with sensing of target BL; (b) majority BLs having same polarity of target data help sensing of target BL.

the sensing of the target BL can be interfered with, as shown in Figure 4.17a. This interference occurs through the already mentioned coupling relationship in the cell array. On the other hand, when the polarity of the majority of BLs is the same as that of the target BL, this coupling effect in the cell array could give assistance to the sensing of the weak target BL data, as shown in Figure 4.17b. As a result, this data polarity determines the type of sensing noise.

From the type of data polarity, four kinds of data patterns can be defined [44,50]. These representative data patterns determine both the best sensing noise and the worst sensing noise. This will be called a solid data pattern in which all the BL data have the same polarities. In addition, the island pattern is formed when the minority of the BLs has opposite data polarity to the majority of the BLs. Because margin failure occurs most frequently while sensing the island data pattern, the sensing noise needs to be improved in the island data pattern, even though solid data could be sacrificed.

4.2.3 RELATION BETWEEN REFRESH TIME AND SENSING NOISE IN ACCORDANCE WITH DATA PATTERN

Generally, the DRAM refresh time is determined by the cell leakage characteristics. However, as cell data patterns vary, the refresh time (tREF) of a specific cell transistor also shows different values and trends. Figure 4.18 illustrates two representative kinds of data patterns, which comprise (a) all one data bits and (b) only one data bit with background data bits of all zero. We measured chip 1, chip 2, and chip 3

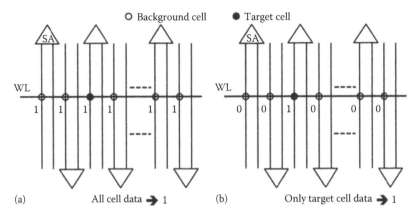

FIGURE 4.18 Two representative kinds of data patterns, which comprise (a) all one data bits and (b) only one data bit with background data bits of all zero.

TABLE 4.1

Chip Information Related to Cell Type, Capacitance between Bit Line and Word Line, and Cell Leakage Characteristics for Chips 1, 2, 3, and 4

	Chip 1	Chip 2	Chip 3	Chip 4
Cell type	Recessed	Recessed	Recessed	Recessed
Fabrication technology	54 nm	54 nm	54 nm	54 nm
BIT-WL cap. [a/u]	1	1	0.1	1
Cell leakage	Better	Worse	Worst	Best
Cell leakage screen	Without	Without	Without	With

fabricated with 54 nm technology. Especially, chip 3 has a different type of cell structure, that is, buried word-line scheme [46], as shown in Table 4.1.

Figure 4.19a shows the dependency of tREF on these data patterns for several types of DRAM chips fabricated with different technologies. The figure shows not only the variation of tREF but also the different dependencies on data patterns for several DRAM chips. We found that the x-axis in Figure 4.19a represents BLSA offsets, which are dependent on data patterns. In addition, we discovered that various cell leakage characteristics determined the slope of this graph.

tREF is determined not only by cell leakage but also by BLSA offset. To clarify the meaning of the x-axis in Figure 4.19a, we measured the BLSA offset according to the data pattern for three kinds of DRAM chips, as shown in Figure 4.19b. By changing the quantity of charge stored in the cell capacitor, we can examine the sensing failure voltage, which is the BLSA offset [47]. When a BLSA is fixed and its own offset is constant, the condition of cell data patterns causes the offset to vary due to the sensing noise in the cell array. This also affects the tREF. The condition of data patterns determines the strength of coupling noise between the bit-line (BL) and word-line (WL) in the cell array [47,50]. Figure 4.20 illustrates the noise mechanism

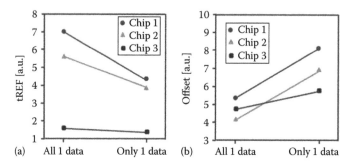

FIGURE 4.19 (a) Dependency of tREF on two kinds of data patterns for several DRAM chips fabricated with different technologies; (b) measured BLSA offset according to data pattern for three kinds of DRAM chips.

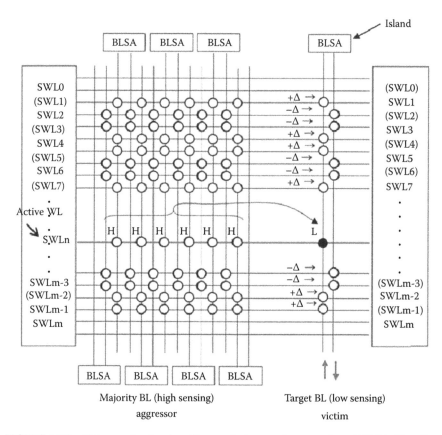

FIGURE 4.20 Noise mechanism in cell array during sensing operation of BLSA. Coupling capacitance between BL and WL is the main origin of sensing noise.

in a cell array during the sensing operation of a BLSA. The majority of BL becomes the aggressor and induces noise in the WL. This occurred noise in m pieces of WL induces secondary noise in the target BL. This noise effect can be reduced by shrinking the capacitance between the BL and WL (C_{BL-WL}). In chip 3, C_{BL-WL} is 10 times smaller than the other 2 DRAM chips [46], so chip 3 has improved offset variation of data patterns.

If a specific BLSA including a target cell is selected, then the tREF variation according to data patterns is determined by the relationship between its own offset variation and cell leakage characteristics, as shown in Figure 4.21. The remaining charge in the cell for successful sensing becomes the BLSA offset including data pattern noise. The cell-discharging curve determines the tREF difference between the data bit pattern (all one) with a small offset and the data bit pattern (only one) with a large offset. For the cell with superior leakage (curve A in Figure 4.21) when the offset variation is the same, its tREF variation must be larger than the worse leakage cell (curve B in Figure 4.21).

Figure 4.22 shows the tREF variation according to offset change. Chip 3, which has the smallest variation of tREF, shows the best offset variation and the worst cell leakage characteristics. On the other hand, chip 1, which has the largest variation of tREF, shows the best cell leakage characteristics. In order to confirm our explanation, we measured the slope of the tREF variation for several cells with various refresh times in three kinds of chips. Chip 1 and chip 2 seemed to demonstrate a different dependency of tREF on offset variation, as shown in Figure 4.22.

However, Figure 4.23 shows that chips 1 and 2 have precisely the same trend of tREF slope versus offset variation for various refresh times, which were measured for a tREF probability range of $[1 \times 10^{-4}, 1][\%]$. From our analysis, we concluded

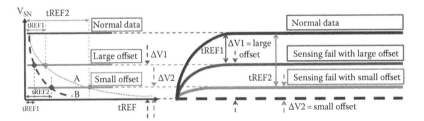

FIGURE 4.21 tREF variation according to data patterns is determined by the relationship between its own offset variation and cell leakage characteristics. A and B curves represent cell discharging voltage for cells showing a better and worse leakage, respectively. Because data pattern determined offset by sensing noise, it is one of the factors affecting refresh time. When fixing offset according to data pattern, discharging curve determined the refresh time variation (tREF2 – tREF1), which is smaller in the cell with worse leakage, representing curve B. ΔV1 and ΔV2 denote large offset and small offset voltage, respectively. At the same time, they denote sensing failure voltages in the corresponding large and small offsets. tREF1 and tREF2 denote refresh time under condition of large offset and small offset, respectively. Thus, normal data become ΔV1 and ΔV2 after tREF1 and tREF2, resulting in data sensing failure, respectively. Therefore, tREF2-tREF1 means refresh time variation according to data pattern determining offset.

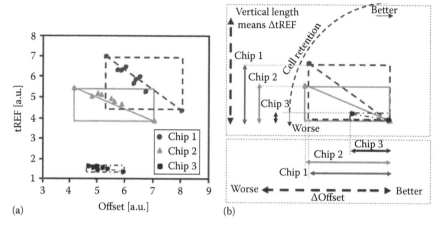

(a)   (b)

FIGURE 4.22 (a) tREF variation according to offset change. (b) Relationship between cell leakage characteristics and tREF variation. (b) x-axis denotes offset variation according to data pattern and y-axis denotes tREF variation (see explanation in Figure 4.21); because chip 1 has better refresh characteristics, it shows larger refresh time variation compared to chip 2. In particular, chip 3 shows best refresh time variation because it has the worst leakage characteristics and the best offset variation according to data pattern.

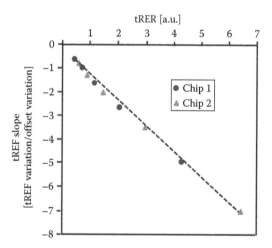

FIGURE 4.23 Chips 1 and 2 have precisely the same trend of tREF slope versus offset variation for various refresh times, which were measured for a tREF probability range of $(1 \times 10^{-4}, 1)$ (%).

that it is not necessary to improve cell leakage characteristics for cells with average tREF; we only need to improve offset variation according to data pattern, in order to reduce tREF variation.

In the tREF distribution range $[1 \times 10^{-7}, 1 \times 10^{-3}][\%]$, we measured the slope dependency for three kinds of chips. We found that chips 1 and 2 revealed an extraordinary trend in this range, as shown in Figure 4.24. In this range, chips 1 and 2 comprised

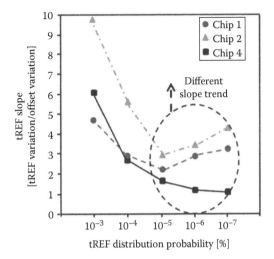

FIGURE 4.24 Chips 1 and 2 reveal an extraordinary trend in range (1×10^{-7}, 1×10^{-3}) (%) compared with chip 4. This phenomenon provides an intuitive explanation of tREF dependency on cell leakage.

cells with a gate-induced drain leakage mechanism such as trap-assisted tunneling (TAT) [30,39]. The main distribution groups [1×10^{-4}, 1] % usually comprise cells with junction leakage by the SRH mechanism and showed the trend of decreasing slope variation. However, the tail distribution group showed an increasing trend after a decreasing one. Because the cells with TAT were eliminated in chip 4 based on redundancy cells, this shows a trend of continuously decreasing slope variation.

Figure 4.25 shows the measured offsets of three different kinds of cells, which share the same BL. Cell 3 shows an extraordinary offset, even though all cell transistors are sensed by the same BLSA. This is because the TAT leakage current occurred

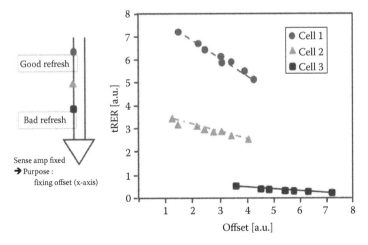

FIGURE 4.25 25 Offsets measured in three kinds of cell sharing same BL. Although they have the same BLSA offset, they show different offsets because of different cell leakage current.

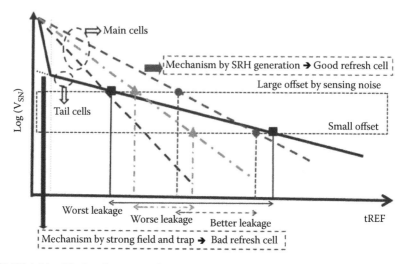

FIGURE 4.26 Discharging curve of cell potential for main and tail cells, respectively. While the main cells show a similar straight discharging curve as denoted by the dotted lines, the tail cells show a different curve as denoted by the black solid line. Leakage mechanism in tail cell is trap assisted tunneling (TAT). TAT leakage current occurs during early stage of retention, so its tREF variation can be larger than other SRH leakage cells, even though main cells with better SRH leakage must show larger tREF variation. tREF variation for three kinds of cells is indicated on x-axis as worst leakage (black), worse leakage (red), and better leakage (blue).

for a very short duration (less than 10 ns) during offset measurement. Figure 4.26 illustrates the discharging curve of cell potential for the main and tail cells, respectively. This curve explains the extraordinary trend of the tREF slope.

4.2.4 How to Improve Sensing Margin

We have offered the definition of a sensing margin in the DRAM chips. There are several important elements for the sensing margin. Therefore, as the sensing margin problems become serious, it is necessary to guarantee the sensing margin enough to succeed in the sensing operation. The easiest solution to enough sensing margin is surely to obtain a large ratio of cell capacitance to bit-line capacitance. But, as the DRAM technology develops further, it is more difficult to keep the same amount of cell capacitance to the past technology. DRAM industries have not been concerned about the sensing noise in the cell array. Therefore, it remains to improve the sensing noise in the cell array by using a new BLSA.

4.2.4.1 Offset Compensation Sense Amplifier

Figure 4.27 illustrates the (b) proposed BLSA scheme, named H-SA (HYNIX-Sense Amplifier), compared with (a) conventional BLSA. The remarkable difference is in whether the driving signal is separated or not. The H-SA has two kinds of pull-up driving lines, that is, UP_T, UP_B, and pull-down driving lines, that is, DN_T, DN_B. Majority BL data polarity determines the choice of driving lines used for data sensing, as shown in the Figure 4.28. The majority of BLs with high data are

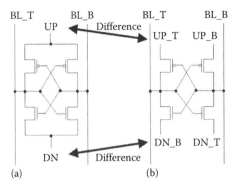

FIGURE 4.27 Illustration of (b) proposed BLSA, named H-SA, compared with (a) conventional BLSA. The main difference is in the separation of pull-up driving lines and pull-down driving lines.

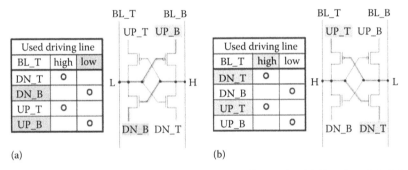

FIGURE 4.28 (a and b) Choice of driving lines used for sensing bit line data. BL_B indicates the reference bit line, whereas BL_T denotes bit line having stored data.

developed by UP_T and DN_T, as shown in Figure 4.28b. On the other hand, UP_B and DN_B are used when the H-SAs sense the majority of BLs of low data, as shown in Figure 4.28a.

The principle of H-SA is illustrated in Figure 4.29. As mentioned previously, when the majority of BLs data is high, the H-SA almost uses the UP_T and DN_T. This means that there is a large amount of current flow in the path of UP_T and DN_T. Therefore, the power drop should be large in these driving lines, as shown in Figure 4.29. But, in the path of UP_B and DN_B, there is small power drop due to a small amount of current flow. This difference between the T and B lines becomes the amount of offset compensation. In H-SA, a charge-shared voltage can be determined as the expression including the offset compensation term, as shown in Figure 4.30. Especially, the island zero pattern shows the offset compensation term of positive value, so the charge-shared voltage should be larger by the amount of the compensation term. As a result, H-SA always shows an improved sensing margin in the island data patterns, even though there is a sacrifice we are willing to endure in the solid data pattern. In Figure 4.30, wave forms of UP and DN illustrate the difference in potential between separated driving lines, which means an offset compensation term, when the data polarity of the majority of bit-lines is high.

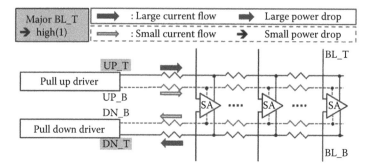

FIGURE 4.29 Situation when majority BLs pose high data. The pull-up and pull-down drivers provide the H-SAs with power using UP_T and DN_T, respectively. This leads to a large amount of current flow in path of the UP_T and DN_T driving lines, resulting in a large power drop.

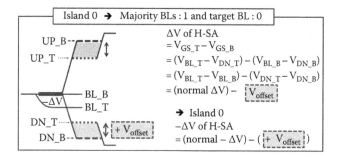

FIGURE 4.30 Illustration denoting the amount of offset compensation in the H-SA. The difference in potential between UP_T and UP_B (or DN_T and DN_B) shows the amount of offset compensation.

H-SA makes use of the voltage drop phenomenon from the current flow in the resistance in favor of suppressing the sensing noise. Therefore, the magnitude of the resistance in the current path becomes the most important factor in noise compensation.

Figure 4.31 illustrates three kinds of H-SA, which are named (a) semi-H-SA, (b) H-SA 1 and 2 in accordance with the existence of connecting metal for decreasing a too large potential gap between the T and B lines. Both H-SA 1 and 2 are fabricated with 68 nm technology, whereas the semi-H-SA is only simulated. The big difference between (a) semi H-SA and (b) H-SA types is in whether the T and B lines share the contact resistance, or not. In the case of semi-H-SA, the difference in the potential between the T and B lines should be determined by only the amount of driving line resistance. Therefore, it is expected that the difference in potential will be too small to compensate for the total amount of sensing noise.

Figure 4.32 shows the simulated potentials of the UP_T, DN_T, and the UP_B, DN_B signals in the early stage of sensing an island data pattern for three kinds of H-SA types. These H-SA types show their own potential difference between the T and B lines, which means that the amount of noise compensation can be controlled

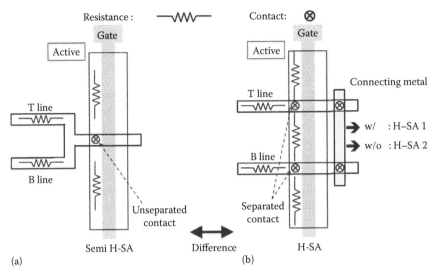

FIGURE 4.31 Three kinds of H-SA types. The main difference between (a) semi H-SA and (b) H-SA is in whether T and B driving lines share contact resistance. H-SA 1 and H-SA2 is differentiated in accordance with the existence of connecting metal.

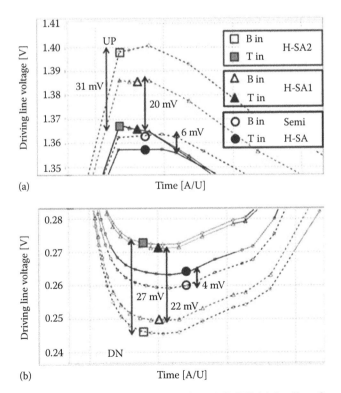

FIGURE 4.32 Simulated voltage results of (a) UP and (b) DN driving lines for semi H-SA, H-SA 1, and H-SA 2.

by several sense amplifier (SA) driver types. In the case of the H-SA 2, there is ~30 mV difference between the T line and the B line. H-SA 1 and semi-H-SA show ~20 and ~5 mV difference, respectively. If this difference is used in the voltage as a compensation method for the sensing noise found in the data patterns, the total amount of the BLSA offset almost disappears.

Figure 4.33 shows the amount of the measured sensing noise in the fabricated DRAM chips including the conventional BLSA, H-SA 1, and H-SA 2. There are four representative data patterns, each with distinctive BLSA offsets. In the conventional BLSA, the offset due to the sensing noise in the island pattern is ~30 mV larger than that found in a solid pattern.

However, in the H-SA 1, island and solid patterns show almost the same offset, so the difference is 5 mV in the one data pattern and 12 mV in the zero data pattern. As a result, the H-SA 1 displays around 18.5% of the total sensing noise measured in conventional SA, as shown in Figure 4.34. Although this voltage drop effect helps the offset in the island pattern data, it has a disadvantage in the offset in the solid pattern data. However, it is more important to guarantee that the maximum offset value is small enough to be able to sense any kind of data pattern. Therefore, the mostly negligible offset in the solid data pattern is not worth consideration.

However, the sensing noise in the solid data pattern can be much larger (by around 13–30 mV) than that in the island data pattern in the H-SA 2, which has a slightly different driver shape than does the H-SA 1. The H-SA 2 shows too large an opposite noise to compensate for the data pattern noise, as shown in Figure 4.34. Therefore, the solid data pattern noise is larger than the island pattern noise in this type. This is not the solution for minimizing the data pattern noise, because of a large solid

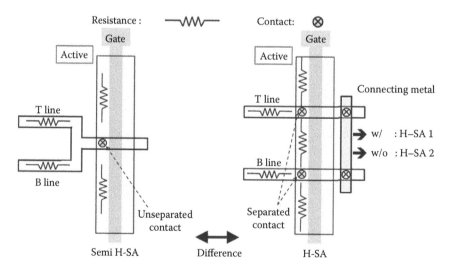

FIGURE 4.33 Measured sensing noise for the several BLSA types. The potential gap between island and solid patterns denotes sensing noise in accordance with data pattern. Normal BLSA shows the sensing noise of ~30 mV.

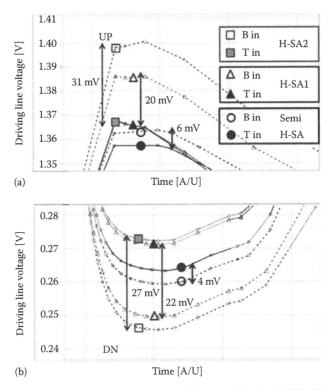

(a)

(b)

FIGURE 4.34 Improved sensing noise for the H-SA types, showing 18.5% of the total noise of normal BLSA in H-SA 1. However, H-SA 2 shows a flipped sensing noise, which means that the sensing noise in the solid pattern data is worse than that in the island pattern.

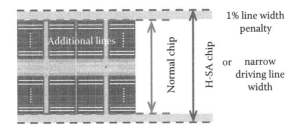

FIGURE 4.35 Chip photo illustrating the line width penalty in this scheme.

pattern noise. Therefore, it is necessary to precisely control the difference in voltage between the T and B driving lines, in accordance with four kinds of data patterns.

The proposed chip has an area penalty for the additional driving lines. It is about less than 1% of the total chip size, as shown in Figure 4.35. This result is due to the additional UP and DN driving lines. In the fabricated 68 nm DRAM chip, it pays the penalty of the narrow line width.

REFERENCES

1. D. Kahng and M. M. Atalla, Silicon-silicon dioxide field induced surface devices, in *IRE Solid-State Device Research Conference*, Carnegie Institute of Technology, Pittsburgh, PA, 1960.

2. G. E. Moore, Progress in digital integrated electronics, *IEDM Tech. Dig.* 21, 11–13, 1975.

3. *International Technology Roadmap for Semiconductors*, Semiconductor Industry Association, CA, 2003.

4. T. H. Ning, Silicon technology directions in the new millennium, in *IEEE 38th International Reliability Physics Symposium*, San Jose, CA, pp. 1–6, April 2000.

5. K. Itho, Y. Nakagome, S. Kimra, and T. Watanabe, Limitations and challenges of multi-gigabit DRAM chip design, *IEEE J. Solid-State Circ.*, 32, 624–634, May 1995.

6. K. Kim, C. Hwang, and J. G. Lee, DRAM technology perspective for gigabit era, *IEEE Trans. Electron Dev.*, 45, 598–608, March 1998.

7. K. Itho, K. Sasaki, and Y. Nakagome, Trends in low-power RAM circuit technologies, *Proc. IEEE*, 83, 524–543, April 1995.

8. D. J. Frank, R. H. Dennard, E. Nowak, P. M. Solomon, Y. Taur, and H. P. Wong, Device scaling limits of Si MOSFETs and their application dependencies, *Proc. IEEE*, 89, 259–288, March 2001.

9. K. Itho, T. Watanabe, S. Kimura, and T. Sakata, Reviews and prospects of high-density DRAM technology, in *Proceedings of the IEEE International Semiconductor Conference (CAS2000)*, Sinaia, Romania, 1, 13–22, 2000.

10. J. H. Comport, DRAM technology: Outlook and challenge, in *IEEE 6th International Conference on VLSI and CAD '99*, Seoul, Korea,1999, pp. 182–186.

11. S. H. Lee, J. Lee, Y. Ahn, D. Ha, G. Koh, T. Chung, and K. Kim, Novel cell transistor using a localized channel and field implantation (LOCFI) technology for improving the data retention time, *J. Korea Phys. Soc.*, 40, 630–635, April 2002.

12. S. Amakawa, K. Nakazato, and H. Mizuta, A new approach to failure analysis and yield enhancement of very large-scale integrated systems, in *Proceedings of 32th European Solid-State Device Research Conference*, Florence, Italy, pp. 147–150, 2002.

13. A. Hiraiwa, M. Ogasawara, N. Natsuaki, Y. Itoh, and H. Iwai, Local-field-enhancement model of DRAM retention failure, *IEDM Tech. Dig.*, San Francisco, CA, 157–160, 1998.

14. T. Hamamoto, S. Sugiura, and S. Sawada, On the retention time distribution of dynamic random access memory (DRAM), *IEEE Trans. Electron Dev.*, 45, 1300–1309, June 1998.

15. L. Selmi, D. Esseni, and P. Palestri, Towards microscopic understanding of MOSFET reliability: The role of carrier energy and transport simulations, *IEDM Tech. Dig.*, 333–336, 2003.

16. C. Lee, J.-M. Yoon, C.-H. Lee, J. C. Park, T. Y. Kim, H. S. Kang, S. K Sung et al., Enhanced dara retention of damascene-finFET DRAM with local channel implantation and <100> fin surface orientation engineering, *IEDM Tech. Dig.*, 61–64, 2004.

17. H. S. Kim, D. H. Kim, J. M. Park, Y. S. Hwang, M. Huh, H. K. Hwang, N. J. Kang et al., An outstanding and highly manufacturable 80 nm DRAM technology, *IEDM Tech. Dig.*, 2003, 411–414, 2003.

18. J. W. Lee, Y. S. Kim, J. Y. Kim, Y. K. Park, S. H. Shin, S. H. Lee, J. H. Oh et al., Improvement of data retention time in DRAM using recessed channel array transistors with asymmetric channel doping for 80 nm feature size and beyond, in *ESSDERC*, Leuven, Belgium, 449–452, 2004.

19. J. Y. Kim, C. S. Lee, S. E. Kim, I. B. Chung, Y. M. Choi, B. J. Park, J. W. Lee et al., The breakthrough in data retention time of DRAM using Recess-channel-array transistor (RCAT) for 88 nm feature size and beyond, in *Symposium on VLSI Technical Digest*, 11–12, 2003.

20. J. Y. Kim, H. J. Oh, D. S. Lee, D. H. Kim, S. E. Kim, G. W. Ha, H. J. Kim et al., S-RCAT (Sphere-shaped-recess-channel-array transistor) technology for 70 nm DRAM feature size and beyond, in *Symposium on VLSI Technical Digest*, 2005, 34–35, 2005.

21. S.-W. Chung, S.-D. Lee, S.-A. Jang, M.-S. Yoo, K.-O. Kim, C.-O Chung, S. Y. Cho et al., Highly Scalable Saddle-Fin (S-Fin) transistor for Sub 50 nm DRAM technology, in *Symposium on VLSI Technical Digest*, 2006, 147–148, 2006.

22. D.-H. Lee, B.-C. Lee, I.-S. Jung, T. J. Kim, Y.-H. Son, S.-G. Lee, Y.-P. Kim, S. Choi, U-I. Chung, and J.-T. Moon, Fin-channel-array transistor (FCAT) featuring sub-70 nm low power and high performance DRAM, *IEDM Tech. Dig.*, 407–410, December 2003.

23. M. J. Lee, J. H. Cho, S. D. Lee, J. H. Ahn, J. W. Kim, S. W. Park, Y. J. Park, and H. S. Min, Partial SOI type isolation for improvement of DRAM cell transistor characteristics, *IEEE Electron Dev. Lett.*, 26(5), 332–334, May 2005.

24. M. J. Lee, C.-K. Baek, S. Jin, I.-Y. Chung, Y. J. Park, and H. S. Min, A new recessed FinFET with R-shaped side channel (RFinFET) for DRAM cell applications, in *IEEE Silicon Nanoelectronics Workshop*, Honolulu, HI, pp. 147–148, June 2006.

25. L. Chang, Y.-K. Choi, D. Ha, P. Ranade, S. Xiong, J. Bokor, C. Hu, and T.-J. King, Extremely scaled silicon nano-CMOS devices, *Proc. IEEE*, 91, 1860–1873, November 2003.

26. B. Yu, H. Wang, A. Joshi, Q. Xiang, E. Ibok, and M.-R. Lin, 15 nm gate length planar CMOS transistor, *IEDM Tech. Dig.*, 937–939, 2001.

27. T. Ghani, K. Mistry, P. Packan, S. Thompson, M. Stettler, S. Tyagi, and M. Bohr, Scaling challenges and device design requirements for high performance sub-50 nm gate length planar CMOS transistors, in *Symposium on VLSI Technical Digest*, 174–175, 2000.

28. S. Jin, J.-H. Yi, J. H. Choi, D. G. Kang, Y. J. Park, and H. S. Min, Modeling of retention time distribution of DRAM cell using Monte-Carlo method, *IEDM Tech. Dig.*, 399–402, December 2004.

29. S. Jin, J.-H. Yi, J. H. Choi, D. G. Kang, Y. J. Park, and H. S. Min, Prediction of data retention time distribution of DRAM by physics-based statistical simulation, *IEEE Trans. Electron Dev.*, 52(11), 2422–2429, November 2005.

30. S. Jin, M. J. Lee, J.-H. Yi, J. H. Choi, D. G. Kang, I.-Y. Chung, Y. J. Park, and H. S. Min, A new direct evaluation method to obtain the data retention time distribution of DRAM, *IEEE Trans. Electron Dev.*, 53(9), 2344–2350, September 2006.

31. J. P. John, V. Ilderem, C. Park, J. Teplik, K. Klein, and S. Cheng, A low voltage graded-channel MOSFET (LV-GCMOS) for sub 1-Volt microcontroller application, in *Symposium on VLSI Technical Digest*, 178–179, June 1996.

32. B. Cheng, V. R. Rao, and J. C. S. Woo, Exploration of velocity overshoot in a high-performance deep sub-100 nm SOI MOSFET with asymmetric channel profile, *IEEE Electron Dev. Lett.*, 20, 538–540, October 1999.

33. J. W. Lee, Y. S. Kim, J. Y. Kim, Y. K. Park, S. H. Shin, S. H. Lee, J. H. Oh et al., Improvement of data retention time in DRAM using recessed channel array transistors with asymmetric channel doping for 80 nm feature size and beyond, in *Solid-State Device Research Conference*, 449–452, September 2004.

34. M. Darwish, J. Lentz, M. Pinto, P. Zeitzoff, T. Krutsick, and H. Vuong, An improved electron and hole mobility model for general purpose device simulation, *IEEE Trans. Electron Dev.*, 44(9), 1529–1538, September 1997.

35. D. M. Caughey and R. E. Thomas, Carrier mobilities in silicon empirically related to doping and field, *Proc. IEEE*, 55, 2192–2193, 1967.

36. D. B. M. Klaassen, A unified mobility model for device simulation—I, *Solid-State Electron.*, 35, 953–960, 1992.

37. D. B. M. Klaassen, A unified mobility model for device simulation–II, *Solid-State Electron.*, 35, 961–967, 1992.

38. G. Vincent, A. Chantre, and D. Bois, Electrical field effect on the thermal emission of traps in semiconductor junction, *J. Appl. Phys.*, 50(8), 5484–5487, August 1979.
39. G. A. M. Hurkx, D. B. M. Klaassen, and M. P. G. Knuvers, A new recombination model for device simulation including tunneling, *IEEE Trans. Electron Dev.*, 39(2), 331–338, February 1992.
40. W.-S. Lee, S.-H. Lee, C.-S. Lee, K.-H. Lee, H.-J. Kim, J.-Y. Kim, W. Yang, Y.-K. Park, J.-T. Kong, and B.-I. Ryu, Analysis on data retention time of nano-scale DRAM and its prediction by indirectly probing the tail cell leakage current, *IEDM Tech. Dig.*, 395–398, December 2004.
41. Y.-S. Chun, B.-J. Park, G.-T. Jeong, Y.-S. Hwang, K.-H. Lee, H.-S. Jeong, T.-Y. Jung, and K. Kim, A new DRAM cell technology using merged process with storage node and memory cell contact for 4 Gb DRAM and beyond, *IEDM Tech. Dig.*, San Francisco, CA, 351–354. 1998.
42. J. M. Park, Y. S. Hwang, D. W. Shin, M. Huh, D. H. Kim, H. K. Hwang, H. J. Oh et al., Novel robust cell capacitor (Leaning Exterminated Ring type Insulator) and new storage node contact (Top Spacer Contract) for 70 nm DRAM technology and beyond, in *Symposium on VLSI Technical Digest*, pp. 34–35, 2004.
43. K. Kim and M.-Y. Jeong, The COB stack DRAM cell at technology node below 100 nm-scaling issues and directions, *IEEE Trans. Semicond. Manuf.*, 15(2), 137–143, 2002.
44. M. J. Lee and K. W. Park, A mechanism for dependence of refresh time on data pattern in DRAM, *IEEE Electron Dev. Lett.*, 31(2), 168–170, February 2010.
45. M. J. Lee, S. Jin, C.-K. Baek, S.-M. Hong, S.-Y. Park, H.-H. Park, S.-D. Lee et al., A proposal on an optimized structure with experimental studies on recent devices for the DRAM cell transistor, *IEEE Trans. Electron Dev.*, 54(12), 3325–3335, December 2007.
46. T. Schloesser, F. Jakubowski, J. Kluge, A. Graham, S. Slesazeck, M. Popp, P. Baars et al., 6F2 buried wordline DRAM cell for 40 nm and beyond, *IEDM Tech. Dig.*, 2008, 33.4.1–33.4.4, December 2008.
47. T. Sekiguchi, K. Itoh, T. Takahashi, M. Sugaya, H. Fujisawa, M. Nakamura, K. Kajigaya, and K. Kimura, A low-impedance open-bitline array for multigigabit DRAM, *IEEE JSSC*, 37(4), 487–498, April 2002.
48. J. H. Ahn, T. H. Kim, S. M. Park, S. H. Wang, and H.-G. Lee, Bidirectional matched global bit line scheme for high density DRAMs, in *Symposium on VLSI Technical Digest*, 91–92, 1993.
49. N. C. C. Lu, H. H. Chao, and W. Hwang, Plate-noise analysis of an on-chip generated half-VDD biased-plate PMOS cell in CMOS DRAMs, *IEEE JSSC*, 20(6), 1272–1276, 1985.
50. M. J. Lee, K. M. Kyung, H. S. Won, M. S. Lee, and K. W. Park, A bitline sense amplifier for offset compensation, Solid-state Circuits Conference Digest of Technical Papers (ISSCC), *2010 IEEE International*, San Francisco, CA, pp. 438–439, February 2010.

5 Concepts of Capacitorless 1T-DRAM and Unified Memory on SOI

Sorin Cristoloveanu and Maryline Bawedin

CONTENTS

5.1 INTRODUCTION

While the scaling of MOS transistors is going on, the miniaturization of the DRAM storage capacitor reaches a critical limit. A novel and responsible strategy consists of attempting to suppress the capacitor. Silicon-on-Insulator (SOI) technology offers the opportunity to store the charges directly in the floating body of a MOSFET, which is also used to read the memory states. These memories, named floating-body 1T-DRAMs, use only one transistor and take advantage of floating-body effects that are usually regarded as parasitic phenomena.

In the last decade, several competing 1T-DRAM variants have been proposed: partially depleted (PD) or fully depleted (FD), planar, vertical or semivertical (FinFET), single gate or double gate, etc. In this chapter, the most promising 1T-DRAM structures will be reviewed by focusing on MSDRAM, ARAM, and Z^2-RAM concepts. The device architecture, scaling issues, and different options for memory programming and reading will be addressed.

The "unified" memory is another advanced and exciting paradigm. The SOI MOSFET is still the ideal candidate as it features two (or more) independent gates: each gate can be assigned distinct tasks (program, store, or read the charge). Solutions are proposed for combining, within a single SOI transistor, volatile and nonvolatile memory (NVM) functionalities as well as for reaching multiple memory states (≥ 4).

5.2   1T-DRAM OPERATING MECHANISMS

The body of SOI transistors is fully isolated (Figure 5.1) providing a suitable storage volume. 1T-DRAMs benefit from floating-body effects and coupling mechanisms that are often viewed as undesirable properties. In all 1T-DRAM versions, state "1" (high drain current) reflects an excess of majority carriers in the body, which causes the potential and the drain current (I_1) to increase (Figure 5.1). Reciprocally, state "0" features lower current (I_0) and higher threshold voltage due to the removal of majority carriers from the body. In SOI n-MOSFETs, the threshold voltage strongly depends on the amount of majority carrier charge in the body. The holes injected into or extracted from the body produce a substantial body potential increase (Figure 5.2a) or decrease (Figure 5.2b). This yields a threshold voltage decrease or increase, hence a drain current enhancement or reduction.

PD SOI 1T-DRAMs are single-gate operated (with grounded back gate $V_{G2} = 0$). Excess carriers are stored in the neutral region of the body. In FD SOI, there is no neutral region, and back-gate biasing is required to accommodate the majority carriers in the accumulated back channel (negative V_{G2} in Figure 5.1). The front and back interfaces are electrostatically coupled, which enables sensing the front-channel drain current for memory reading. The magnitude of the *front* inversion current reflects the condition at the *back* channel: depletion (memory state "0") or accumulation (state "1"). 1T-DRAMs can be classified according to the mechanism serving to generate the majority carriers:

- *Impact ionization.* A large drain–voltage bias V_D is applied while the front interface is in inversion mode ($V_D > |V_{G1} - V_{TH1}|$). The electric field at the pinch-off region is sufficiently high to generate electron–hole pairs by impact ionization. The holes move toward the back interface and are

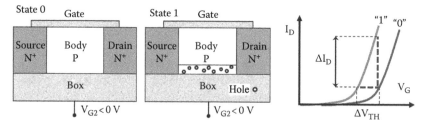

FIGURE 5.1   Schematic configuration of SOI n-MOSFET used as 1T-DRAM memory cell and current–voltage characteristics in "0" and "1" states.

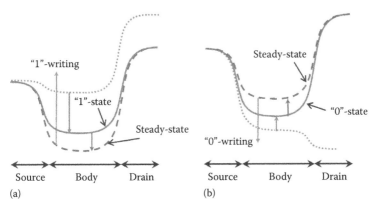

FIGURE 5.2   Front-surface potential variations during (a) the "1"-state and (b) the "0"-state writing and reading in a 1T-DRAM. The "1"-state writing is performed by impact ionization or B2B tunneling and the "0"-state by drain forward biasing or capacitive coupling. Although the potential evolves between programming and reading, the final potential of "0"-state during reading is lower than that in "1"-state (solid lines) resulting in an increase in the threshold voltage and a lower drain current.

accumulated in the PD body [1–5]. In FD devices (thinner silicon film thickness), a negative back-gate bias holds the hole charge inside the body. The impact ionization allows fast 1-state writing but leads to high power consumption and reliability issues (hot-carrier degradation).

- *Bipolar junction transistor (BJT).* The floating-body SOI MOSFET uses the source N^+, body P, and drain N^+ regions respectively as emitter, base, and collector [1]. To turn on the BJT, a hole current has to be generated in the floating base in order to increase the base (or body) potential. A negative V_D pulse forward biases the body/drain (here emitter) junction, and a negative V_{GI} bias is applied to keep the excess of generated majority carriers inside the base. As the source/body junction is reversed biased, the electrons flowing from the drain to the source trigger the electron/hole pair generation by impact ionization at the body/source junction; the generated holes are stored inside the body under the gate, which is negatively biased. According to a more effective BJT programming variant [6–8], positive pulses are applied on the gate and drain terminals in order to enhance the body potential increase. The gate switches from a negative bias (required for the majority carrier storage) to zero (or a less negative bias), increasing the body potential and turning on the source/body (emitter) junction. Holes are generated by impact ionization in the body through the reverse-biased body/drain junction. Programming and reading in BJT mode are fast and power efficient but require high drain voltage.
- *Band-to-band (B2B) tunneling.* The majority carriers (holes) are created at the gate-to-drain (or source) overlap region [9–12]. Negative gate and positive drain pulses are applied to enhance the local electric field and enable the band bending in the gate overlap region. At the programming onset (negative gate voltage, $V_{GI} \ll V_{FBI}$), the body potential decreases by

dynamic gate coupling and becomes negative. The holes are gradually filling the body and cannot escape through the source because the body potential is negative. Therefore, the carrier storage and the deeper body potential
well are more efficient compared with other programming methods. When
the gate returns to a less negative value (hold and read), the hole concentration becomes larger than that at equilibrium. A temporary excess drain
current (overshoot) is observed during the reading. As the MOSFET lies in
off-state mode during the programming, this generation mechanism is very
attractive from a power consumption viewpoint.

- *Gate tunneling current.* The direct tunneling current through ultrathin
  gate oxides is used to inject the majority carriers into the body [13]. In
  the specific p-MOSFET configuration of Figure 5.3c, there is an N^+ body
  contact left floating. The polysilicon gate covering the body contact is N^+
  doped in order to improve the injection of electrons (here the majority
  carriers) by tunneling from the polysilicon conduction band into the body.
  As electron tunneling is more efficient, a p-MOSFET was used instead of
  the standard n-MOSFET for the 1T-DRAM application. To program the
  "1"-state, a negative gate bias is applied, and the resulting electron tunneling current induces the body potential decrease. As the n-type body potential becomes negative, a higher drain current can be observed. The main
  advantages of this method are low bias and power consumption reduction
  as compared with the 1-state programming by impact ionization.
- *Photo-generation.* If optical interconnections succeed in CMOS technology, they could also serve for memory devices. 1T-DRAM programming
  assisted by optically generated holes was proposed and demonstrated
  experimentally [14]. The "1"-state programming method is based on the
  BJT conduction, which is triggered by a source of light. The gate voltage
  pulse used to switch on the bipolar latch is replaced by the carrier photo-
  generation. During the memory operations (read, hold, program), the
  gate voltage is kept at a constant negative value. Although the worst case
  (disturbance for "0"-state holding) has not been investigated yet, the proposed programming method is an interesting candidate for electro-optic
  hybrids in memory applications.

The programming of the 0-state requires the hole removal by

1. Forward biasing the drain– or source–body junction.
2. Dynamic coupling between front and back gates in FD, where the sudden
   increase in the gate bias increases the body potential above equilibrium and
   naturally turns on the junctions.

The 1T-DRAM reading is nondestructive because the drain voltage is low (except
for BJT method) and does not alter the memory states.

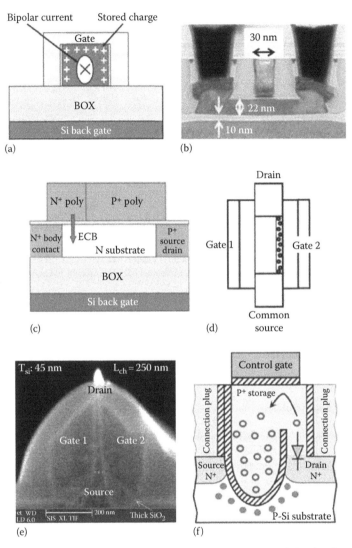

FIGURE 5.3 1T-DRAM variants: (a) triple-gate Z-RAM (From Okhonin, S. et al., *Int. Electr. Dev. Meet.*, 925–928, 2007), (b) 45 nm node FD 1T-DRAM (From Avci, U.E. et al., *IEEE International SOI Conference*, 29–30, 2008), (c) 1T-DRAM programmed by gate tunneling current (From Guegan, G. et al., *Sol. State Dev. Mater.*, 2010), and (d and e) vertical double-gate 1T-DRAM with independent gates and common source (From Jeong, H. et al., *IEEE Trans. Nanotechnol.*, 6, 352, 2007; Ertosun, M.G. et al., *IEEE Electron Dev. Lett.*, 29, 615, 2008). (f) Floating junction gate (FJG) cell. (From Wang, P.F. and Gong, Y., *IEEE Electron Dev. Lett.*, 29, 1347, 2008.)

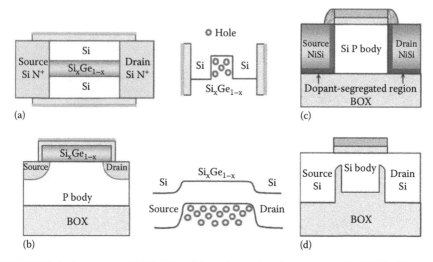

FIGURE 5.4 Innovative 1T-DRAMs with engineered body structure: (a) Si/SiGe/Si body stack and (b) convex channel structure filled with SiGe. The diagrams on the right show the potential well used for majority carrier storage. 1T-DRAMs with engineered source/drain terminals: (c) dopant-segregated Schottky contacts and (d) block-oxide junctions.

5.3   1T-DRAM ARCHITECTURES

Figures 5.3 through 5.7 present the protagonists of the 1T-DRAM family. These devices combine special architectural features with the programming methods exposed earlier.

5.3.1   Z-RAM VARIANTS

The first generation of 1T-DRAMs, called Z-RAM, was demonstrated on PD-SOI MOSFETs with 100 nm design rules. Impact ionization and forward drain biasing were initially used for programming the "1" and "0" states, respectively [1]. Improved performance was achieved with the second generation where the BJT effect and the capacitive coupling proved to be more efficient [6]. Underlapped gate structure was implemented for reaching longer retention time (70 ms at 85°C). The advantage is that the maximum electric field is lowered, hence reducing the B2B leakage current during holding and reading. Devices with 55 nm gate length and 80 nm thick Si film, tested in cell array configuration, showed a threshold voltage shift larger than $\Delta V_{TH} = 500$ mV and a very high I_1/I_0 current ratio (10^4–10^6) [7].

The Z-RAM principles, namely, the BJT action for programming and reading [6], were extended to FD SOI and FinFET/SOI devices. Figure 5.3a shows a Triple-Gate Z-RAM with 50 nm gate length and 11 nm fin width [8]. Advanced Z-RAMs, fabricated with 45 nm node CMOS, demonstrated that intentional asymmetry of the source and drain terminals is beneficial for longer retention and lower-voltage operation.

5.3.2   Fully Depleted 1T-DRAM

Early variants used the impact ionization and the forward drain bias techniques, together with a negative back-gate voltage, for programming the "1" and "0" states. The memory was read in strong inversion ($V_{G1} > V_{TH1}$, Figure 5.1). To enhance the retention time (25–100 ms) and the threshold voltage shift ($\Delta V_{TH} = 420$ mV), the CMOS/SOI process included LDD, moderate channel doping (3×10^{17} cm^{-3}), and negatively biased P-doped field plate located below the BOX [4]. The device of Figure 5.3b featured a high-K gate dielectric, a 22 nm thick film, and a thin (10 nm) BOX suitable for further lowering the back-gate voltage [5].

5.3.3   1T-DRAMs with Nonconventional MOSFET Architectures

While any standard SOI MOSFET can be operated in 1T-DRAM mode, structural modifications of the transistor can bring enhanced retention time and memory window. More or less exotic transistor architectures have been proposed, as discussed in the following. In order to become viable on the memory market, there is a tradeoff between performance gain and the complexity of the technology needed for such nonconventional devices.

Vertical 1T-DRAM (Figure 5.3d and e), with double-gate or gate-all-around configurations, has been demonstrated [15–17]. In a vertical transistor, the channel length is no longer an area-limiting factor, hence the goal of reaching 4F^2 feature size for ultimate integration density seems feasible. The volatile memory operation has been investigated with 2-D numerical simulations as well as with fabricated prototype devices. Majority carriers were generated either by impact ionization or by B2B tunneling. For a sensing margin of 40 µA/µm, the retention time at room temperature was 4 ms, needing further improvements. An interesting solution is to make use of the independent biasing of the two lateral gates (Figure 5.3e).

The floating junction gate (FJG) device [18] has a floating gate with "U-shape" connected to the drain via a gated p–n diode (Figure 5.3f). The U-shape gate increases the storage volume and extends the channel length. "1"-state is programmed by injecting positive charges into the floating gate through the gated p–n diode. The resulting potential increase enhances the electron density in the channel. High-positive drain voltage and negative MOS gate voltage are applied. The hole current is enhanced, thanks to B2B tunneling in the reversed-biased p$^+$/n$^+$ junction. "0"-state is written by forward biasing the diode, which removes holes from the gate. Numerical simulations predict 300 µA/µm sensing margin and 6 s retention time.

The quantum well (QW) 1T-DRAM variant uses an engineered body adding in the Si film a thin-layer semiconductor with narrower bandgap (SiGe). The SiGe layer serves as storage well for holes (Figure 5.4a). According to simulations, this QW memory has the capability to efficiently store the holes closer to the front gate (as compared with the storage at the film–BOX interface), so improving the V_{TH} shift, current sense margin, and retention time.

Another *QW* [19] has a *convex channel* in the SiGe layer located just underneath a raised gate oxide (Figure 5.4b). During the "1"-state programming, the holes are stored in the SiGe layer. As the junction potential barrier is lowered, holes easily diffuse through source and drain filling the SiGe region. If the bandgap is further reduced, a deeper potential well will be formed in the heterostructure for benefit in sensing margin and retention time.

An alternative strategy for increasing the retention time aims at reducing the leakage current through the junctions. The 1T-DRAM with *band-gap engineered source and drain* was proposed to form a deeper potential well in the body [20]. SiC source and drain make the body-to-junction potential barrier significantly higher. The potential well being deep, due to valence band offset, more holes can be stored in the body. The programming mechanisms are impact ionization for writing "1"-state and forward-biased drain–body junction current for "0"-state. Based on 2-D simulations, the sensing margin may reach 100 μA/μm, which is three times larger than that for conventional all-silicon 1T-DRAM. The hole leakage during the "1"-state programming was predicted to be two to three orders of magnitude lower.

Dopant-segregated Schottky contacts and FinFET configurations were also envisioned for improving the hole storage (Figure 5.4c) [21]. Dopant segregation with partial S/D silicidation enables the holes to flow into the body, which would be impossible with plain-metal Schottky contacts. By assuming a minimum sense margin of 100 μA/μm, the simulated retention time reached 70 ms. A competing variant, shown in Figure 5.4d, features self-aligned block oxide, which again attenuates the leakage-induced degradation of the stored charge, thanks to the reduced junction depth [22]. The processing seems to be complex so a more technology-friendly approach would be an SOI MOSFET with thinned junctions (as opposed to the current trend of raised source and drain terminals in CMOS/SOI).

5.3.4 MSDRAM MEMORY CELL

MSDRAM principles exploit the MSD hysteresis effect (Figure 5.5a), which appears in regular FD SOI MOSFETs [10]. A moderately inverted back channel is formed by applying an appropriate positive bias on the back gate ($V_{G2} \geq V_{TH2}$). The front gate V_{G1} is swept from strong accumulation ($V_{G1} < -4$ V) to nearly 0 V. The drain voltage can be low ($V_D \sim 100$–200 mV). For high negative V_{G1}, B2B tunneling occurs and efficiently fills the front channel with holes. The transistor is at equilibrium, and high current flows at the back channel when reading state "1" ($V_{G1} \approx -3$ V). By contrast, for reverse V_{G1} scan from 0 to -3 V, the device does not reach equilibrium and operates in deep depletion. Since there are no holes immediately available to fill the front channel, the body potential drops, temporarily suppressing the back-channel current (state "0").

MSDRAMs have two distinct advantages: very wide memory window ($-4 < V_{G1} < -2$ V) and outstanding current ratio I_1/I_0 exceeding six orders of magnitude. The MSDRAM takes full advantage of double-gate operation mechanisms and features low-power consumption and superior reliability. Customized source and drain architectures and thin BOX allow enhancing the retention time while reducing the programming voltage (≤ 2 V) and back-gate bias. For example, front-gate overlap increases B2B tunneling rate during programming whereas back-gate underlap reduces the parasitic B2B hole generation during "0"-state hold. A single-gate

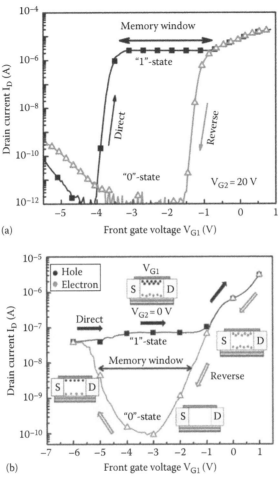

(a)

(b)

FIGURE 5.5 (a) Measured drain current I_D vs. increasing (direct scan) and decreasing (reverse scan) front-gate voltages ($V_{G2} = 30$ V, $V_D = 0.1$ V, $T_{BOX} = 400$ nm, L = 1.5 μm). (b) Current hysteresis in MSDRAM cell fabricated on ONO BOX. The back gate is grounded, hence the MSD hysteresis effect is entirely governed by the nonvolatile charge in the buried dielectric. A positive charge stored in the nitride is sufficient to invert the back channel, leading to single-gate MSDRAM operation.

MSDRAM version (Figure 5.5b) is possible by replacing the positive bottom gate bias ($V_{G2} \geq 0$) with a nonvolatile positive charge trapped in the BOX [23].

The MSDRAM scalability was demonstrated by 2-D simulations of 25–50 nm channel lengths. Figure 5.6a shows 14 s retention time for $I_1/I_0 = 10$ and fast programming time (5 ns). Recent measurements confirm that MSD effect is maintained in small-area MOSFETs (0.1 μm²) [24]. Additional experiments revealed the superiority of the MSD programming mechanisms (B2B tunneling and capacitive coupling) compared with conventional programming based on impact ionization and forward drain biasing (Figure 5.6b).

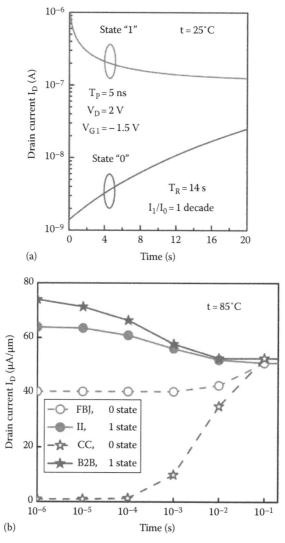

(a)

(b)

FIGURE 5.6 (a) Simulated short-channel MSDRAM: drain current I_D vs. time during the "0"- and "1"-states hold. $V_{G2} = 0.25$ V and $V_D = 0.1$ V; 40 nm thick Si film and 50 nm long channel. The front- and back-gate oxides are 3 and 6 nm thick, and the body doping is 10^{16} cm^{-3}. Fourteen-second retention time for high sensing margin ($I_1/I_0 = 10$) is achieved at 25°C. During "1"-state writing by B2B tunneling, V_D and V_{G1} are equal to 2 and -1.5 V with a programming time T_P of 5 ns. (b) Experimental retention at 85°C comparing conventional programming methods (impact ionization II and forward-biased drain junction FBJ) with MSD programming (B2B tunneling and capacitive coupling CC) in an FDSOI cell ($L_G = 0.35$ μm, $T_{si} = 55$ nm). (From Hubert, A. et al., *Solid-State Electron.*, 53, 1280, 2009.)

5.3.5 ARAM MEMORY CELLS

The 1T-DRAMs presented earlier all require the coexistence of electrons and holes in a thin body. But in ultrathin SOI films (T_{si} < 10 nm, needed for digital CMOS/SOI circuits), the so-called super-coupling effect forbids the formation of accumulation and inversion layers facing each other [25]; the corresponding electric field ($\geq 2\Phi_F/T_{si}$) would indeed be too large for the body or gate oxide to sustain it. In sub-10 nm thick films, the body potential tends to become quasi-flat (even for V_{GF} > 0 and V_{GB} < 0), being controlled by the "stronger" gate, so that only electrons *or* only holes are available in the body.

ARAM concept solves the super-coupling rule by physically isolating, with a middle oxide (MOX), the holes and electrons in two dedicated semibodies [26]. The upper semibody is used for majority carrier storage and the lower semibody for electron current sense (Figure 5.7). The MOX isolation offers the advantage of maintaining electron and hole layers very close to each other (on both sides of the dielectric layer) even in ultrathin SOI films (<10–15 nm) fit for CMOS scaling. To write "1" (Figure 5.7a), the excess of holes is generated, as in regular 1T-DRAMs, by impact ionization or B2B tunneling in the upper semibody. The positive hole charge induces a dynamic increase in the upper semibody potential, which enables by electrostatic coupling an electron current to flow in the lower semibody (state "1"). By contrast, if no holes are stored, the lower semibody is FD, and no current flows (state "0," Figure 5.7b).

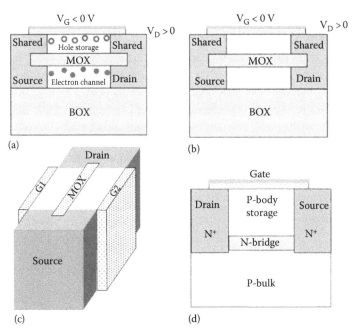

FIGURE 5.7 Configuration of ARAM cells: (a) In state "1," the holes stored in the upper semibody enable a large electron current in the lower semibody. (b) In state "0," both semibodies are fully depleted, and the current is zero. MOX and semibodies are typically 3–5 nm thick. (c) ARAM implementation in FinFETs. (d) A2RAM device (without MOX) in bulk or SOI.

The ARAM structure and operation remind the HRAM memory device proposed earlier in GaN/AlGaN heterostructures [27]. ARAM operates with a single gate, which makes simpler the programming and reading signals. ARAM architecture is also more realistic than the floating-junction gate (FJG, Figure 5.2f) in terms of scaling and process complexity. The proof of concept has been extended to DG and FinFETs (Figure 5.7c) but still needs to be verified at the experimental level. Nonvolatile functionality can be added by using oxide/nitride/oxide (ONO) stack for the MOX dielectric. The MOX is easily deposited on the initial Si film (lower semibody), after which epitaxial growth of the upper semibody layer completes the ARAM structure.

A second-generation device (A2RAM) was devised in order to suppress the MOX. The body is now composed of an FD P-layer on top of an N^+ layer (Figure 5.7d). The source and drain are short-circuited by the N^+ bridge. The top semibody stores the holes whereas the N^+ bridge serves for current sensing. When holes are accumulated at the surface (due to the negative V_{G1}), they screen the vertical electric field so that the electron current flows through the undepleted bridge. Conversely, if the top P-body is temporarily depleted of holes, the gate field is no longer screened and fully depletes the bridge, naturally canceling the drain current. Retention time over 100 ms was predicted at 85°C in optimized 22 nm node SOI cells. Preliminary measurements have validated the A2RAM concept on both SOI and bulk-Si wafers [28].

5.3.6   Z²-RAM Memory Cell

The Z^2-FET is an FD PIN diode [29], where the body is partially covered by the gate (Figure 5.8a). Although the diode is forward-biased, the current is initially blocked by gate-induced barriers. The front and back gates are biased such as to form two potential barriers preventing the injection of electrons from N^+ drain and of holes from P^+ source into the body. The gate biasing actually emulates a virtual PNPN thyristor configuration despite the body is undoped. The current remains low until V_D increases enough to slightly lower the electron injection barrier. Electrons injected

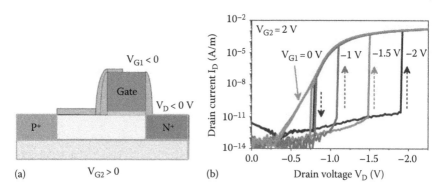

FIGURE 5.8 (a) Schematic view of the Z^2-FET transistor and 1T-DRAM memory. (b) Typical $I_D(V_D)$ characteristics showing gate-controlled hysteresis.

from N$^+$ contact into the channel flow to the P$^+$ source where they reduce the hole barrier. A few holes able to flow from source to drain further reduce the electron barrier. This positive feedback turns the device on and completely eliminates the injection barriers. The feedback mechanism results in a strong $I_D(V_D)$ hysteresis (Figure 5.8b), the amplitude of which is gate controlled and useful for capacitorless memory. Alternatively, the feedback can be triggered by slightly increasing the gate bias (at constant V_D). In this case, the device shows steep transition between off and on states: for $\Delta V_G < 1$ mV, the current changes by eight orders of magnitude [30]. These unrivaled features explain the device name: Z^2-FET for *zero* subthreshold swing and *zero* impact ionization (unlike a standard thyristor).

To program the "1" or "0" states, holes are stored or not stored under the negatively biased gate. Memory reading merely consists of discharging this stored charge [29]. In "0"-state, there is no discharge current, hence the diode remains blocked (negligible read current). In "1"-state, the discharge current $\Delta Q_G/\Delta t$ is sufficient to turn on the Z^2-FET and the read current is high. The read pulse should be very fast (1 ns or less) in order to take advantage of a minimum amount of stored charge ΔQ_G for triggering the device. It is worth underlying this key advantage: the memory state is defined by the transient current $\Delta Q_G/\Delta t$, not by the stored charge ΔQ_G as in other 1T-DRAMs, so that the number of holes to be stored is no longer a critical limitation. Besides excellent speed capability, the Z^2-FET memory shows scalability down to 30 nm gate length and offers long retention, $\sim$1 V operating voltage, and regenerative (nondestructive) reading.

5.3.7 UNIFIED MEMORY

An ideal memory is expected to comply with three requirements: high density, high speed, and nonvolatility. Such "universal memory" has not been conceived yet, so these three different targets have been pursued independently: DRAM, SRAM, and flash. A more realistic paradigm shift from "scaling" (more Moore) to "multifunction" (more than Moore) is envisioned by assigning the transistor different memory tasks. The "unified memory" (URAM) aims at combining the functionalities of NVM and volatile DRAM in a single transistor.

The URAM concept consists of implementing an NVM charge-trapping layer within the 1T-DRAM (Figure 5.9a). Several materials can serve for the NVM core region, including nanocrystals or ONO stacks. For example, electrons are injected and trapped in the silicon nitride layer like in a standard SONOS memory. The programming and erasing steps are performed with either Fowler–Nordheim or hot carrier injection through a thin tunnel oxide above the MOSFET channel. For volatile memory operation, the floating body is the storage volume as in 1T-DRAMs. This URAM idea is attractive and demonstrated by preliminary results [31–33]. Excess holes, generated by impact ionization for the 1T-DRAM "1"-state programming, were accumulated in the body region with lower potential. For "0"-state, the excess holes were swept out of the body by forward biasing the junctions. The FinFET shown in Figure 5.4c featured Schottky contacts and specific gate stack. This particular URAM cell was designed to speed up the flash memory programming while attenuating the FinFET short-channel effects.

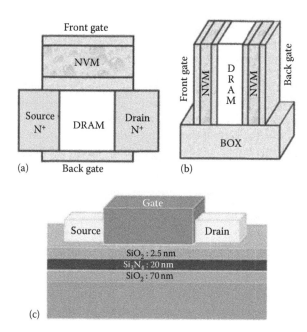

FIGURE 5.9 Concepts of unified memory combining the nonvolatile (NVM) functionality of the ONO flash and the volatile functionality of the 1T-DRAM in (a) double-gate planar MOSFET with ONO gate, (b) FinFET with two ONO gate stacks for multi-bit storage, and (c) FinFET with ONO buried insulator.

Additional FinFETs with ONO BOX and implanted junctions have been fabricated and tested in URAM configuration (Figure 5.9c) [34]. The buried nitride layer stores the NVM charge, which is detected by the current flowing at the front gate. The physical separation of the back and front interfaces, respectively used for programming and reading, avoids the disturbance of the stored charge during reading. Combined with enhanced scalability and CMOS compatibility, this feature is a clear benefit of ONO FinFETs. According to the polarity of the back-gate voltage, electrons ($V_{G2} > 0$) or holes ($V_{G2} < 0$) can be injected into the buried ONO layer by Fowler–Nordheim tunneling. The front-channel current reflects the variation of the stored charge via the "vertical" coupling effect, as shown in Figure 5.10a. Reducing the fin width below a critical size alters the sensitivity because the lateral gates tend to govern the back-surface potential ("lateral" coupling), gradually masking the effect of the trapped nonvolatile charge. However, in short FinFETs, the fringing field from source/drain through the BOX and substrate ("longitudinal" coupling) opposes the lateral gate action and restores the impact of the nonvolatile ONO charge. As a result, in shorter devices, the memory margin is extended [34].

An alternative programming method uses the drain bias, so eliminating the need for substrate bias (Figure 5.10b). For $V_D = +3$ V (with $V_{G2} = V_{GF} = 0$), holes are injected and trapped into the ONO. Reciprocally, for $V_D = -3$ V, electrons are trapped. The drain current at the front channel is very sensitive to the type and amount of charge stored (Figure 5.10b), and the retention time exceeds 10 years. When V_D is

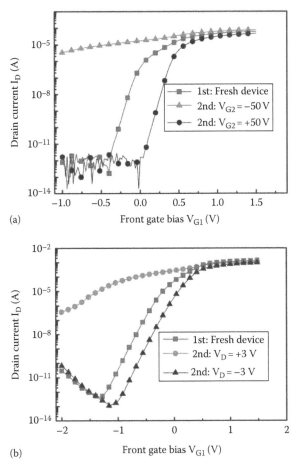

FIGURE 5.10 Transfer characteristics of unified memory combining the nonvolatile (NVM) functionality of the ONO flash and the volatile 1T-DRAM functionality in FinFETs with buried ONO stack. Carrier trapping in the nitride layer is achieved by (a) back-gate biasing (Fowler–Nordheim tunneling) or (b) drain biasing.

used for programming, the charge is actually localized near the drain, leading to a net difference between $I_D(V_D)$ characteristics measured in direct mode (source to drain) and reverse mode (drain to source). This asymmetry opens the avenue to "multiple bit" URAM. It has been experimentally demonstrated that positive *or* negative charge can be trapped near the drain *or* near the source. The four NVM states are easily discernible: the charge polarity is deduced from the current magnitude (increased or decreased level), whereas the charge localization is unambiguously resolved from the difference between direct and reverse currents [34].

Planar SOI-like MOSFETs have also been fabricated and tested for URAM feasibility. The transistor featured ONO BOX and Si back gate (ground plane), both localized underneath the transistor body [23]. When the ONO layer is not charged, multiple dynamic memory states are programmable by varying the back-gate bias

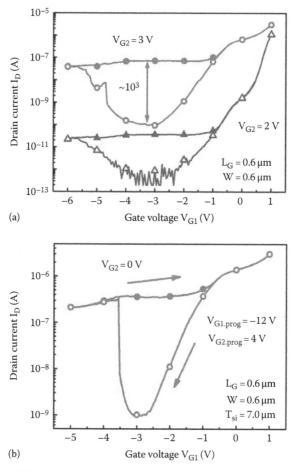

FIGURE 5.11 MSDRAM hysteresis in FD DG ONO MOSFET, where the nitride layer was integrated at the bottom gate. Direct and reverse scans of the front gate voltage are performed for (a) different back-gate voltages with uncharged ONO layer and (b) grounded back gate with the nitride layer positively charged.

(Figure 5.11a). A positive charge trapped in the ONO acts like a positive back-gate voltage. It was demonstrated that the MSD effect subsists even at $V_{G2} = 0$ (Figure 5.11b), which means that the MSDRAM becomes virtually single-gate-operated. In addition, the MSD current level depends on the polarity and concentration of nonvolatile stored charges, confirming capability for multi-bit operation.

The combination of NVM and 1T-DRAM features within the same cell is promising for higher integration density and reduced cost per bit, but is still challenging in several respects. For example, interference between the programming voltages of 1T-DRAM and NVM must be avoided [35,36]. Also the threshold voltage shift resulting from volatile and nonvolatile storage must be decorrelated. The reading procedure has to be revised in particular for the multi-bit operation.

5.4   CONCLUSIONS

The floating-body capacitorless DRAM memory is a very attractive device. There are several competing 1T-DRAM options, the principles of which have been discussed and critically compared. While brilliant, some of these solutions are not likely to be adopted by the technology. B2B tunneling appears to be very efficient for "1"-state programming, and capacitive coupling is suitable for "0"-state. 1T-DRAM volatile capability can be enriched by adding nonvolatile charge storage. Preliminary demonstrations with SOI transistors and FinFETs indicate that unified and multibit memory cells are manufacturable. The main concerns are the retention time for volatile operation in very short cells and the addressing/discrimination of volatile and NVM states.

ACKNOWLEDGMENT

This work has been encouraged and supported by the European Union programs (EUROSOI+, Reach 22) AMNESIA project (ANR 2011 JS03 001 01) and the WCU project of NRF (Korea).

REFERENCES

1. S. Okhonin, M. Nagoga, J.M. Sallese, P. Fazan, A capacitor-less 1T-DRAM cell, *IEEE Electron Device Letters*, **23**, 85–87 (2002).
2. C. Kuo, T.J. King, C. Hu, A capacitorless double gate DRAM technology for sub-100-nm embedded and stand-alone memory applications, *IEEE Transactions on Electron Devices*, **50**, 2408–2416 (2003).
3. T. Shino, T. Ohsawa, T. Higashi, K. Fujita, N. Kusunoki, Y. Minami, M. Morikado et al., Operation voltage dependence of memory cell characteristics in fully depleted floating-body cell. *IEEE Transactions on Electron Devices*, **52**, 2220–2226 (2005).
4. T. Hamamoto, Y. Minami, T. Shino, N. Kusunoki, H. Nakajima, M. Morikado, T. Yamada et al., A floating-body cell fully compatible with 90-nm CMOS technology node for a 128-Mb SOI DRAM and its scalability, *IEEE Transactions on Electron Devices*, **54**, 563–571 (2007).
5. U.E. Avci, I. Ban, D.L. Kencke, P.L.D. Chang, Floating body cell (FBC) memory for 16-nm technology with low variation on thin silicon and 10-nm box, *IEEE International SOI Conference*, New Paltz, NY, 29–30 (2008).
6. S. Okhonin, M. Nagoga, E. Carman, R. Beffa, E. Faraoni, New generation of Z-RAM. *International Electron Devices Meeting*, 925–928 (2007).
7. K.W. Song, H. Jeong, J.W. Lee, S.I. Hong, N.K. Tak, Y.T. Kim, Y.L. Choi et al., 55 nm capacitor-less 1T-DRAM cell transistor with non-overlap structure, *International Electron Devices Meeting,* 1–4 (2008).
8. S. Okhonin, M. Nagoga, C.W. Lee, J.P. Colinge, A. Afzalian, R. Yan, N. Dehdashti Akhavan et al., Ultra-scaled Z-RAM cell, *IEEE International SOI Conference*, New Paltz, NY, 157–158 (2008).
9. E. Yoshida, T. Tanaka, A capacitorless 1T-DRAM technology using gate-induced drain-leakage (GIDL) current for low-power and high-speed embedded memory, *IEEE Transactions on Electron Devices*, **53**, 692–697 (2006).
10. M. Bawedin, S. Cristoloveanu, D. Flandre, A capacitorless 1T-DRAM on SOI based on dynamic coupling and double-gate operation. *IEEE Electron Device Letters*, **29**, 795–798 (2008).

11. S. Puget, G. Bossu, C. Fenouiller-Beranger, P. Perreau, P. Masson, P. Mazoyer, P. Lorenzini, J.M. Portal, R. Bouchakour, T. Skotnicki, FD-SOI floating body cell eDRAM using gate-induced drain-leakage (GIDL) write current for high speed and low power applications, *International Memory Workshop(IMW '09)*, Monterey, CA, 1–2 (2009).

12. T. Tanaka, E. Yoshida, T. Miyashita, Scalability study on a capacitorless 1T-DRAM: From single-gate PD-SOI to double-gate fin DRAM, *Electron Devices Meeting, 2004. IEDM Technical Digest. IEEE International*, 919–922 (2004).

13. G. Guegan, P. Touret, G. Molas, C. Raynaud, J. Pretet, A novel capacitor-less 1T-DRAM on partially depleted SOI pMOSFET based on direct-tunneling current in the partial n+ poly gate, *Solid State Devices and Materials*, (2010).

14. D.I. Moon, S.J. Choi, J.W. Han, Y.K. Choi, An optically assisted program method for capacitorless 1T-DRAM, *IEEE Transactions on Electron Devices*, **57**, 1714–1718 (2010).

15. H. Jeong, K.W. Song, I.H. Park, T.H. Kim, Y.S. Lee, S.G. Kim, J. Seo et al., A new capacitorless 1T-DRAM cell: Surrounding gate MOSFET with vertical channel (SGVC Cell), *IEEE Transactions on Nanotechnology*, **6**, 352–357 (2007).

16. H.K. Chung, H. Jeong, Y.S Lee, J.Y. Song, J.P. Kim, S.W. Kim, J.H. Park, J.D. Lee, H. Shin, B.G. Park, A capacitor-less 1T-DRAM cell with vertical surrounding gates using Gate-Induced Drain-Leakage (GIDL) current, *IEEE Silicon Nanoelectronics Workshop*, Honolulu, Hawaii, 1–2 (2008).

17. M.G. Ertosun, H. Cho, P. Kapur, K.C. Saraswat, A nanoscale vertical double-gate single-transistor capacitorless DRAM, *IEEE Electron Device Letters*, **29**, 615–617 (2008).

18. P.F. Wang, Y. Gong, A Novel 4.5 F2 capacitorless semiconductor memory device, *IEEE Electron Device Letters*, **29**, 1347–1348 (2008).

19. M.G. Ertosun, P. Kapur, K.C. Saraswat, A highly scalable capacitorless double gate quantum well single transistor DRAM: 1T-QW DRAM, *IEEE Electron Device Letters*, **29**, 1405–1407 (2008).

20. T. Poren, H. Ru, W. Dake, Performance improvement of capacitorless dynamic random access memory cell with band-gap engineered source and drain, *Japanese Journal of Applied Physics*, **49**, 04DD02 (2010).

21. S.J. Choi, J.W. Han, S. Kim, H. Kim, M.G. Jang, J.H. Yang, J.S. Kim et al., High speed flash memory and 1T-DRAM on dopant segregated Schottky barrier (DSSB) FinFET SONOS device for multi-functional SoC applications, *International Electron Devices Meeting*, 1–4 (2008).

22. Y.M. Tseng, J.T. Lin, Y.C. Eng, S.S. Kang, H.J. Tseng, Y.C. Tsai, B.T. Jheng BT, B.H. Lin, A new process for self-aligned silicon-on-insulator with block oxide and its memory application for 1T-DRAM, *International Conference on Solid-State and Integrated-Circuit Technology*, 1154–1157 (2008).

23. K.H. Park, M. Bawedin, J.H. Lee, Y.H. Bae, J.H. Lee, S. Cristoloveanu, Fully depleted double-gate MSDRAM cell with additional nonvolatile functionality, *Solid-State Electronics*, **67**, 17–22 (2012).

24. A. Hubert, M. Bawedin, S. Cristoloveanu, T. Ernst Dimensional effects and scalability of Meta-Stable Dip (MSD) memory effect for 1T-DRAM SOI MOSFETs, *Solid-State Electronics*, **53**, 1280–1286 (2009).

25. S. Eminente, S. Cristoloveanu, R. Clerc, A. Ohata, G. Ghibaudo, Ultra-thin fully-depleted SOI MOSFETs: Special charge properties and coupling effects, *Solid-State Electronics*, **51**, 239–244 (2007).

26. N. Rodriguez, F. Gamiz, S. Cristoloveanu, A-RAM memory cell: Concept and operation, *IEEE Electron Device Letters*, **31**, 972–974 (2010).

27. M. Bawedin, M.J. Uren, F. Udrea, DRAM concept based on the hole gas transient effect in a AlGaN/GaN HEMT, *Solid-State Electronics*, **54**, 616–620 (2010).

28. N. Rodriguez, S. Cristoloveanu, F. Gamiz, Novel capacitorless 1T-DRAM cell for 22-nm node compatible with bulk and SOI substrates, *IEEE Transactions of Electron Devices*, **58**(8), 2371–2377 (2011).

29. J. Wan, C. Le Royer, A. Zaslavsky, S. Cristoloveanu, A compact capacitor-less high-speed DRAM using field effect-controlled charge regeneration, *IEEE Electron Device Letters*, **31**, 179–181 (2012).

30. J. Wan, S. Cristoloveanu, C. Le Royer, A. Zaslavsky, A feedback silicon-on-insulator steep switching device with gate-controlled carrier injection, *Solid-State Electronics*, **76**, 109–111, (2012).

31. J.W. Han, S.W. Ryu, C. Kim, S. Kim, M. Im, S.J. Choi, J.S. Kim et al., A Unified-RAM (URAM) cell for multi-functioning capacitorless DRAM and NVM, *International Electron Devices Meeting*, 929–932 (2007).

32. D.I. Bae, B. Gu, S.W. Ryu, Y.K. Choi, Multiple data storage of URAM (Unified-RAM) with Multi Dual Cell (MDC) method, *Silicon Nanoelectronics Workshop*, Honolulu, Hawaii, 1–2 (2008).

33. J.W. Han, S.W. Ryu, S. Kim, C.J. Kim, J.H. Ahn, S.J. Choi, J.S. Kim et al., A bulk FinFET Unified-RAM (URAM) cell for multifunctioning NVM and capacitorless 1T-DRAM, *IEEE Electron Device Letters*, **29**, 632–634 (2008).

34. S.J. Chang, M. Bawedin, W. Xiong, J.H. Lee, S. Cristoloveanu, FinFlash with buried storage ONO layer for flash memory application, *Solid-State Electronics*, **70**, 59–66 (2012).

35. J.W. Han, S.W. Ryu, S.J. Choi, Y.K. Choi, Gate-Induced Drain-Leakage (GIDL) programming method for soft-programming-free operation in Unified RAM (URAM), *IEEE Electron Device Letters*, **30**, 189–191 (2009).

36. S.W. Han, C.J. Kim, S.J. Choi, D.H. Kim, D.I. Moon, Y.K. Choi, Gate-to-source/drain nonoverlap device for soft-program immune Unified RAM (URAM), *IEEE Electron Device Letters*, **30**, 544–546 (2009).

6 A-RAM Family
Novel Capacitorless 1T-DRAM Cells for 22 nm Node and Beyond

Francisco Gamiz, Noel Rodriguez, and Sorin Cristoloveanu

CONTENTS

6.1 INTRODUCTION

Semiconductor memory market represents more than 30% of the total semiconductor market in the world; in particular, dynamic random access memory (DRAM) industry revenues are around 35 billions of U.S. dollars every year, and it is expected to continue to grow in the coming years. In addition, emerging silicon and package technologies will further drive lower cost and new applications. However, the difficulty of scaling and developing new technologies and investments to build new factories is increasing at about the same rate as the memory bit growth in the world. At the same time, the industry is becoming aware that we are facing physical and electrical scaling limitations. As we close in on scaling limits, the use of new materials, new structures, manufacturing processes, and circuit design, mostly related to the capacitor scalability, will become unavoidable. In this

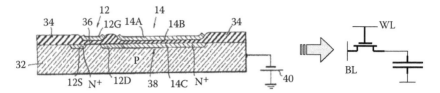

FIGURE 6.1 Schematics of a standard DRAM cell. (After Dennard, R., Field effect transistor memory, U.S. Patent No. 3,387,286.)

chapter, we will review some of the solutions recently introduced to solve the problem, and we will discuss in detail one of these solutions, the A-RAM family.

The profitable semiconductor memory business is strongly based on a very simple concept, the 1-transistor–1-capacitor (1T–1C) DRAM cell introduced by Robert Dennard (IBM) many years ago [1]. The device cell comprises an access transistor and a storage capacitor as shown in Figure 6.1. To hold the "1" binary state, a certain amount of charge is stored in the capacitor, while the capacitor is left uncharged to represent the "0" state. The basic principle of operation has remained unaltered for more than three decades [1], although the cell has evolved to complicated three-dimensional structures, including high-k materials and corrugated capacitor surfaces [2]. The demand of DRAM constitutes one of the most vivid markets of the semiconductor industry. The basic cell has evolved during the years, even more aggressively than high-performance systems, as can be observed in Figure 6.2, where we represent the number of transistors per die in integrated circuits of memory and in microprocessors. As seen, the number of transistors in memory chips quadruplicates every year, as a consequence of the scaling of the transistor, but also thanks to the use of innovative structures to build the storage capacitor. However, serious constraints are at present threatening the DRAM

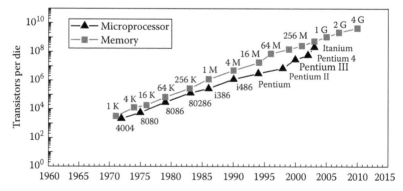

FIGURE 6.2 Evolution of the number of transistors per die in microprocessors and in DRAM memory chips.

survival, as industry continues pushing the dimensions of the semiconductor devices toward the decananometer range [3]:

I. The cell capacitor should be able to store enough charge to allow the discrimination, immune to noise, between the two memory states. Increasing or maintaining the capacitance size is in contradiction to scalability because of the high area consumption.

II. The charge must be transferred with a minimum delay from and into the capacitor.

III. Both the capacitor and the transistor junction should have a very low leakage current to avoid a degradation of the retention time.

The most challenging constraint is the minimum size of the capacitor: 25/30 fF per cell is required, regardless of the technological node, to safely distinguish between the "1" and "0" states in logic circuits. In order to preserve this minimum capacitance, we have (1) to keep the area of the capacitor, what it is against the essence of the scaling; (2) to develop complex processes to produce corrugated surfaces [4]; or (3) to use dielectrics with very high permittivity to build the capacitor, but frequently these materials are not stable or compatible with CMOS process. Several solutions adapted to build the capacitor of the cell are shown in Figure 6.3. Figure 6.3a shows a stacked capacitor in the front end of the chip, while Figure 6.3b shows trenched capacitors in the back end [5].

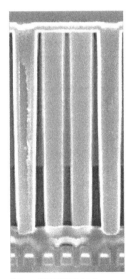
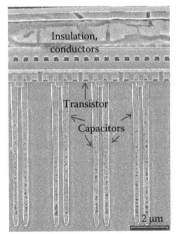

(a)   1T1C DRAM
       stacked capacitor

(b)   1T1C DRAM
       trenched capacitor

FIGURE 6.3 (a) TEM cross section of a DRAM cell using stacked capacitors. (After Lewis, J., Scaling the challenge of memory at 45 nm and below, http://chipdesignmag.com/print.php?articleId=1695?issueId=0, Chip Design Magazine, 2002.) (b) SEM cross section through a 64 Mbit DRAM using trenched capacitors. (University of Kiel, http://www.tf.uni-kiel.de/matwis/amat/semitech_en/kap_3/backbone/r3_1_2.html.)

These solutions are becoming very complicated; they are high-cost processes and not easily CMOS compatible. If memory makers continue to rely on a 1T + 1C (one transistor, one capacitor) design, at some point, the capacitor will become so tall or so deep that it will not be possible to cost-effectively manufacture it in volume. Already, more than 30% of the DRAM manufacturing cost is accounted purely for the capacitor. This percentage will increase as smaller and smaller process geometries are used.

The external storage element is a limitation for the DRAM survival. The step forward that each technology node requires in terms of technological, material, and design advances represents a paramount challenge for the integration of the capacitor inside the cells. This is the main reason of the increasing research activity during the last decade around new memory cells, free of capacitors. 1T-DRAM is a vast concept that includes a set of memories intended to be potential substitutes of the standard DRAM technology. All of them have a common feature: they avoid using any external storage element. The memory cell is composed of a single device (transistor or transistor-like) where the information is stored, that is, the same device is used to store the information and to read it. Within the 1T-DRAM family, a large collection of devices has been accommodated: from single transistors to more complicated multi-gate or thyristor-like structures [6]. In this chapter, we are going to focus on a particular set belonging to the 1T-DRAM family: the so-called floating-body (FB) 1T-DRAM cells.

6.2 FLOATING-BODY MEMORIES

FB memories take advantage of, in principle, a harmful effect that appears in partially depleted (PD)-SOI transistors: the kink effect [7]. If we represent the output characteristics (I_D–V_D curve) of a PD-SOI transistor as in Figure 6.4, we observe that at a particular drain voltage, a change in the slope of the curve occurs. This increase in the drain current is produced by the accumulation of holes in the body of the transistor: when the drain voltage is large enough, impact ionization (II) occurs at the drain edge of the channel, generating electron–hole pairs. While electrons are drifted to the drain, holes are pushed to the isolated body of the transistor where they cannot escape due to the presence of the buried oxide (BOX) and the reversed channel–drain and channel–source P–N junctions. The accumulation of holes in the body of the transistor increases the body potential, decreasing at the same time the threshold voltage of the transistor, which produces the current increase and the change in the slope. Kink effects worsen the differential drain conductance of the device and are strongly dependent on the operating speed, which affects the performance of analog circuits. For an amplifier, the gain at low frequency is substantially degraded with the kink effect. While kink effect or FB effect presents a problem in logic and analog applications of PD-SOI devices, it could be exploited as the underlying principle for alternative 1T-DRAM cells (known as floating-body DRAMs, FB-DRAM). For this reason, the effect is sometimes called the Cinderella effect in the context of these technologies, because it transforms a disadvantage into an advantage [8].

6.2.1 PARTIALLY DEPLETED FB/1T-DRAM

How can we take advantage of FB effect to build a memory cell? Let's consider a PD-SOI transistor (as the one shown in Figure 6.4).

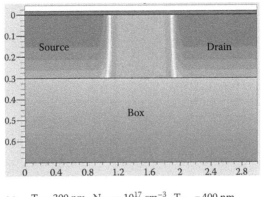

(a) $T_{Si} = 300$ nm $N_{A,Si} = 10^{17}$ cm^{-3} $T_{Box} = 400$ nm

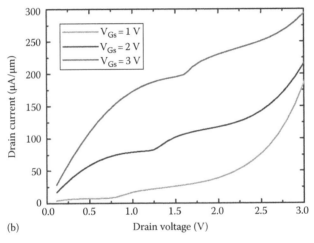

(b)

FIGURE 6.4 (a) Cross section of a partially depleted SOI (PD-SOI) transistor with 1 μm of channel length. (b) Calculated I_D–V_D curves for the device in (a) showing the floating-body effect. Simulations have been performed using Silvaco ATLAS (http://www.silvaco.com).

Initially we applied a low voltage to the drain, $V_{DS} = 0.1$ V, and a ramp voltage to the gate, for example, $V_{GS} = 0$–2 V (Figure 6.5a). We obtained then the typical I_D–V_G curve for the transistor. As the drain voltage is low, no II phenomenon appears at the drain edge of the channel, and the population of holes in the body is quite low. At that point, if we apply a high voltage pulse to the drain for some time, the II produced at the drain edge of the channel triggers the accumulation of holes in the body of the transistor as seen in the third column of Figure 6.5b, which modifies the threshold voltage of the transistor. If we apply now the same voltage ramp to the gate, we obtain a higher drain current as shown in Figure 6.5b. The difference in the drain current for the same value of drain and gate voltages allows us to define a "0" state and a "1" state. To come back to the initial situation, that is, the "0" state, we need to eliminate the hole excess in the body of the transistor. To do so, we could apply, for example, a negative bias pulse to the drain. Doing so, we are forward-biasing the drain-body P–N junction, and the excess holes scape through the drain (Figure 6.5c).

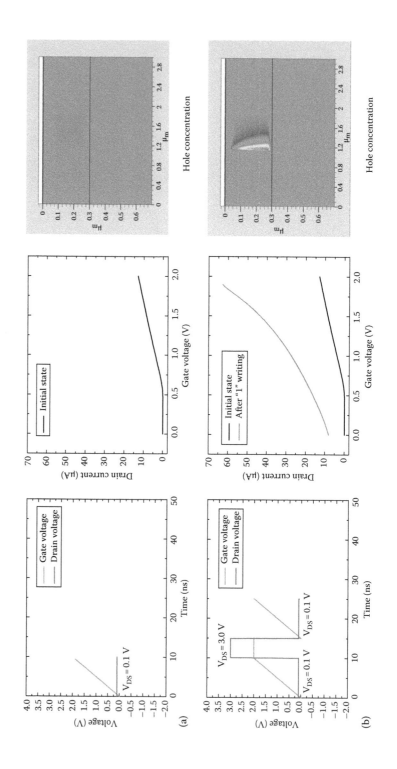

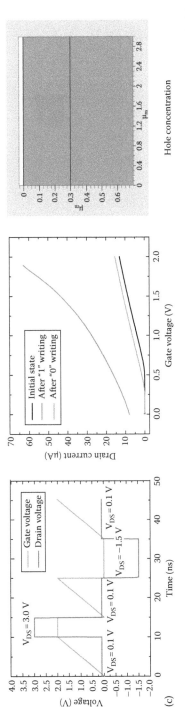

FIGURE 6.5 Evolution of the current (second column) and the hole concentration in the body (third column) in the device of Figure 6.4a under the bias pattern (V_D and V_G) shown in the first column. (a) Without hole accumulation in the floating body of the transistor (before writing "1" state), (b) With hole accumulation in the floating body of the transistor (after writing "1" state), and (c) Without hole accumulation in the floating body of the transistor (after writing "0" state). Evolution of the current (second column) and the hole concentration in the body (third column) in the device of Figure 6.4a under the bias pattern (V_D and V_G) shown in the first column.

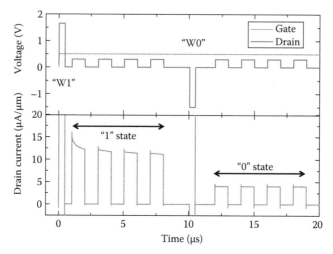

FIGURE 6.6 Simulation results for the operation of a PD-1T-RAM. The picture shows the bias pattern (top) and the driven current (bottom). For simplicity, the gate bias is maintained constant ($V_G > V_T$). The floating body is initially charged with holes generated by impact ionization (W1), and the cell state is read four times by using a small drain bias. At $t = 10$ μs, the cell is purged (write "0" state: W0) and read again four times reflecting difference with respect to the previous "1" states. $L = 1$ μm, $T_{Si} = 300$ nm, $T_{ox} = 3$ nm, $T_{BOX} = 400$ nm, $N_A = 10^{17}$ cm^{-3}.

Figure 6.6 summarizes the operation of a PD-SOI transistor as 1T-DRAM cell. The top side of the figure shows an example of the bias pattern used to demonstrate the 1T-DRAM functionality of the PD-SOI MOSFET. In the bottom side, the driven current is monitored. Initially, the holes are injected in the FB by II due to the large current driven by the device ("W1" in Figure 6.6): the highly energetic electrons at the drain edge of the channel knock valence electrons out of their bound state to a state in the conduction band, creating electron–hole pairs. Electrons are evacuated through the drain, while holes are accumulated into the neutral body of the silicon film. The hole overpopulation of the body of the device leads to a decrease in the threshold voltage and therefore a transitory increase in the drain current. The cell can be purged of charge by forward-biasing the drain-to-body junction (negative drain bias, "W0" in Figure 6.6). In this process, holes are evacuated from the FB through the channel-to-drain P–N junction. If the cell state is read again, the current level remains in the stable level (lower current).

6.2.2 FULLY DEPLETED FB/1T-DRAM

Although, theoretically, PD-FB/1T-DRAM cells scale much better than 1T–1C-DRAM cells because of the lack of the storage capacitor, if we want to scale further these devices, we have to reduce the silicon thickness (Ref. 22); in such a case, PD devices become fully depleted (FD) devices, where we do not have an floating body. However, the storage of holes inside the body of the transistor can also be

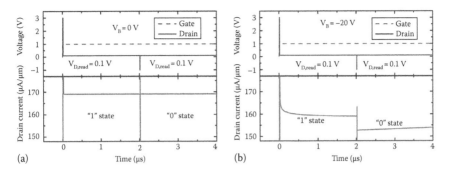

FIGURE 6.7 Simulation results for the operation of an FD-1T-RAM. The picture shows the bias pattern (top) and the driven current (bottom). For simplicity, the front-gate bias is maintained constant ($V_G > V_T$). (a) The substrate is biased at 0 V, and due to the lack of floating-body effects, the memory effect is not manifested. (b) A back-gate bias is applied, creating a potential well where holes can be stored. The difference between the drain currents in "1" and "0" states represents the memory effect. L = 200 nm, T_{Si} = 70 nm, T_{ox} = 1 nm, N_A = 10^{16} cm^{-3}.

achieved in FD-SOI transistors by applying a negative bias to the back-gate (below the BOX). The negative bias creates an electrostatic potential well where holes can be accumulated leading to a storage node.

To illustrate this effect, we show, in Figure 6.7, simulation results corresponding to a relatively thin (film thickness T_{Si} = 70 nm, BOX thickness T_{BOX} = 50 nm) SOI transistor with (Figure 6.7a) and without (Figure 6.7b) negative substrate bias. Because of the undoped thin body, the unbiased device ($V_B = 0$) lacks from FB effect, and the holes generated by II recombine so quickly that the difference between states is not noticeable (Figure 6.7a).

By contrast, if a negative back-gate bias is applied to the device (emulating a second gate with the substrate and BOX), the potential well allows the accumulation of holes (Figure 6.7b), that is, the potential well acts as the storage node for the positive carriers. Holes can be injected and removed in the same way as in Figure 6.6. Note then in this case, both "1" and "0" current levels become unstable: for the "1" state, the potential well becomes overpopulated with holes, and equilibrium is recovered by recombination (current overshoot); purging the cell by forward-biasing the drain to body junction leads an underpopulation of holes; for the "0" state, the back interface is driven into depletion, but the holes are restored by junction leakage, gate-induced drain leakage (GIDL) and thermal generation, tending to corrupt the "0" state and convert it to the "1" state.

Different groups have tried to exploit the concept of FD-1T-DRAM, proposing different structures and different mechanisms to write "1" and "0" states, attempting to improve the states margins, retention times, and power consumption during the operation of the cell [9–21].

6.2.3 SCALING LIMITS OF FB/1T-DRAM

One of the most questioned drawbacks of FB/1T-DRAM has been their scalability. The scalability of FD-SOI transistors requires a decrease in the film thickness in

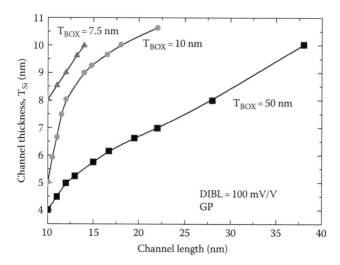

FIGURE 6.8 Silicon film thickness as a function of the channel length in an FD-SOI transistor to have a DIBL (drain-induced barrier lowering) equal to 100 mV/V. This condition means a good control of the short channel effects (SCEs) and roughly implies a relationship $L_{ch} > 4 \, T_{Si}$ between the channel length and the silicon film thickness. This constraint is relaxed when a thinner buried oxide (BOX) thickness is used.

order to have a good control of short channel effects (SCEs). Figure 6.8 shows the silicon thickness needed to have a good SCE control (measured as a drain-induced barrier lowering of 100 mV/V as a function of the channel length) [22].

As shown in Figure 6.8 ($T_{BOX} = 50$ nm), for a channel length of 35 nm, we need a silicon thickness of $T_{Si} = 9$ nm, but if we reduced the channel to 20 nm, we need a silicon thickness of 6 nm, that is, the film thickness of the device should be around four times thinner than the gate length (this condition can be relaxed if we use thinner BOXs, a ground plane, or double-gate devices). This "scaling rule" entails that FB-DRAMs should be compatible with body thicknesses below 10 nm in order to be competitive in future technology nodes, which represents a challenge for the FB/1T-DRAM family. Several studies have revealed severe degradation in the readout current margin between states of 1T-DRAMs when the body thickness decreases below 30 nm [23]. This limitation, also known as super-coupling effect [24], is basically an electrostatic result: the thinner the body, the more difficult to generate the potential difference in order to accommodate a high concentration of electrons at one surface and a high concentration of holes at the opposite surface of the same silicon layer.

In order to beat this paramount limitation while maintaining enough performance in terms of sensing margins and retention time despite the scaling of the film thickness, several architectures and material combinations have been envisaged. The idea behind all of them is the effective separation of the stored carriers and the sensing carriers by creating dedicated volumes (potential wells) inside the transistor body (multibody devices).

6.3 MULTIBODY MEMORIES: ADVANCED-RAM FAMILY

Different structures have been proposed during last years sharing the concept of multibody memories, that is, the electrostatic and physical separation of stored carriers and sensing carriers:

1. The single transistor quantum well 1T-DRAM [25]
2. Convex channel 1T-DRAM [26]
3. A-RAM family [27,28]

6.3.1 ADVANCED-RAM MEMORY CELL

In 2010, researchers at the University of Granada in Spain and Minatec in Grenoble (France) proposed a totally new concept of 1T-DRAM cell, called advanced RAM or A-RAM [29], using some of the 1T-DRAM fundamentals but featuring a novel architecture and electrostatic properties. A-RAM cell has been designed to physically separate majority and minority carriers even in ultrathin FD SOI layers [30] by eliminating the super-coupling effect.

The essence of the A-RAM is an FD-SOI transistor, which features two ultrathin semibodies physically isolated by a middle oxide (MOX) but sharing the source and drain regions (Figure 6.9). When this device is operated as a memory cell, the top semibody is used to store for majority carriers (holes), while the bottom semibody serves to sense the logic state of the device through an electron current. The MOX constitutes the key element of the A-RAM cell: electron and hole populations can be brought very close to each other, unlike in an ultrathin single-body 1T-DRAMs [24]: the low-k insulator amplifies the electrostatic potential difference in the transistor body. The A-RAM structure can be fabricated by local oxidation (MOX) of the bottom semibody, followed by epitaxial regrowth of both the upper semibody and source/drain regions. The shorter the device, the better the quality of the epitaxial layer.

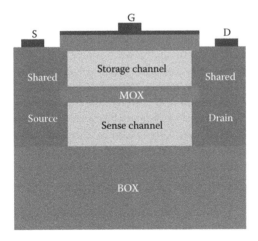

FIGURE 6.9 Schematic configuration of an A-RAM cell.

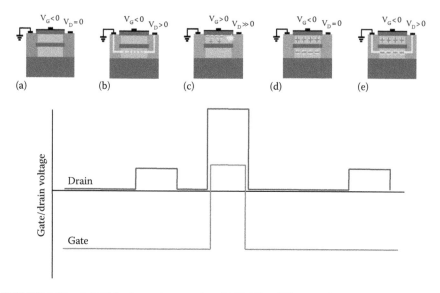

FIGURE 6.10 A-RAM schematic operation: (a) Holding "0" state. (b) Reading "0" state by increasing drain bias: no current flow. (c) Writing "1" state: gate voltage is increased over the threshold voltage to switch on the channel, and a high drain voltage is used to produce impact ionization in the drain edge of the channel. (d) Holding "1" state: hole overpopulation in the top semibody. (e) Reading "1" state: current flow between the source and the drain.

The SON process can also be considered: growth of a sacrificial SiGe layer, Si epitaxy, and SiGe etch leaving a cavity to be refilled with the MOX dielectric [31].

If there is no charge accumulated in the top semibody (storage channel, holes), the electron concentration in the bottom semibody (sense channel) remains extremely low, defining state "0" (Figure 6.10a). Source and drain regions are electrically isolated: even raising the drain voltage results in a negligible current (Figure 6.10b). Memory state "1" is programmed by charging the top semibody with holes (Figure 6.10c) via II or band-to-band (BTB) tunneling [32]. A volume inversion electron channel is then activated, via electrostatic coupling, in the ultrathin bottom semibody, establishing electrical continuity between source and drain regions. In "retention" phase, negative gate bias and zero drain voltage are used to maintain the potential well for the holes in the storage channel (Figure 6.10d). If the drain voltage is increased, a substantial current flows through the transistor (Figure 6.10d).

Up to date, A-RAM operation has been validated only by numerical simulations. Poisson and continuity equations were solved in transient mode by accounting for the most relevant mechanisms (BTB tunneling, II, and generation–recombination processes) using Silvaco ATLAS. Figure 6.11a shows a possible bias sequence for writing and reading the transistor states; the sensed drain current is shown in Figure 6.11b.

At the starting time (t = 0), the "0" bit has been written, and therefore the upper semibody of the cell is discharged. The cell state, read by slightly increasing the drain voltage (0.1 V), shows a negligible drain current. Next, "1" state is programmed by II (a gate voltage pulse is embedded in a drain voltage pulse). This makes sure that

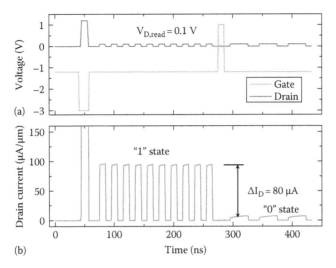

(a)

(b)

FIGURE 6.11 Electrical simulation results for the operation of A-RAM cells performed with Silvaco ATLAS (L_G = 50 nm, T_{total} = 20 nm, T_{MOX} = 4 nm, T_{Si} = 8 nm). (a) Bias sequence for writing and reading memory states and (b) sensed drain current for the bias pattern shown in (a).

when the gate voltage decreases back to the negative retention value, hole recombination is inhibited. During the gate bias transient, the lateral field remains high enabling more holes to be generated. In Figure 6.11, "1" state has been sensed 10 times, showing large enough current (80 µA/µm) with no degradation of the current level.

Reading is nondestructive: the refresh is needed only to compensate for temporal charge derive. The "0" state is written by pulsing the gate bias to a positive value (typically, 1 V for T_{ox} = 1.5 nm). Holes are rapidly eliminated (ns) by the junctions that become forward-biased due to the sudden increase in the body potential (with respect to the source and drain potentials). The electron concentration in the bottom semibody decreases, and the readout drain current returns to a subthreshold value. This capacitive coupling enables the "0" state writing without applying any drain bias (typical procedure in 1T-DRAMs is to forward-bias the drain–channel junction to eliminate the charge). The temporal evolution of the "1" and "0" states has been simulated for a device with a total thickness of 20 nm. MOX thickness was considered to be 4 nm (Figure 6.12). The negative gate bias controls the concentration of holes. If the amount of generated holes is larger than the value that can be accommodated in the potential well supported by the gate bias, temporary hole recombination occurs leading to an exponential current decay in "1" state as shown in Figure 6.12; for longer periods of time, equilibrium is achieved, and the current reaches a saturated constant value. The slow derive of the "0" state is caused by the progressive repopulation of the upper body with holes resulting from thermal generation, GIDL, and leakage current in the source and drain junctions. After 0.1 s, the margin between states is maintained, showing a large difference between "0" and "1" levels. The drain voltage must be kept reasonably low when testing the cell state; otherwise, II in the sense channel may be triggered destroying the information. For a given total device thickness, increasing the MOX thickness (i.e., thinner semibodies)

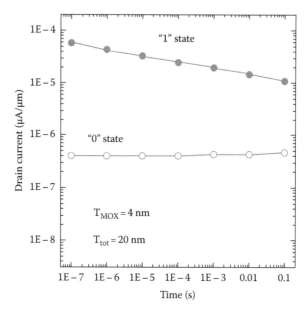

FIGURE 6.12 Evolution of the (solid symbols) "1" and (open symbols) "0" states with time. V_{GI}(Retention) = –2 V, V_D(Reading) = 0.1 V. T_{tot} = 20 nm, T_{MOX} = 4 nm; T_{ox} = 2 nm; T = 300 K, L = 45 nm.

benefits the cell "1" state and slightly improves retention time. The full device optimization implies a tradeoff between coupling, electrostatic potential difference, and quantum limitations (threshold voltage increase and phonon confinement in ultrathin bodies) [27].

In summary, A-RAM cell exploits a dedicated body partitioning for charge storing and current sensing in an FD-SOI transistor. The cell features easy discrimination of "1" and "0" states, defined by the concentration of majority carriers stored in the upper body. Numerical simulations demonstrate competitive programming and retention times, which result from the physical separation of the two types of carriers. The A-RAM is operated in single-gate mode and achieves low-power operation and enhanced scalability. The A-RAM concept is versatile, its architecture being amenable to structural (FinFET) and doping (P/N) modifications.

6.3.2 SECOND GENERATION ADVANCED-RAM (A2RAM)

As an alternative to the physical isolation (MOX), the electrical isolation of the two types of carriers could also be carried out by a vertical P–N junction, originating a new cell architecture named A2RAM as illustrated in Figure 6.13 (which is compatible with SOI substrates (a) and with Si-bulk substrates (b)) [33]. To do so, a conventional MOSFET is modified by connecting the n+ source and drain regions through a buried N-type layer (named N-bridge), underneath the P-channel.

Figure 6.14 shows an example of a realistic doping profile at the middle of the device, perpendicular to the gate, and obtained from process simulations with ATHENA [34]: a 10^{18} cm^{-3} N-type layer is located in-between the low-doped P-type

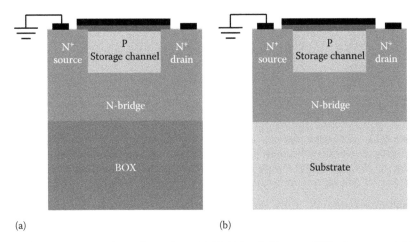

FIGURE 6.13 Schematic configuration of A2RAM cell (a) on an SOI substrate and (b) on bulk substrate. A buried N-layer (N-bridge) connects the source and the drain.

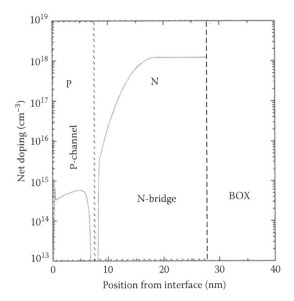

FIGURE 6.14 Vertical net doping profile at the middle of the channel on the SOI substrate from process simulator Silvaco ATHENA demonstrating fabrication feasibility of the A2RAM.

body and the P-type substrate. The P–N stack was formed with high-vacuum chemical vapor deposition to epitaxially deposit the two layers of silicon at 900°C. The total device thickness that can be achieved is very thin (typically in the range of 15–30 nm). Since the source and the drain have the same doping polarity as the N-bridge, the source and drain regions are, in principle, electrically short-circuited through the bridge. The basic idea for memory cell operation is to suppress/enable

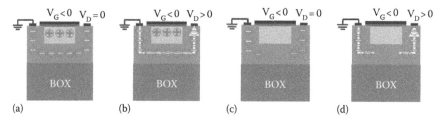

FIGURE 6.15 Schematic memory cell operation: (a) Holding "1": the P-body is charged, screening the gate field. (b) Reading "1": majority carrier (electron) current flows through the N-bridge. (c) Holding "0": the P-body is discharged (deep depletion), and the N-bridge is FD. (d) Reading "0": the current through the N-bridge is negligible.

this short-circuiting by fully depleting/no-depleting the N-bridge through the accumulation/emptying of holes in the upper channel somehow.

The top P-body is used as a storage node, whereas the lower body (i.e., a relatively high-doped N-bridge) serves for current sense discriminating the two memory states. The simplified sequence of states is shown in Figure 6.15. The upper body (P-type) is charged with holes generated either by BTB tunneling [32] or by II [35] (in the MOS transistor). These holes can be retained in the top body with negative gate bias (see Figure 6.15a). In this situation, the vertical electric field, originated from the negative gate bias, is screened by the positive hole charge of the P-body and has a minor effect on the majority carriers of the N-bridge. The bridge is partially-depleted, and a small drain bias leads to an electron current, I_1, flowing through the bridge (no current flows through the P-body) reading the "1" state (Figure 6.15b). When the top body is discharged of holes (state "0"), the gate field is no longer screened, and the N-bridge is FD (Figure 6.15c). The lack of holes in the N-bridge causes a very low current, I_0, if the drain bias is increased (Figure 6.15d).

The differences with conventional FB/1T-DRAMs are threefold:

1. The drain current, defining the cell state, is due to electrons (majority carriers) flowing or not in the volume of the bridge.
2. The use of an insulator substrate is optional.
3. The super-coupling effect is suppressed because the coexistence of electrons and holes in the same silicon slab is ensured by the vertical P–N junction.

We have used numerical simulations to demonstrate the functionality, as the memory cell, of the A2RAM device; Poisson and continuity equations were solved self-consistently in 2-D.

First, we studied the steady-state operation of the cell: the combination of the parallel P and N channels results in unconventional I_D–V_G curves (Figure 6.16). In steady state, the N-bridge is always nondepleted regardless of the gate voltage. For negative V_G, the potential difference is basically absorbed by the accumulated holes in the upper P-region: the more negative the gate voltage, the more holes accumulate screening the field. The current is weakly dependent on V_G. For positive gate bias, the current flow comes from the parallel combination of the majority carriers

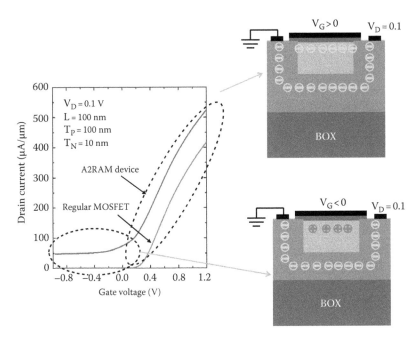

FIGURE 6.16 Comparison of I_D–V_G characteristics of A2RAM device and conventional FD-SOI MOSFET. For $V_G > V_{TH}$, the drain current in the A2RAM cell is the superposition of the inversion channel current of the top MOSFET and the current of the majority carriers of the N-bridge. For $V_G < V_{TH}$, the MOSFET channel is off, and only the N-bridge contributes to the drain current.

(electrons) of the N-bridge (which behaves like a resistor) and the minority carriers (electrons) of the top MOSFET. As the gate bias is increased, the top inversion channel becomes dominant. Notice that this behavior is different from that of a depletion-mode NMOSFET, where the conduction can be effectively cut at a certain negative gate bias.

The transient behavior is analyzed in Figure 6.17:

1. From A to B, a positive voltage is applied to the gate. The upper channel becomes inverted with electrons, and the behavior of the device is similar to the behavior of a MOSFET transistor.
2. From B to C, a negative voltage is suddenly applied to the gate. The channel becomes depleted of electrons, and as there are no sources of holes, the upper channel becomes empty of carriers. The negative electric field induced by the negative gate voltage also depletes the N-bridge, and as a consequence, there is no current at all at the device.
3. If the gate voltage is decreased to even more negative values (C–D), BTB tunneling starts to appear in the source–channel and drain–channel overlapped regions. This process injects holes into the channel that screen the negative electric field induced by the gate. As the negative gate electric field is now weaker, the N-bridge becomes partially undepleted and

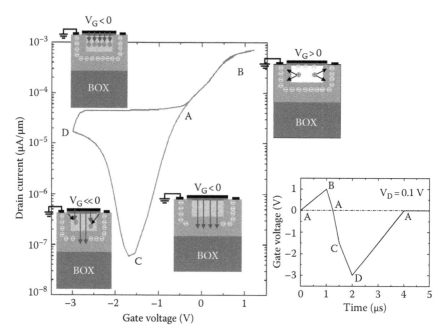

FIGURE 6.17 Transient behavior of the A2RAM device of Figure 6.16: (a) From A to B, a positive voltage is applied to the gate. The upper channel becomes inverted with electrons, and the behavior of the device is similar to the behavior of a MOSFET transistor. (b) From B to C, a negative voltage is suddenly applied to the gate. The channel becomes depleted of electrons, and the upper channel becomes empty of carriers. The negative electric field induced by the negative gate voltage also depletes the N-bridge, and as a consequence, there is no current at all at the device. (c) If the gate voltage is decreased to even more negative values (C–D), band-to-band tunneling injects holes into the channel, and the N-bridge becomes partially undepleted. (d) If the gate voltage is reduced to zero (D–A), band-to-band tunneling stops, no more holes are injected in the channel, and the drain current becomes constant.

   the drain current starts to increase. The greater the hole injection, the higher the drain current.

4. If the gate voltage is reduced to zero (D–A), BTB tunneling stops, no more holes are injected in the channel, and the drain current becomes constant.

As observed in Figure 6.17, under negative gate voltages, there is a current window for the same values of gate and drain voltages, which, depending on the population of holes in the upper channel, allows the definition of two memory states:

1. "1" state: Upper channel is populated with holes that screen the negative electric field. Drain current flows through the buried N-bridge.
2. "0" state: Upper channel is fully depleted of carriers, that is, there are no carriers at all in it. The negative electric field induced by the negative gate voltage depletes the N-bridge, and no current flows between drain and source. This state is a transient effect. After a long time, thermal carrier generation will restore the hole population in the channel, and the "0" state will disappear.

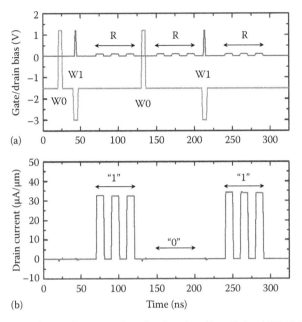

FIGURE 6.18 Waveforms demonstrating the functionality of the A2RAM cell: (a) Bias pattern for gate and drain. (b) Drain current. Writing is performed by BTB tunneling. $L = 22$ nm, $\varphi_m = 4.5$ eV, $T_{ox} = 2$ nm, $T_{P-body} = 8$ nm, and $T_{N-bridge} = 10$ nm.

Figure 6.18 shows the memory operation of the cell. The gate is biased well below the threshold voltage of the transistor to generate an accumulation of holes in the channel in steady-state conditions, which will determine the current level of the "1" state. Initially, the cell is purged by writing a "0" state. To do so, a positive voltage pulse is applied to the gate. This pulse forward-biases the channel–source and channel–drain P–N junctions, and holes are evacuated from the channel. If the gate suddenly comes back to the negative value, the channel becomes empty of carriers.

We propose two alternative mechanisms to write the "1" states (restore the hole population in the P-body): BTB tunneling [36] by means of an overbias pulse of the retention gate voltage, or II by applying a positive gate voltage in the gate activating the MOSFET. Nevertheless, BTB tunneling is best suited for low-power embedded applications since the writing current is typically several orders of magnitude lower than that with the II mechanism (during the writing time, there is an additional contribution of current coming from the MOSFET, which is not present when using the BTB mechanism). Waveforms demonstrating the cell functionality are shown in Figure 6.18 using the BTB alternative. As observed, the "0" state corresponds to zero drain current.

Figure 6.19 shows the transient evolution of the two states in a 22 nm-channel length cell under continuous reading condition. In spite of the use of a very low drain voltage to read the state of the cell, parasitic BTB tunneling (GIDL) at the drain edge of the channel generates holes in the channel, degrading the "0" state, which is unstable. In addition to BTB tunneling, other mechanisms contribute to the stability

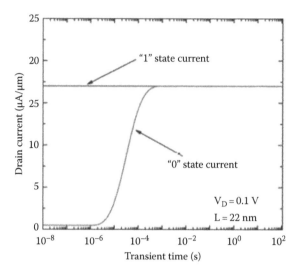

FIGURE 6.19 Evolution of the "1" and "0" state currents versus the transient time under continuous reading ($V_D = 0.1$ V, $V_G = -1.5$ V, T = 85°C).

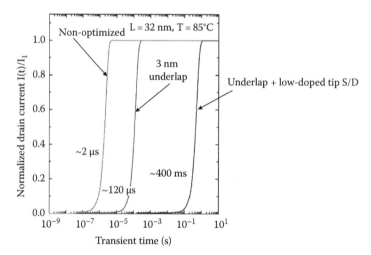

FIGURE 6.20 Evolution of the "0" state current at 85°C for a 32 nm channel length device ($T_{Si} = 18$ nm), showing a retention time around 2 μs. This retention time can be improved to 120 μs by using an underlap between gate oxide and drain and source extensions or even to 400 ms if low-doped tips are added to drain and source.

of the "0" state, such as junction leakage, thermal generation, or even II if the drain reading bias becomes large. Figure 6.20 shows the evolution of the "0" state current at 85°C for a 32 nm channel length device ($T_{Si} = 18$ nm), showing a retention time around 2 μs. This retention time can be improved to 120 μs by using an underlap between gate oxide and drain and source extensions or even to 400 ms if low-doped tips are added to drain and source. A2RAM cells have successfully been fabricated on both SOI and bulk-Si substrates.

6.4 CONCLUSIONS

The FB 1T-DRAM family has grown exponentially during the last decade aiming to develop innovative memory cells and replace the standard 1T+1C DRAM, which, after more than 30 years of unchallenged success, shows signs of exhaustion. Each particular approach has its own advantages and drawbacks in terms of CMOS compatibility, scalability, power dissipation, and performance. Some of them have stand-up featuring promising characteristics, but no one is yet competitive enough to beat the DRAM supremacy. One of these promising alternatives, fully compatible with CMOS technology, is the A-RAM family. This new concept of 1T-DRAM cell features N/P body partition, which enables the physical separation of the hole storage and sense electron current. The hole concentration controls the partial or full depletion of the N-body, modulating the sense majority carrier current. The cell is compatible with single-gate operation and ultimate scaling down to the 22 nm node. As confirmed by recent measurements, the A2RAM cell features attractive performance (long retention, wide memory window, simple programming, nondestructive reading, and low-power operation) for embedded systems on bulk and SOI substrates.

ACKNOWLEDGMENTS

This work was supported in part by the Junta de Andalucia under Research Project TIC2010–6902, and by the Spanish Government under Research Projects TEC2011–28660 and BIOTIC-20F12/37.

REFERENCES

1. R. Dennard, Field effect transistor memory, U.S. Patent No. 3,387,286.
2. H. Suname, T. Kure, N. Hashimoto, K. Itoh, T. Toyabe, and S. Asai, A corrugated capacitor cell (CCC). *IEEE Transactions on Electron Devices* 1984; 31(6):746–753.
3. K. Kim, Perspectives on giga-bit scaled DRAM technology generation, *Microelectronics Reliability* 2000; 40(11):191–206.
4. H. Watanabe and I. Honma, Stacked capacitor having a corrugated electrode U.S. Patent No. 5.835.337, 1998; H. Watanabe and I. Honma, Stacked capacitor having a corrugated electrode, U.S. Patent No. 6.022.772, 2000.
5. J. Lewis, Scaling the challenge of memory at 45 nm and below, http://chipdesignmag.com/print.php?articleId=1695?issueId=0, *Chip Design Magazine*, 2002.
6. F. Nemati and J.D. Plummer, A novel thyristor-based SRAM cell (T-RAM) for high-speed, low-voltage, giga-scale memories, *IEDM Technical Digest* 1999, pp. 283–289.
7. T. Ouisse, G. Ghibaudo, J. Brini, S. Cristoloveanu, and G. Borel, Investigation of floating body effects in silicon-on-insulator metal-oxide-semiconductor field-effect transistors. *Journal of Applied Physics* 1991; 70(7):3912–3919.
8. R. Tom Halfhill, Z-RAM shrinks embedded memory, *Microprocessor Report*, Reed Electronics Group, October 2005; 19, pp. 30–33.
9. H.J. Wann and C. Hu, A capacitorless DRAM cell on SOI substrate. *IEDM Technical Digest*, December 5–8, 1993, pp. 635–638.
10. S. Okhonin, M. Nagoga, J. Sallese, and P. Fazan, A SOI capacitor-less 1T-DRAM concept. In: *Proceedings of the 2001 IEEE International SOI Conference*, Durango, CO, 2001, pp. 153–154.

11. S. Okhonin, M. Nagoga, J.M. Sallese, and P. Fazan, A capacitor-less 1T-DRAM cell. *IEEE Electron Device Letters,* February 2002; 23(2):85–87.

12. T. Hamamoto, Y. Minami, T. Shino, N. Kusunoki, H. Nakajima, M. Morikado, T. Yamada et al., A floating-body cell fully compatible with 90-nm CMOS technology node for a 128-Mb SOI DRAM and its scalability. *IEEE Transactions on Electron Devices* 2007; 54:563–571.

13. I. Ban, U.E. Avci, U. Shah, C.E. Barns, D.L. Kencke*, and P. Chang, Floating body cell with independently-controlled double gates for high density memory. *IEDM Technical Digest* 2006, pp. 1–4.

14. M. Bawedin, S. Cristoloveanu, J.G. Yun, and D. Flandre, A new memory effect (MSD) in fully depleted SOI MOSFETs. *Solid-State Electronics* 2005; 49(n ± 9):1547–1555.

15. F. Assaderaghi, J. Chen, R. Solomon, T. Chan, P. Ko, and C. Hu, Time dependence of fully depleted SOI MOSFET's subthreshold current. In *Proceedings of IEEE International SOI Conference*, October 1991, Vail, CO, pp. 32–33.

16. M. Bawedin, S. Cristoloveanu, and D. Flandre, A capacitor-less 1T-DRAM on SOI based on double gate operation. *IEEE Electron Device Letters* 2008; 29(n ± 7):795–798.

17. A. Hubert, M. Bawedin, G. Guegan, T. Ernst, O. Faynot, and S. Cristoloveanu, SOI 1T–DRAM cells with variable channel length and thickness: Experimental comparison of programming mechanisms. *Solid-State Electronics*, November/December 2011; 65–66C:256–262.

18. S. Okhonin, M. Nagoga, E. Carman, R. Beffa, and E. Faraoni, New generation of Z-RAM. *IEDM Technical Digest* 2007, pp. 925–928.

19. G. Giusi, M.A. Alam, F. Crupi, and S. Pierro, Bipolar mode operation and scalability of double-gate capacitorless 1T-DRAM cells, *IEEE Transactions on Electron Devices*, August 2010; 57(8):1743–1750.

20. D.-Il Moon, S.-J. Choi, J.-W. Han, S. Kim, Y.-K. Choi, Fin-width dependence of BJT-based 1T-DRAM implemented on FinFET, *IEEE Electron Device Letters*, September 2010; 31(9):909–911.

21. Z. Lu, N. Collaert, M. Aoulaiche, B. De Wachter, A. De Keersgieter, W. Schwarzenbach, O. Bonnin et al., A novel low-voltage biasing scheme for double gate FBC achieving 5s retention and 10^{16} endurance at 85°C, *IEDM Technical Digest* 2010, pp. 12.3.1–12.3.4.

22. O. Faynot, F. Andrieu, C. Fenouillet-Béranger, O. Weber, P. Perreau, L. Tosti, L. Brevard et al., Planar FDSOI technology for sub 22 nm nodes, *IEDM Technical Digest*, 2010, pp. 3.2.1–3.2.4.

23. U.E. Avci, I. Ban, D.L. Kencke, P.L.D. Chang, Floating body cell (FBC) memory for 16-nm technology with low variation on thin silicon and 10-nm BOX. In *IEEE International SOI Conference,* New Paltz, NY, 6–9 October 2008, pp. 29–30.

24. S. Eminente, S. Cristoloveanu, R. Clerc, A. Ohata, and G. Ghibaubo, Ultra-thin fully depleted SOI MOSFETs: Special charge properties and coupling effects. *Solid-State Electronics* 2007;51(2):239–244.

25. M.G. Ertosun, P. Kapur, and K.C. Saraswat, A highly scalable capacitorless double gate quantum well single transistor DRAM: 1T-QW DRAM. *IEEE Electron Device Letters* 2008; 29:1405–1407.

26. M.H. Cho, C. Shin, and T.J.K. Liu, Convex channel design for improved capacitorless DRAM retention time. In *Simulation of Semiconductor Processes and Devices (SISPAD '09), The 2009 on International Conference,* San Diego, CA, 2009, pp. 1–4.

27. N. Rodriguez, F. Gamiz, and S. Cristoloveanu, A-RAM memory cell: Concept and operation, *IEEE Electron Device Letters* 2010;31(9):972–974.

28. N. Rodriguez, S. Cristoloveanu, and F. Gamiz, Capacitor-less A-RAM SOI memory: Principles, scaling and expected performance, *Solid-State Electronics* 2011; 59(1):44–50.

29. N. Rodriguez, F. Gamiz, and S. Cristoloveanu, Point memoire RAM a un transistor, Patent FR09/52 452.
30. N. Rodriguez, S. Cristoloveanu, and F. Gamiz, A-RAM: Novel capacitorless DRAM memory. In *Proceedings of IEEE International SOI Conference*, Foster City, CA, 2009, pp. 1–2.
31. M. Jurczak, T. Skotnicki, M. Paoli, B. Tormen, J. Martins, J. Regolini, D. Dutartre et al. Silicon-on-Nothing (SON) an innovative process for advanced CMOS, *IEEE Transactions on Electron Devices,* November 2000; 47(11):2179–2187.
32. E. Yoshida and T. Tanaka, A capacitorless 1T-DRAM technology using gate-induced drain-leakage (GIDL) current for low-power and high-speed embedded memory, *IEEE Transactions on Electron Devices,* April 2006; 54(4):692–697.
33. N. Rodriguez, F. Gamiz, and S. Cristoloveanu, Novel capacitorless 1T-DRAM cell for a 22-nm node compatible with bulk and SOI substrates, *IEEE Transactions on Electron Devices* 2011, ED-58:2371–2377.
34. *ATHENA User's Manual.* Santa Clara, CA: Silvaco, 2012.
35. R. Ranica, A. Villaret, C. Fenouillet-Beranger, P. Malinge, P. Mazoyer, P. Masson, D. Delille et al., A capacitor-less DRAM cell on 75 nm gate length, 16 nm thin fully depleted SOI device for high density embedded memories, *IEDM Technical Digest* 2004, pp. 277–280.
36. R. Ranica, A. Villaret, P. Malinge, P. Mazoyer, D. Lenoble, P. Candelier, F. Jacquet et al., A one transistor cell on bulk substrate (1T-bulk) for low-cost and high-density eDRAM. In *VLSI Symposium on VLSI Technology, 2004*, Digest of Technical Papers, Honolulu, Hawaii, June 2004, 427, pp. 128–129.

Part III

Novel Flash Memory

7 Quantum Dot-Based Flash Memories

Tobias Nowozin, Andreas Marent,
Martin Geller, and Dieter Bimberg

CONTENTS

7.1 INTRODUCTION

Some of the main technological achievements in the past decades are the capabilities to process and store increasing amounts of data. Such capabilities represent the foundation of the modern information society and depend on the success of the semiconductor industry to increase the performance of micro- and nanoelectronic devices. The key strategy since the invention of the integrated circuit has been "simply" downscaling the feature size, which has been coined as *Moore's law*. Downscaling will lead to feature sizes of just 10 nm as projected by the ITRS for 2020 [1]. Obviously, the feature size is reaching dimensions, in which quantum mechanics starts to dominate the physical properties of the underlying material. A growing number of difficulties in realizing such small structures are expected, and very little progress based just on downscaling is feasible beyond that limit. Therefore, considerable research effort is devoted to the search for alternative memory technologies.

One promising approach is the use of self-organized nanomaterials in future memories. In particular, self-organized quantum dots (QDs) based on III–V materials provide a number of advantages, as billions of them can be fabricated simultaneously with a high area density (10^{10}–10^{12} cm^{-2}) without any lithography in a bottom-up approach. In addition, they show extremely fast carrier capture and relaxation in the sub-picosecond range [2]. By properly designing the barrier/QD material combination, the retention time of charges in the QDs can be tuned, potentially up to 10^6 years at room temperature [3]. Hence, a QD-based memory could exhibit ultrasmall size, long retention times, and a fast write/read access.

7.2　CONVENTIONAL CHARGE-BASED SEMICONDUCTOR MEMORIES (DRAM AND FLASH)

The market for semiconductor memories is divided mainly between two memories: the dynamic random access memory (DRAM) and the flash memory. Being essentially different in design, both have their advantages and disadvantages in terms of performance. The DRAM is fast with a typical access time of 10 ns, but volatile with retention times in the range of some 10 ms, requiring periodical refresh of the information, consuming energy. In contrast, the flash is nonvolatile (with a retention time of more than 10 years), but suffers from a slow write time of some microseconds.

The basic cell structures of the DRAM and flash are schematically shown in Figure 7.1. The standard DRAM cell (Figure 7.1a) consists of one transistor and one capacitor (1T–1C) per cell [4]. The information bit is stored in the form of charges on the capacitor plates while the transistor controls the write and read access to the cell. The transistor itself is activated via a bit-line and word-line array. The main shortcoming of this design is charge leakage in the capacitor, which leads to short retention times.

The flash cell (Figure 7.1b) consists of a metal–oxide–semiconductor field-effect transistor containing a storage node that acts as additional gate (floating gate) [4]. The floating gate is isolated by a dielectric (such as SiO_2 or HfO_2), which acts as a barrier to restrain the charge carriers within the storage node. Depending on the charge state of the floating gate, the channel charge varies and the turn-on voltage of the transistor shifts. A typical barrier height of about 3.2 eV leads to retention times of >10 years. Writing and erasing charges from the floating gate are slow as the carriers have to overcome the barrier by either hot-electron injection or Fowler–Nordheim tunneling.

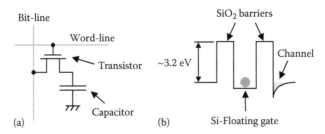

FIGURE 7.1　Schematics of conventional semiconductor memories: (a) 1T–1C cell of the DRAM and (b) flash memory cell.

This writing/erasing mechanism also leads to the successive destruction of the barriers and low endurance of the flash.

7.3 MEMORY BASED ON SELF-ORGANIZED QUANTUM DOTS

The QD-based flash memory (QD-flash) [5,6] is a charge-based memory in which charge carriers are confined within a heterostructure made out of III–V compound semiconductors with large band offsets. There are five key advantages of the QD-flash as compared to the conventional flash memory based on the Si/SiO$_2$ material system:

- *The barrier height can be designed.* By combining various materials, the band structure of the device can be designed. The properties of the memory can be tailored to be either very fast with infrequent refresh, or slower but nonvolatile, or anywhere in between. The wide variety of different material combinations is a decisive advantage as compared to the Si/SiO$_2$ material system.
- *Defect-free interfaces.* Although technologically advanced and abundant, the Si/SiO$_2$ material system is far from being perfect. Defects and dangling bonds at the Si/SiO$_2$ interface are a major issue and present a limit when the structures are scaled down to the nanoscale. In contrast, III–V interfaces can be ideal: free of any defect, atomically almost abrupt, even if moderate strain is present due to lattice mismatch.
- *Voltage tunability.* The effective barrier height in the QD-flash memory is voltage-tunable and can be reduced to zero, which should allow very fast write times on the sub-picosecond time scale.
- *Writing/erasing does not damage the structure.* Unlike in the flash memory, where extremely large fields are necessary [7], the QD-flash does not require high fields, which increases the endurance and the reliability.
- *Hole-based charge storage can be used.* Hole-confining type-II systems (such as material combinations based on GaSb) create huge confining potentials for holes while being repulsive for electrons. This way, storage times of 10^6 years might be realized [3], a prerequisite for nonvolatility.

The following sections will give an overview of the QD-flash memory concept, starting with some basics about III–V semiconductors and heterostructures (Section 7.3.1), self-organized QDs (Section 7.3.2), followed by the QD-flash memory concept (Section 7.3.3), and what has been achieved so far experimentally concerning storage times (Section 7.3.4), charge detection (Section 7.3.5), write times (Section 7.3.6), and a first fully functional demonstrator (Section 7.3.7). The last section (7.3.8) will discuss the problems that still need to be solved.

7.3.1 III–V SEMICONDUCTOR COMPOUNDS AND HETEROSTRUCTURES

Molecular beam epitaxy (MBE) and metal–organic chemical vapor deposition (MOCVD) are epitaxial growth techniques to deposit defect-free semiconductor heterostructures with atomically abrupt interfaces. In a device based on a

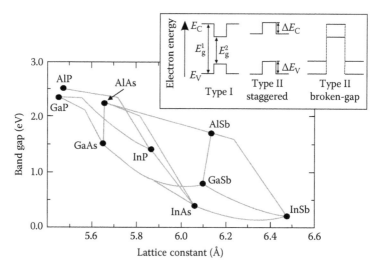

FIGURE 7.2 Band gaps of III–V compound semiconductors and their ternary alloys vs. the lattice constants at zero temperature. The inset shows three possible interfaces of double heterostructures. (After Vurgaftman, I. et al., *J. Appl. Phys.*, 89(11), 5815, 2001.)

heterostructure, the optical and electronic properties can be tailored by selecting chemically different materials and geometric dimensions for the individual layers.

The band gap of the most important III–V semiconductors and their ternary alloys is shown in Figure 7.2 vs. their lattice constants. In a heterostructure, besides the band gap, the relative positions of the valence and conduction bands are of critical importance, which is referred to as the band alignment. The band gap and the band alignment determine the band offsets of the valence and conduction bands. Three different band alignments are distinguished (see inset of Figure 7.2). If two semiconductors are aligned as type-I, the valence and conduction bands of the semiconductor with the smaller band gap lie completely within the band gap of the other semiconductor (i.e., InAs/GaAs). Both carrier types, electrons and holes, are localized in the narrower gap material. In a type-II staggered heterojunction, the band gaps of the two materials show only partial overlap (i.e., GaSb/GaAs), while for a type-II broken-gap alignment, the band gaps of the two semiconductors do not overlap at all.

The band gaps of many III–V semiconductors lie within the range of the photon energy of UV or visible light and the near infrared and have hence found numerous applications as LEDs, semiconductor lasers, and detectors.

7.3.2 SELF-ORGANIZED QUANTUM DOTS

Self-organized QDs are low-dimensional heterostructures, which confine electrons and/or holes within all three spatial directions [8]. As the dimensions of the QDs are below the de Broglie wavelength of the charge carriers, discrete energy levels form within the confinement potential, a property that has coined the term *artificial atom* for QDs.

Semiconductor QDs are typically grown by MBE or MOCVD in the Stranski–Krastanow [9] growth mode. Dome- or pyramid-shaped QDs form on an initially two-dimensional (2-D) layer as a result of self-organization effects, which take place due to total energy minimization of lattice-mismatched semiconductors at the heterointerface [10]. The resulting ensembles of small islands are coherent and defect-free and can be very regular and homogeneous in size, shape, and composition (called self-similarity). A typical QD has a diameter of some 10 nm, a height of about 2–5 nm and area densities of 10^{10}–10^{12} cm^{-2} can be obtained. By changing the growth parameters, the size and shape of the QDs can be modified, enabling to tailor the optical and electronic properties.

7.3.3 MEMORY CELL BASED ON SELF-ORGANIZED QDS

Due to the large confinement of charge carriers within the QDs, they can be used as storage units in a memory device. Schematics of the band structure for a hole-based QD-flash memory and its basic operation principles are shown in Figure 7.3 [5,6,11]. The idea is to embed a layer of self-organized QDs in a modulation-doped field-effect transistor (MODFET). The QDs are used to store the information, while the MODFET is used to control the charge state of the QDs and to perform the memory operations. In the vicinity of the QD layer, a 2-D carrier gas is placed, which is used for the read-out of the stored information.

In order to store a logic "1" (here defined as QDs filled with charge carriers), an emission barrier is needed to prevent the charge carriers from unwantedly leaving the QDs. The emission barrier is formed by the binding potential of the QDs (Figure 7.3a). To store a logic "0" (empty QDs), a capture barrier is needed to prevent charge carriers from being captured into the QDs. The capture barrier is formed by the band bending of the Schottky diode, which acts as gate in the MODFET structure.

The write process is shown in Figure 7.3b. When a gate bias is applied in forward direction, the capture barrier can be completely eliminated. The holes get captured by the QDs and relax down to the lowest hole state. This carrier capture and relaxation process is extremely fast for QDs and was predicted to be on a sub-picosecond time scale at room temperature [2,12]. Hence, the great advantage of the QD-flash concept as compared to the conventional flash memory is its ability to completely eliminate the capture barrier, which can be tuned by the gate voltage.

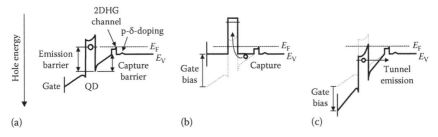

FIGURE 7.3 Schematic depiction of the basic memory operation of the QD-flash based on hole storage. (a) Storage, (b) writing, and (c) erasing operations.

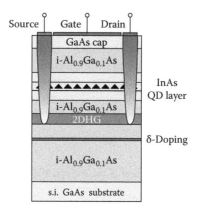

FIGURE 7.4 Schematic device structure of the QD-flash based on hole storage. A layer of self-organized QDs is inserted into a MODFET structure and acts as additional gate. The charge state of the QDs is detected via a measurement of the conductance of the 2-DHG.

The erasing principle is shown in Figure 7.3c. A gate bias in reverse direction leads to a larger electric field, which increases the band bending. This narrows the triangular emission barrier and increases its transparency, hence increasing the tunneling probability. The holes tunnel out of the QDs into the valence band continuum.

The read-out of the information stored in the QDs is done by a conductance measurement in the 2-DHG. Holes inside the QDs decrease the carrier density in the 2-DHG due to the field effect and lower the mobility due to increased scattering (see Section 7.3.5). Both effects can be directly measured as a reduction of the conductance of the 2-DHG.

The structure of a QD-flash for holes is shown in Figure 7.4. A layer of self-organized QDs is inserted into a MODFET structure. Underneath the QD layer, a quantum well or a heterointerface facilitates the formation of a 2-DHG. The holes are provided by δ-doping, while the rest of the device is nominally undoped. The 2-DHG is contacted via two ohmic source/drain contacts, and a Schottky contact acts as control gate.

7.3.4 STORAGE IN SELF-ORGANIZED QDS

The storage time in self-organized QDs is limited by the carrier emission and capture processes. The key parameter is the barrier height yielding the localization energy of the QDs. The localization energy depends on both the III–V material combination and the size of the QDs. In general, a larger localization energy leads to a longer storage time. Hence, material combinations must be searched for, which have a large band offset either in the conduction or preferably the valence band, where carriers have a larger effective mass.

Three basic carrier emission mechanisms exist and are observed in QDs: tunneling, thermally assisted tunneling, and thermal activation (see insets of Figure 7.8). For pure tunneling emission, the charge carriers tunnel directly through the triangular barrier. For thermally assisted tunneling, the charge carriers are activated to an intermediate energy level thermally for $\Delta E \sim kT$ and successively tunnel out of the QDs through a barrier that is now effectively smaller and narrower than that for the initial level. For thermal activation, the charge carriers are emitted from the QDs across the barrier. The process with the highest emission rate will dominate. For a memory, the device must be designed in such a way that thermal activation is the limiting factor to the storage time. For pure thermal emission, the emission rate e_a is [13]

$$e_a = \gamma T^2 \sigma_\infty \exp\left(\frac{-E_a}{kT}\right) \qquad (7.1)$$

where
E_a is the activation energy
T is the temperature
k is the Boltzmann constant
σ_∞ is the capture cross section for $T = \infty$
γ is a temperature-independent constant

Based on Equation 7.1, deep-level transient spectroscopy (DLTS) [13,14] is used to determine the localization energy for various QD material systems. Table 7.1 shows the measured localization energies and the resulting storage times at room temperature. Indeed, QDs with a larger localization energy have a longer storage time for the same carrier type. Figure 7.5 shows the storage times of various material systems vs. the measured localization energies on a semilogarithmic scale [3]. As a rule of thumb, the storage time increases by about one order of magnitude for every 50 meV additional localization energy. An extrapolation yields a localization energy of ~1.14 eV to reach 10-year storage time. Using 8-band-k·p theory [15,16], the localization energy for QDs of various material combinations can be calculated. Based on the extrapolation of the experimental data in Figure 7.5, the storage time can

TABLE 7.1
Experimental and Theoretically (Marked with an Asterisk) Predicted Localization Energies and Storage Times in Various QD Heterostructures

Material System	Charge Carrier Type	Localization Energy	Storage Time at 300 K
InAs/GaAs [34]	Electron/hole	290 meV/210 meV	~200 ns/~0.5 ns
GaAs$_{0.4}$Sb$_{0.6}$/GaAs [30,31]	Hole	450 meV	1 µs
InAs/Al$_{0.6}$Ga$_{0.4}$As [32]	Hole	560 meV	5 ms
InAs/Al$_{0.9}$Ga$_{0.1}$As [3]	Hole	710 meV	1.6 s
GaSb/GaP [33]	Hole	~1.4 eV*	>10^6 years*
GaSb/AlAs [3]	Hole	~1.4 eV*	>10^6 years*

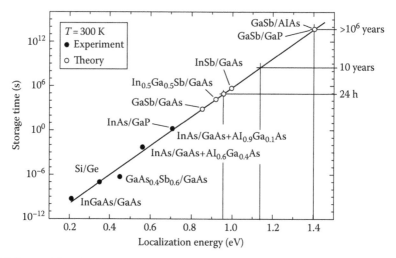

FIGURE 7.5 Experimental (black data points) and theoretical (white data points) carrier storage times vs. the localization energy of different QD heterostructures. An extrapolation allows to predict the storage times based on the localization energies calculated by 8-band-k·p theory.

be predicted. A very promising material system is based on antimony (Sb), which has type-II properties with exclusive hole localization, and the difference of the band gaps goes almost completely into the valence band offset. When combining GaSb with either GaP or AlAs, a localization energy of ~1.4 eV should be reached, much beyond 1.14 eV and sufficient for nonvolatility.

7.3.5 CHARGE DETECTION IN QDS USING A 2-DEG/2-DHG

The QD layer couples to an adjacent 2-D carrier system, a mechanism that is used to detect the logic state in the QDs. Charges inside the QDs can couple in two ways to the 2-D carrier gas [17]. In the Drude model of a 2-D gas, the conductance is [18]

$$\sigma = n_{2\text{-D}} \frac{e^2}{m^*} \tau \qquad (7.2)$$

where
   $n_{2\text{-D}}$ is the area density of the charges in the 2-D gas
   e is the elementary charge
   m^* is the effective mass
   τ is the inverse of the sum of the individual scattering rates of all processes involved (following Matthiessen's rule)

The first coupling mechanism influences the scattering time τ. The charges inside the QDs act as scattering centers for the free moving charges inside the 2-D gas and hence contribute to the total scattering rate $1/\tau$. The QDs can be described like remote impurities that lead to a scattering rate in the 2-D gas of [18]

$$\frac{1}{\tau} \approx \frac{\pi \hbar N n_{QD}}{8 m^* (k_F |d|)^3} \tag{7.3}$$

where
  N is the number of charges per QD
  n_{QD} is the area density of the QDs
  m^* is the effective mass
  k_F is the wave vector at the Fermi energy
  d is the distance of the 2-D system to the QD layer

The scattering rate induced by the charges inside the QDs must be higher than that of any other scattering mechanism in order to be the control parameter of the conductance.

As phonon population increases with increasing temperature, the coupling by scattering can be neglected at room temperature [19,20].

The second coupling mechanism influences the area density $n_{2\text{-D}}$ of the carriers. The 2-D gas is depleted of free carriers by the field effect induced by the charges in the QDs. In a simple model where the gate contact, the QDs, and the 2-D channel are capacitively coupled, the effect of charges inside the QDs on the channel charge, commonly expressed as threshold voltage shift ΔV_{th}, which is necessary to keep the channel charge constant, can be approximated as [21,22]

$$\Delta V_{th} \approx \frac{N n_{QD}}{C_{\text{gate/QD}}} \tag{7.4}$$

where $C_{\text{gate/QD}} = (\varepsilon_s \varepsilon_0 A)/d$ is the capacitance between the QD layer and the gate contact with the permittivity ε_s, the vacuum dielectric constant ε_0, the gate area A, and the distance d between the gate and the QD layer. It is assumed that the QD layer distance to the channel is much smaller than the distance to the gate. The field effect is independent of the temperature if the carriers remain localized at all temperatures.

Due to the localization of the carriers within two dimensions, the 2-D gas is also perfectly suited to detect single- and many-particle states inside the QDs [23,24], which has been demonstrated up to temperatures of ~77 K [25]. If the Fermi function is steep enough (low temperature) and the gate voltage is swept in such a way that the system can relax into thermal equilibrium for each voltage step, the charge carriers in the 2-D gas tunnel into the QDs. This lowers the carrier density within the 2-D gas and increases the scattering. Both effects can be detected via the conductance. Whenever the Fermi level passes a peak (valley) in the density of states of the QD ensemble, the conductance will decrease (increase) as more charge carriers from the 2-D gas tunnel to the QDs and do not contribute to the source/drain current anymore (i.e., lowering $n_{2\text{-D}}$ in Equation 7.2). Figure 7.6 shows the capacitance–voltage

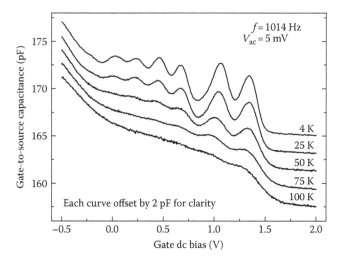

FIGURE 7.6 Gate-source capacitance–voltage curves of a MODFET structure with embedded QDs. The 2-DHG is used to detect the many-particle density of states of the QD ensemble. Each peak represents a many-particle hole state in the QD ensemble.

curve of a hole-based MODFET structure with an embedded layer of QDs, similar to the one shown in Figure 7.4. Whenever the tunneling probability is increased between the QDs and the 2-DHG, at voltages where the Fermi level is aligned with a peak in the density of states within the QD ensemble, the capacitance rises. Due to many-particle effects, such as Coulomb repulsion (Coulomb blockade) and exchange interaction, the peaks represent the many-particle density of states within the QD ensemble. Individual hole levels could be distinguished up to a temperature of about ~77 K. Above that temperature, the Fermi function in the 2-DHG detector broadened and the peaked structure vanished. Nevertheless, the charge state in the QDs can still be detected at room temperature (although the energy levels cannot be discriminated individually anymore). With a slightly adapted capacitance model, the individual-level splittings of the many-particle hole states were extracted [25].

7.3.6 WRITE AND ERASE TIMES IN QDS

One figure of merit for a memory is the time that is needed to write the information to the memory cell. We separate now the processes of writing and erasing of information. Writing means charging the QDs, while erasing stands for discharging the QDs.

In a QD-based memory, the carrier capture and relaxation times to the QDs limit the possible write times. Electron capture has been investigated by interband-pump intraband-probe measurements [12], while hole capture has been studied in DLTS experiments [2]. Both studies revealed a capture and relaxation within picoseconds at room temperature, three orders of magnitude faster than that in a DRAM cell. This extremely fast carrier capture should enable fast write times in a QD-based memory (<ns), which are independent of the storage time.

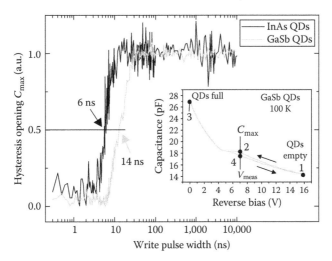

FIGURE 7.7 Write time limits for InAs/GaAs and GaSb/GaAs QDs. The inset shows the C–V hysteresis of the GaSb QD sample with the individual steps in the measurement cycle.

To study the limits of the write time in QD-based memories, a pn-diode structure with a layer of QDs embedded into the p-doped region was studied [26]. Figure 7.7 shows the results for a diode with type-I InAs/GaAs QDs and a diode with GaSb/GaAs QDs. The measurement principle is the following and is illustrated in the inset of Figure 7.7. It shows the C–V hysteresis curve of the GaSb QDs, in which the memory effect due to the carrier confinement in the QDs can clearly be seen as hysteresis. Around a bias voltage of 0 V, the QDs are above the valence band of the surrounding GaAs and are filled with holes, whereas at higher reverse bias, the QDs are below the Fermi level and the holes are emitted, leading in equilibrium to empty QDs. The measurement cycle is the following: first, the diode is set to high reverse bias (~16 V) to discharge the QDs (step 1), then the larger capacitance is measured at the measurement voltage V_{meas} (step 2). After that, a short write pulse to 0 V is applied to fill the QDs with holes (step 3), and at last, the lower capacitance is measured at V_{meas} (step 4). The difference in the two measured capacitance values gives the maximum hysteresis opening ΔC_{max}. For each measurement cycle, the write pulse width is successively reduced from 10 μs to 300 ps, resulting each in a measurement point for ΔC_{max}. At some point in the measurement, the write pulse width is too short to completely fill the QDs with holes, and ΔC_{max} will drop. This drop can be seen in both curves in Figure 7.7. We define the write time limit as the point where ΔC_{max} has decreased to 50%. This gives a value of 6 ns for the InAs/GaAs QDs and 14 ns for the GaSb/GaAs QDs. Hence, a write time in the range of DRAM values is demonstrated. These write time limits do not yet reveal the intrinsic properties of the QDs (capture and relaxation of carriers), but are controlled by the RC parasitics of the diode structure, which had not been designed for high-speed measurements. With an optimized device design, much faster write times, close to the physical limit, should be possible.

The erase process is essentially different, as the underlying process is not capture and relaxation but tunneling, where the barrier shape and the barrier height have to be taken into account. To study the erase process, the same pn-diode structure that

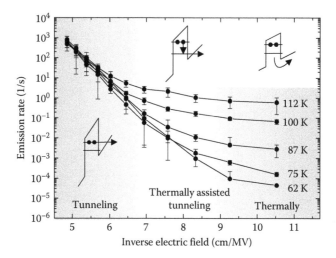

FIGURE 7.8 Emission rate (corresponding to the erase time) vs. the inverse electric field. A transition from pure tunneling via thermally assisted tunneling to thermal emission can be observed. The linear dependence of the tunneling rate on the inverse field allows to extrapolate the field needed for a desired erase time in the pure tunneling regime.

was used in the write time measurements, was studied at different temperatures and different bias voltages. The measurement principle was the same as for the write time measurements, with the reversed sequence of the steps in the duty cycle. Figure 7.8 shows the emission rates vs. the inverse electric field of a sample with GaSb QDs embedded in a pn-diode. In this depiction, a pure tunneling emission process appears as a straight line [27]. A clear transition from pure thermal emission via thermally assisted tunneling to pure tunneling emission can be observed when the reverse bias voltage and hence the electric field are increased (decreasing inverse field). From the linear dependence observed in the tunneling regime, the electric field that would be necessary for 10 and 1 ns erase time limits can be extrapolated, and we get ~500 and ~650 kV/cm, respectively. These values are small as compared to the 8–10 MV/cm, which are used in conventional flash memories [7], but are expected to increase as soon as the carrier localization energy in the QDs is increased.

7.3.7 LOW-TEMPERATURE DEMONSTRATOR

To demonstrate the feasibility of the QD-flash memory concept, a first prototype was fabricated as a demonstrator [6,11]. A schematic structure is shown in Figure 7.9. It uses InAs/GaAs QDs to store holes. The design is based on a GaAs-MODFET structure, in which a 2-DHG is formed inside a QW in the vicinity of the QD layer. First, a 1 μm thick GaAs buffer layer is deposited on top of a semi-insulating substrate. Then, a 40 nm thick p-doped layer is grown, followed by a 7 nm GaAs spacer layer. On top of the spacer, an 8 nm wide $In_{0.25}Ga_{0.75}As$ layer is deposited, followed by 20 nm of undoped GaAs and the InAs/GaAs QDs layer. Finally, the structure is completed by 180 nm undoped GaAs. The device is processed into Hallbar mesas with an active gate area of 740×310 μm^2 using conventional chemical

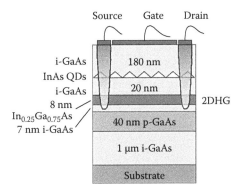

FIGURE 7.9 Schematic depiction of the InAs/GaAs QD MODFET low-temperature demonstrator that was used as proof of principle. The 2-DHG is formed inside an $In_{0.25}Ga_{0.75}As$ QW.

wet-etching techniques. The source/drain contacts are metalized using Ni/Zn/Au and annealed at 400°C for 3 min. The Schottky gate contact was made by Ni/Au.

To study the influence of the holes stored inside the QDs on the conductance of the 2-DHG, the drain current I_D was measured while simultaneously sweeping the gate bias V_G from −0.5 to 1.5 V. The resulting hysteresis curve can be seen in Figure 7.10a. At a gate bias of −0.5 V, the QDs are above the Fermi level and capture holes until they are fully charged. In contrast, at a gate voltage of 1.5 V, the QDs are well below the Fermi level and all holes are emitted. If the sweep time between these two values is smaller than the storage time of the holes in the QDs at a given temperature, a hysteresis curve can be seen. At a temperature of 50 K, the hole storage time in InAs/GaAs QDs is much longer than the sweep time of 1 ms. When the temperature is increased, the relative hysteresis opening (with respect to the upper value) decreases since the storage time becomes shorter. This can be seen in Figure 7.10b, where the effect of the temperature and the sweep time is shown.

To determine the write and erase times of the demonstrator, we used a method similar to the one described in Section 7.3.6, substituting the capacitance measurement by the measurement of the drain current. The write times are shown

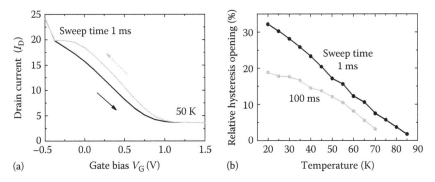

FIGURE 7.10 (a) Drain current hysteresis curve of the demonstrator. (b) Relative hysteresis opening vs. the temperature for two different sweep times.

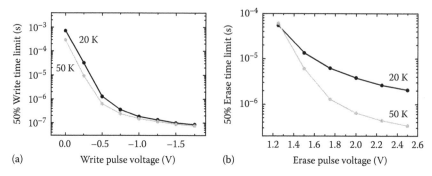

FIGURE 7.11 (a) Write and (b) erase times vs. different pulse voltages for two different temperatures.

in Figure 7.11a for two different temperatures and different write pulse voltages. Starting with a write pulse at 0 V, the write times decrease by three orders of magnitude when increasing the write pulse voltage to −1.75 V, where the Schottky diode of the MODFET structure is extremely forward biased. The write times then stay nearly constant at around 80 ns. This value is in good agreement with the cutoff frequency of the RC parasitics of the device. A dependence of the write time on the temperature could not be observed here since the carrier capture process is expected to be temperature independent. The erase times are shown in Figure 7.11b for different erase pulse voltages and two different temperatures. Here also, a decrease with the pulse voltage can be observed as the band structure is further tilted by the increased electric field, and hence the tunneling probability of the holes in the QDs is increased. The erase time is about 350 ns for an erase pulse of 2.5 V and a temperature of 50 K. In contrast to the carrier capture, the carrier emission process does show a temperature dependence as the underlying emission process can also involve thermally assisted tunneling.

7.3.8 CHALLENGES TO BE SOLVED

The use of self-assembled QDs as storage units has produced extremely promising results so far. The storage time at room temperature was increased by nine orders of magnitude from 0.5 ns up to 1.6 s by using a variety of III–V material systems. Write times in the range of the DRAM access time were demonstrated, and a first demonstrator was fabricated and successfully tested. Still, considerable challenges remain. These are mainly the following:

- *The storage time needs to be increased to reach nonvolatility.* To reach a storage time of 10 years at room temperature, an increase in the localization energy up to at least ∼1.14 eV is necessary (see Figure 7.5). The challenge is here mainly related to the growth of high-quality heterostructures that combine materials with a larger difference in the band gaps than hitherto, facilitating a larger band offset in either the conduction or the valence band. Unfortunately, band gaps are inversely connected to the lattice constant (see Figure 7.2), making it more difficult to combine these materials. The most promising candidates are GaSb on $Al_xGa_{1-x}As$ and GaSb on GaP due to the type-II properties of these heterostructures.

- *The read-out efficiency must be improved.* The transfer characteristics and the hysteresis opening of the demonstrator are still too small to be used in a memory application. The transfer characteristic can be optimized by altering the MODFET design, whereas the hysteresis opening (i.e., threshold voltage shift usually denoted as ΔV_{thr}) can be increased by a stack of QD layers, which increase the number of charges stored in the storage unit. Also, the number of charges that a QD can store increases with increasing localization energy.

- *The write times need to be decreased to reach the physical limit.* To achieve write times closer to the physical limit of carrier capture and relaxation (~ps), the RC parasitics of the device must be optimized. In the MODFET structure, this can be achieved by scaling the gate area and the device size down. If the gate area is scaled down, while the gate aspect ratio is kept constant, the sheet resistance stays constant while the capacitance scales linearly with the gate area, reducing the overall RC value. Thus the write time will be reduced.

- *The erase time needs to be decreased.* In contrast to the write times, the limitations for the erase times are not related to device size or design, but are a result of the tunneling emission. One solution might be to increase the electric field during the erasing process by applying a higher bias voltage. This increases the band bending and narrows the triangular emission barrier, enhancing the tunneling rate [27]. Another option would be the implementation of a resonant tunneling structure where the tunneling rates are enhanced [28], if the energy levels of a superlattice structure are in resonance with each other. This poses a challenge to the band design of the superlattice in order to be able to repeatedly switch between a high tunneling probability and a very low tunneling probability.

- *A fully integrated device design of the QD-flash needs to be developed.* To facilitate QDs as storage units within a commercially available memory device, such as USB sticks and flash cards, a fully integrated device design must be developed similar to the established Si-CMOS technology. This is no fundamental problem, since high-performance computers fully based on III–V technology were demonstrated more than a decade ago.

7.4 CONCLUSION

We have presented a memory concept (QD-flash) based on self-organized QDs as storage units. A QD-flash will be able to realize fast write times, while being at the same time nonvolatile. The key of the concept is the use of III–V compound semiconductor heterostructures that offer a vast flexibility for detailed band structure design. The use of self-organization effects are used in a bottom-up approach to form nanometer-sized structures that act as a confining potential for charge carriers.

These QDs offer extremely fast carrier capture times (~ps) independent of the storage time. Write times in pn-diodes with embedded QDs were measured and are already identical to DRAM access time, at the moment limited only by the RC parasitics of the diode structure. The storage time in QDs scales exponentially with the confining potential (localization energy). We have increased the storage time at room temperature from nanoseconds in InAs/GaAs QDs to seconds in InAs/Al$_{0.9}$Ga$_{0.1}$As QDs. From an extrapolation based on 8-band-k·p calculations, material combinations with larger localization energy, reaching longer storage times, were predicted. Heterostructures based on Sb in combination with Al$_x$Ga$_{1-x}$As or GaP are found to be very promising candidates to reach nonvolatility. A 2-DHG used as charge detector proved to be a very sensitive detector, with which even many-particle hole states in QDs could be detected up to a temperature of ~77 K. A first demonstrator memory was fully functional, and storage of holes inside the QDs was shown up to a temperature of ~80 K with write times of ~80 ns and erase times of ~350 ns.

Despite enormous progress achieved so far, big challenges still exist. Structures exhibiting longer storage times need to be epitaxially grown with a high material quality, the read-out efficiency needs to be increased, the write times have to be decreased down to the physical limits by downscaling the device size, the erase times have to be decreased, and the successful integration of the QD-flash concept to a technology similar to CMOS has to be accomplished.

ACKNOWLEDGMENTS

We recognize contributions of and fruitful discussions with many others: Andrei Schliwa, Erik Stock, Johannes Gelze, Annika Högner, Franziska Luckert, Konstantin Pötschke, Andreas Beckel, and Bastian Marquardt. This work was partly funded by the DFG, contract number BI284/29-1, in the framework of the NanoSci-E+ project QD2D, contract numbers BI284/30-1 and GE2141/1-1, of the European Commission, and the VIP project HOFUS of the BMBF.

REFERENCES

1. International Technology Roadmap for Semiconductors (ITRS), Executive Summary, Edition 2011.
2. M. Geller, A. Marent, E. Stock, D. Bimberg, V.I. Zubkov, I.S. Shulgunova, and A.V. Solomonov, Hole capture into self-organized InGaAs quantum dots, *Appl. Phys. Lett.*, 89(23), 232105, 2006.
3. A. Marent, M. Geller, A. Schliwa, D. Feise, K. Pötschke, D. Bimberg, N. Akcay, and N. Öncan, 10[sup 6] years extrapolated hole storage time in GaSb/AlAs quantum dots, *Appl. Phys. Lett.*, 91(24), 242109, 2007.
4. R. Waser, *Microelectronics and Information Technology*. Berlin, Germany: Wiley-VCH, 2003.
5. M. Geller, A. Marent, and D. Bimberg, Speicherzelle und Verfahren zum Speichern von Daten (Memory cell and method for storing data), International patent EP 2097904 BI, June 15, 2011.
6. A. Marent, T. Nowozin, J. Gelze, F. Luckert, and D. Bimberg, Hole-based memory operation in an InAs/GaAs quantum dot heterostructure, *Appl. Phys. Lett.*, 95, 242114, 2009.

7. R. Bez, E. Camerlenghi, A. Modeli, and A. Visconti, Introduction to flash memory, *Proc. IEEE*, 91(4), 489–502, 2003.

8. D. Bimberg, M. Grundmann, and N. N. Ledentsov, *Quantum Dot Heterostructures.* Chichester, England: John Wiley & Sons, 1998.

9. I. N. Stranski and L. Krastanow, Zur Theorie der orientierten Ausscheidung von Ionenkristallen aufeinander, Sitzungsber. Akad. Wiss. Wien, Math.-Naturwiss. K1, Abt. 2B, vol. 146, p. 797, 1938.

10. V. A. Shchukin and D. Bimberg, Spontaneous ordering of nanostructures on crystal surfaces, *Rev. Mod. Phys.*, 71, 1125, 1999.

11. A. Marent, T. Nowozin, M. Geller, and D. Bimberg, The QD-flash: A quantum dot-based memory device, *Semicond. Sci. Technol.*, 26, 014026, 2011.

12. T. Müller, F. F. Schrey, G. Strasser, and K. Unterrainer, Ultrafast intraband spectroscopy of electron capture and relaxation in InAs/GaAs quantum dots, *Appl. Phys. Lett.*, 83(17), 3572–3574, 2003.

13. D. V. Lang, Deep-level transient spectroscopy: A new method to characterize traps in semiconductors, *J. Appl. Phys.*, 45(7), 3023, 1974.

14. P. Blood and J. W. Orton, *The Electrical Characterization of Semiconductors: Majority Carriers and Electron States.* London, U.K.: Academic Press, 1992.

15. O. Stier, M. Grundmann, and D. Bimberg, Electronic and optical properties of strained quantum dots modeled by 8-band-k.p theory, *Phys. Rev. B*, 59, 5688, 1999.

16. A. Schliwa, M. Winkelnkemper, and D. Bimberg, Impact of size, shape and composition on piezoelectric effects and the electronic properties of InGaAs/GaAs quantum dots, *Phys. Rev. B*, 76, 205324, 2007.

17. B. Marquardt, A. Beckel, A. Lorke, A. D.Wieck, D. Reuter, and M. Geller, The influence of charged InAs quantum dots on the conductance of a two-dimensional electron gas: Mobility vs. carrier concentration, *Appl. Phys. Lett.*, 99(22), 223510, 2011.

18. J. H. Davies, *The Physics of Low-Dimensional Semiconductors.* New York: Cambridge University Press, 1998.

19. W. Walukiewicz, H. E. Ruda, J. Lagowski, and H. C. Gatos, Electron-mobility in modulation-doped heterostructures, *Phys. Rev. B*, 30(8), 4571–4582, 1984.

20. W. Walukiewicz, Hole mobility in modulation-doped heterostructures: GaAs-AlGaAs, *Phys. Rev. B*, 31(8), 5557, 1985.

21. L. Guo, E. Leobandung, L. Zhuang, and S. Y. Chou, Fabrication and characterization of room temperature silicon single electron memory, *J. Vac. Sci. Technol. B*, 15(6), 2840–2843, 1997.

22. D. Nataraj, N. Ooike, J. Motohisa, and T. Fukui, Fabrication of one-dimensional GaAs channel-coupled InAs quantum dot memory device by selective-area metal-organic vapor phase epitaxy, *Appl. Phys. Lett.*, 87, 193103, 2005.

23. B. Marquardt, M. Geller, A. Lorke, D. Reuter, and A. D. Wieck, Using a two-dimensional electron gas to study non-equilibrium tunneling dynamics and charge storage in self-assembled quantum dots, *Appl. Phys. Lett.*, 95, 022113, 2009.

24. B. Marquardt, M. Geller, B. Baxevanis, D. Pfannkuche, A. D. Wieck, D. Reuter, and A. Lorke, Transport spectroscopy of non-equilibrium many-particle spin states in self-assembled quantum dots, *Nat. Commun.*, 2, 209, 2011.

25. T. Nowozin, A. Marent, G. Hönig, A. Schliwa, D. Bimberg, A. Beckel, B. Marquardt, A. Lorke, and M. Geller, Time-resolved high-temperature detection with single charge resolution of holes tunneling into many-particle quantum dot states, *Phys. Rev. B*, 84, 075309, 2011.

26. M. Geller, A. Marent, T. Nowozin, D. Bimberg, N. Akcay, and N. Öncan, A write time of 6 ns for quantum dot-based memory structures, *Appl. Phys. Lett.*, 92(9), 092108, 2008.

27. T. Nowozin, A. Marent, M. Geller, D. Bimberg, N. Akcay, and N. Öncan, Temperature and electric field dependence of the carrier emission processes in a quantum dot-based memory structure, *Appl. Phys. Lett.*, 94, 042108, 2009.

28. S. M. Sze and K. K. Ng, *Physics of Semiconductor Devices*. Hoboken, NJ: John Wiley & Sons, 2006.

29. I. Vurgaftman, J. R. Meyer, and L. R. Ram-Mohan, Band parameters for III-V compound semiconductors and their alloys, *J. Appl. Phys.*, 89(11), 5815–5875, 2001.

30. L. Müller-Kirsch, R. Heitz, U. W. Pohl, D. Bimberg, I. Häusler, H. Kirmse, and W. Neumann, Temporal evolution of GaSb/GaAs quantum dot formation, *Appl. Phys. Lett.*, 79, 1027–1029, 2001.

31. M. Geller, C. Kapteyn, L. Müller-Kirsch, R. Heitz, and D. Bimberg, 450 meV hole localization energy in GaSb/GaAs quantum dots, *Appl. Phys. Lett.*, 82(16), 2706–2708, 2003.

32. A. Marent, M. Geller, D. Bimberg, A. P. Vasi'ev, E. S. Semenova, A. E. Zhukov, and V. M. Ustinov, Carrier storage time of milliseconds at room temperature in self-organized quantum dots, *Appl. Phys. Lett.*, 89(7), 072103, 2006.

33. D. Bimberg, A. Marent, T. Nowozin, and A. Schliwa, Antimony-based quantum dot memories, *Proc. SPIE*, 7947, 79470L, 2011.

34. M. Geller, E. Stock, C. Kapteyn, R. L. Sellin, and D. Bimberg, Tunneling emission from self-organized In(Ga)As/GaAs quantum dots observed via time-resolved capacitance measurements, *Phys. Rev. B*, 73(20), 205331, 2006.

Part IV

Magnetic Memory

8 Spin-Transfer-Torque MRAM

Kangho Lee

CONTENTS

8.1 INTRODUCTION

8.1.1 MOTIVATION FOR EMBEDDED STT-MRAM: APPLICATION PERSPECTIVES

As silicon industry is moving toward the end of technology roadmap, providing cost-effective and power-efficient system-on-chip memory solutions has become ever more challenging. While there are increasing demands for embedded memory capacity, conventional embedded working memories such as embedded SRAM and DRAM have been facing scalability challenges along with increasing static leakage power. The static leakage power consumption of embedded working memories, particularly in case of high-performance mobile chips, accounts for a substantial portion of total power consumption, which is expected to exacerbate at future technology nodes. Considering that embedded memory occupies more than 50% of the total chip area of commercial state-of-the-art mobile chipsets, it is important to develop an alternative embedded memory technology that can improve energy efficiency and reduce cost without compromising the benefits of conventional working memories.

Novel memory devices such as phase-change RAM, ferroelectric RAM, and resistive RAM have actively been investigated; however it has been challenging to meet two essential requirements for working memories: unlimited endurance and fast read/write speed (10 ns or less). None of the emerging memory technologies except for spin-transfer-torque magnetoresistive random access memory (STT-MRAM) has not demonstrated more than 10^{12} write endurance combined with good scalability and fast read/write operations. This positions STT-MRAM as a promising emerging memory technology that has a potential to replace the conventional working memories.

Embedded STT-MRAM may provide additional benefits, particularly for futuristic low-power wireless applications. There have been growing demands for ultralow-power wireless solutions for implantable medical devices, wireless healthcare monitoring devices, etc. Typically, these applications do not require high-speed operations; however, the power requirement is expected to be very stringent, possibly sub-mW. In this case, static leakage power from conventional memory arrays may occupy a significant portion of the total power consumption. Nonvolatility of STT-MRAM can eliminate a substantial portion of the static leakage power. Small form factor and low cost would be critical factors as well. A typical STT-MRAM bitcell consists of one MTJ cell and one access transistor connected in series (1T–1MTJ). The size of a 1T–1MTJ bitcell can be much smaller than that of embedded SRAM or DRAM bitcells. Since an MTJ module can also be integrated into a CMOS back-end-of-line (BEOL) without substantial process overheads, decreased bitcell size leads to cost reduction. Finally, STT-MRAM can simplify a system architecture. Due to nonvolatility of STT-MRAM, flash memory for code storage can be removed from the system. This also minimizes IO transactions and helps reduce the total cost as well as the form factor.

Depending on target applications, desirable attributes of embedded STT-MRAM can be different. However, the common key challenge for the success of embedded STT-MRAM is to minimize energy per write operation within proper voltage headrooms. This chapter covers magnetoelectric properties of magnetic tunnel junctions (MTJs), memory operations of STT-MRAM bitcells, and recent advances and prospects of STT-MRAM technology.

8.1.2 RECENT INDUSTRIAL EFFORTS FOR MRAM DEVELOPMENT

Before diving into technical details of STT-MRAM, it would be good to briefly review recent industrial efforts for MRAM development. In 1990s, semiconductor industry started MRAM development. It was Freescale Semiconductor that shipped the first 4 Mb MRAM product in 2006. The MRAM module was integrated into a 180 nm CMOS logic platform. Freescale Semiconductor spun off its MRAM business to a new company called Everspin Technologies. By 2008, over 1 million MRAM chips were sold. In 2010, Everspin introduced new 16 Mb MRAM chips. All the MRAM products from Everspin are conventional MRAM based on field-induced switching, not STT-MRAM. Currently, Everspin is the only company shipping MRAM products. MagIC, IBM, and NEC have also been working on conventional MRAM. NEC reported a high-speed 32 Mb MRAM macro suitable for embedded systems in 2009 [1]. However, conventional MRAM has a fundamental scalability problem because scaling down MTJ cells entails a substantial increase in switching fields, and thereby more power consumption. In addition, the programming current to generate the switching field is too high to be integrated into a low-power logic platform at advanced CMOS technology nodes. For these reasons, conventional MRAM has not made a significant impact on memory industry, serving only niche markets.

In contrast, STT-MRAM is a scalable technology. Critical switching current density (J_c) is proportional to the magnitude of the STT effect, which is largely determined by the material property and the film structure of a free layer. Scaling down MTJ cells leads to smaller critical switching current (I_c) because I_c is simply J_c times MTJ area. Hence, for a given J_c, the write power of STT-MRAM scales down as the size of MTJ cells shrinks. Since Sony reported the first chip-level demonstration of STT-MRAM in 2005 [2], semiconductor industry has actively been exploring STT-MRAM technology. MTJ was integrated into a 180 nm CMOS platform. This milestone was followed by Hitachi and Tohoku University that presented a 2 Mb STT-MRAM integrated into a 200 nm CMOS platform in 2007 [3]. The MTJ size was 50×100 nm^2. Operations of 40 ns read and 100 ns write were demonstrated. IBM and MagIC reported statistical behaviors of MTJs using a 4 kb STT-MRAM test chip and suggested that a 64 Mb STT-MRAM chip at the 90 nm node would be feasible [4]. In 2009, Qualcomm and TSMC presented 45 nm STT-MRAM embedded into a standard CMOS logic platform that employs low-power transistors and Cu/low-k BEOL [5]. Grandis reported a 256 kb STT-MRAM integrated into a 90 nm CMOS platform in 2010, demonstrating J_c as low as ~1 MA/cm^2 [6]. Fujitsu demonstrated improved bitcell switching yields using MTJs with reversed MTJ film stacks, so-called top-pinned structures [7]. Samsung investigated feasibility of STT-MRAM as next-generation nonvolatile memory to replace DRAM and NOR Flash, showing that the STT-MRAM bitcell size can be scaled down to sub-30 nm technology node [8]. Hynix and Grandis also reported fully integrated 64 Mb STT-MRAM using modified DRAM processes at the 54 nm technology node [9]. The bitcell size was 14 F^2, and the MTJ size was ~54×108 nm^2.

All the previous works mentioned earlier utilized in-plane MTJs whose magnetization lies in the film plane. However, it is questionable whether deeply scaled in-plane MTJs will be able to serve future CMOS technology nodes (28 nm or beyond). In 2010, Toshiba reported 64 Mb STT-MRAM using perpendicular MTJs [10]. In perpendicular MTJs, the free layer is magnetized perpendicular to the film plane due to strong crystalline anisotropy or surface anisotropy, which provides more rooms for scaling down MTJ. This will be discussed later in Section 7.4.1. IBM and MagIC also presented 4 kb STT-MRAM based on pMTJs, demonstrating superior MTJ performances that may be sufficient to yield a 64 Mb chip [11].

8.2 MAGNETIC TUNNEL JUNCTION: STORAGE ELEMENT OF STT-MRAM

MRAM defines binary states (states "0" and "1") by two discrete resistance values of an MTJ. Figure 8.1 illustrates a typical MTJ film stack that consists of multiple metallic films separated by a thin (~1 nm) MgO tunnel barrier. The layers with

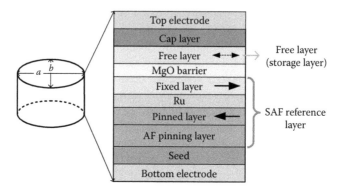

FIGURE 8.1 Illustration of a typical MTJ film structure.

arrows (free, fixed, and pinned layers) are ferromagnetic metals (FMs). Soft fer-romagnetic alloys such as NiFe and CoFeB have been used for the free layer whose moment direction can be switched by external excitations as indicated by the dou-ble-ended dotted arrow. Due to thin-film shape anisotropy, the moment typically resides in the film plane. The cap layer is inserted between the top electrode and the free layer to protect the free layer from the following process steps and/or tune the magnetoelectric properties of the free layer. The layers between MgO and the seed layer are to achieve reliable reference layers (reference to the free-layer moment) so that the fixed-layer moment does not change in the presence of the external excita-tions. The antiferromagnetic pinning layer, typically PtMn or IrMn, is deposited on the seed layer. The moment direction of the pinned layer is biased via the exchange bias effect during magnetic annealing following film depositions. The fixed-layer moment is antiferromagnetically coupled to the pinned layer moment via a nonmag-netic Ru spacer, which is known as interlayer exchange coupling. This scheme, called synthetic antiferromagnetic (SAF) reference layers (typically CoFe/Ru/CoFeB), has been widely adopted since it provides means to achieve reliable reference layers and control magnetostatic coupling between the free layer and the reference layers. The seed layer is to provide smooth surface and preferable crystallographic orientation for subsequent film depositions. All the films can be grown by a physical vapor depo-sition system. However, the MgO tunnel barrier can also be produced by oxidation of a thin Mg layer, which is desirable from a manufacturing standpoint. The MTJ film stacks can be integrated into CMOS BEOL and patterned by either ion milling or reactive ion etching. An individual patterned MTJ cell typically represents 1 bit. In this section, essential physics of MTJs are explained in conjunction with key MTJ performance metrics for STT-MRAM: tunneling magnetoresistance ratio (TMR), thermal barrier (E_B) for data retention, critical switching current (J_c).

8.2.1 MAGNETIZATION DYNAMICS IN FERROMAGNETIC METALS

Strong ferromagnetism commonly observed in FMs such as Co, Ni, Fe, and their alloys originates from spontaneous alignment of microscopic magnetic moments associated with electron motions (orbital motion and spin). The orbital motion of a single electron forms a current loop, exhibiting magnetic dipole moment (M_{ob}) associated with its angular momentum (L). Classical electromagnetism tells us that

$$M_{ob} = \frac{e}{2m_e} L,$$ (8.1)

where
  e is the electron charge
  m_e is the electron mass

The magnetic moment corresponding to the first Bohr orbit, called Bohr magnetron (μ_B), is 0.927×10^{-20} erg/Oe in cgs unit. The electron spin, a purely quantum-mechanical phenomenon, also shows the magnetic moment exactly equal to μ_B. Hence, μ_B is considered a natural unit of electron magnetic moment. In general, electron magnetic moment (m) is given by

$$m = \frac{ge}{2m_e} S = -\gamma S,$$ (8.2)

where
  g is a spectroscopic splitting factor ($g = 1$ for orbital motion and $g = 2$ for spin)
  S is the spin angular momentum
  γ is called the gyromagnetic ratio

Note that γ is positive. Energy felt by an electron in the presence of a time-dependent external magnetic field (B) is $-m \cdot B$ (Zeeman energy), hence the Hamiltonian is given by $\gamma S \cdot B$. The time derivative of the expectation value of S, denoted as $<S>$ in the following equation, can be computed using Schrödinger's equation:

$$\frac{d <S>}{dt} = \frac{1}{i\hbar} <[S, H]> = -\gamma <S> \times B$$ (8.3)

From Equations 8.2 and 8.3, the motion of an electron in magnetic fields can be described by

$$\frac{dm}{dt} = -\gamma m \times B$$ (8.4)

In a typical ferromagnet, electron spins in the 3d orbitals are spontaneously aligned due to strong quantum-mechanical exchange forces among adjacent spins. The exchange energy (E_{ex}) between two spins ($\sigma_{i,j}$) is given by

$$E_{ex} = -2J_{ij}\sigma_i \cdot \sigma_j$$ (8.5)

where J_{ij} is the exchange integral. With the exchange energy for individual electrons considered, the Hamiltonian for a ferromagnet can be approximated as

$$H = \sum_i \gamma S_i \cdot B + \sum_{i,j} -2J_{ij}\sigma_i \cdot \sigma_j \qquad (8.6)$$

Assuming that all the adjacent spins are aligned in the ferromagnet, the magnetization (M) of the ferromagnet can simply be described by

$$\frac{dM}{dt} = -\gamma M \times B \qquad (8.7)$$

where M is defined as the total electron dipole moment per unit volume.

In a static magnetic field, Equation 8.7 tells us that the magnetization precesses around the applied field at an angular frequency of γB (known as the Larmor frequency) as illustrated in Figure 8.2a. However, we know from magnetic hysteresis measurements that with a sufficiently large field, the magnetization becomes saturated to its maximum, saturated magnetization (M_s), and aligned to the field direction. The precession motion alone does not explain this. Adding damping torque can make the magnetization spiral into the field direction after a finite time (order of nanoseconds) as illustrated in Figure 8.2b. Hence, a damping term has been added into Equation 8.7 by introducing phenomenological damping parameter, often called Gilbert damping constant (α), which leads to the following equation, known as Landau–Lifshitz–Gilbert equation:

$$\frac{1}{\gamma}\frac{dM}{dt} = -M \times B - \frac{\alpha}{M} M \times (M \times B) \qquad (8.8)$$

Physical origins of the damping torque have been attributed to energy relaxations due to interactions with s-electrons, spin–orbit interactions, etc. Damping can be characterized from ferromagnetic resonance measurements, and α values reported with typical free-layer materials is ~0.01 or less.

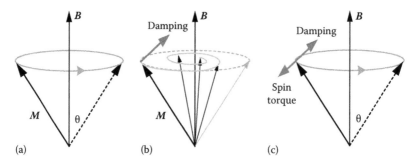

FIGURE 8.2 Precession of magnetization (a) without considering damping, (b) in the presence of damping torque, and (c) with spin torque opposing damping torque.

While energy is dissipated by the damping motion, it may also be possible to transfer energy to electrons in the 3d orbitals (d-electrons) by external excitations. In 1996, Slonczewski [12] and Berger [13] theoretically predicted that spin-polarized currents, primarily carried by electrons in the 4s orbitals (s-electrons), can transfer spin angular momentum of s-electrons to d-electrons, exerting spin torque to d-electrons. Another term is added to Equation 8.8 to account for such current-induced magnetic excitation:

$$\frac{1}{\gamma}\frac{d\boldsymbol{M}}{dt} = -\boldsymbol{M}\times\boldsymbol{B} - \frac{\alpha}{M}\boldsymbol{M}\times(\boldsymbol{M}\times\boldsymbol{B}) + \frac{a_J}{M}\boldsymbol{M}\times(\boldsymbol{M}\times\boldsymbol{n}_s) \tag{8.9}$$

where
$\boldsymbol{n}_s$ is the direction of spin polarization of the incoming current
a_J is given by $\hbar JP/2eM_st$
J is the current density
P is the spin polarization factor
t is the film thickness

The last term can be viewed as spin torque opposing damping torque as illustrated in Figure 8.2c. When the spin-torque excitation balances out damping, the magnetization can precess without being damped. This has been utilized to produce high-frequency oscillators. Also, as the spin-torque excitation becomes large enough to overcome damping, the magnetization can be switched to another energetically favorable orientation. Current-induced magnetization reversal will be explained more in detail in Section 8.2.4.

8.2.2 Tunneling Magnetoresistance Ratio

Due to the spin-dependent tunneling effect, the angle (θ) between the free-layer and the fixed-layer moments determines the MTJ resistance, resulting in the minimum (maximum) resistance when the two moments are parallel (antiparallel) to each other. In general, the MTJ resistance (R) can be described by

$$R = \frac{R_\perp}{1+(TMR/2)\cos\theta} \tag{8.10}$$

where $R_\perp$ is the resistance measured in the perpendicular magnetic configuration ($\theta = \pi/2$). R can have any values between antiparallel-state resistance (R_{ap}) and parallel-state resistance (R_p). The TMR, defined as ($R_{ap} - R_p$)/R_p, is often used as a figure of merit for the read operation. The read operation of STT-MRAM is to sense the difference between MTJ cell resistance and predefined reference cell resistance. Although the reference cell scheme and MTJ resistance distributions also affect the read margin significantly, it is critical to achieve large TMR for high-speed read operations.

Since TMR has been an essential element for MRAM development, it would be beneficial to briefly review the history of TMR development. Since Jullière discovered

TMR (14% at 4.2 K) in an Fe/Ge/Co junction in 1975 [14], many researchers have contributed to the enhancement of *TMR*. Miyazaki et al. at Tohoku University and Moodera et al. at MIT demonstrated the *TMR* ratio over 10% at room temperature, using amorphous AlO_x as a tunnel barrier. Amorphous AlO_x has been used in conventional MRAM, providing the *TMR* ratio of ~70%. The next milestone was the introduction of polycrystalline MgO. Following the theoretical predictions that the TMR ratio can be increased to more than 1000% using Fe/MgO/Fe junctions [15,16], many research groups have reported enhanced TMR using MgO. Recently, TMR ~ 600% at room temperature has been reported using CoFeB/MgO/CoFeB junctions annealed at high temperature (550°C) [17]. More detailed history of TMR development can be found elsewhere [18].

8.2.3 ENERGY BARRIER FOR DATA RETENTION

When the MTJ film is patterned into an elliptical shape, shape anisotropy allows the free-layer moment to have only two energetically favorable directions along the easy-axis as shown in Figure 8.3 (i.e., θ = 0 or π). For a single-domain nanomagnet, the energy barrier (E_B) between these two states in the presence of external magnetic field (H_{ext}) is given by

$$E_B = \frac{M_s H_k V}{2} \left(1 - \frac{H_{ext}}{H_k} \right)^2 \tag{8.11}$$

where
  M_s is the saturation magnetization of the free layer
  V is the free-layer volume
  H_k is the effective uniaxial anisotropy field

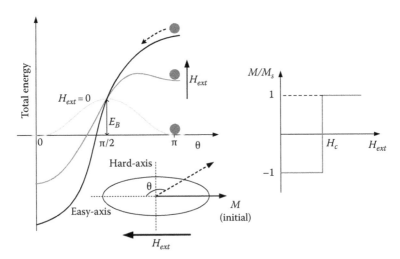

FIGURE 8.3 Changes in energy landscape when an external field is applied along the easy-axis. The magnetization (*M*) rotates coherently as indicated by the magnetization curve.

In Figure 8.3, the magnetization is initially oriented at $\theta = \pi$. When H_{ext} is applied to the opposite direction of the magnetization along the easy-axis, the energy barrier seen from the magnetization state decreases, resulting in switching of the magnetization when this energy barrier becomes comparable to thermal energy. Note that the magnetization remains to be at $\theta = \pi$ with an insufficient magnetic field, hence magnetization switching occurs coherently as illustrated in the magnetization-field curve. The switching field is often called coercivity field (H_c).

Since scaling down MTJ cells leads to volume reduction, it is important to have sufficient $M_s - H_k$ product to meet the target standby data retention requirement and ensure adequate thermal stability for patterned MTJ cells. Nonswitching probability, $F(t)$, of a single MTJ cell is commonly described by the Néel–Brown relaxation time formula, $F(t) = \exp(-t/\tau)$, with the relaxation time constant $\tau = \tau_0 \exp(E_B/k_B T)$. τ_0 is typically assumed to be 1 ns. To retain a single bit for 10 years, this formula requires E_B to be at least $40k_B T$. The E_B requirement increases with increasing memory capacity; however, $60k_B T$ has been suggested to be a good target for reliable operations. The E_B requirement can also be alleviated by adding error-correction-code circuits. However, one should be careful when measuring E_B that represents realistic standby data retention. The most accurate method would be to directly measure dwell times of magnetization states at elevated temperatures and extract E_B using Néel–Brown relaxation time formula. However, for MTJ cells with reasonable E_B, this measurement takes too much time. Assuming that thermally activated magnetization reversal follows the same path as field-driven switching, one can estimate E_B by characterizing dependence of H_c on field pulse width (t_p) or temperature and fitting the data with Sharrock's formula:

$$H_c = H_k \left[1 - \left(\frac{k_B T}{E_B} \ln\left(\frac{t_p}{\tau_0} \right) \right)^c \right] \qquad (8.12)$$

where c can be considered a fitting parameter and typically ranges between 0.5 and 1. This method has been used to estimate E_B of recording media for hard disk and MTJs for STT-MRAM as well as conventional MRAM.

8.2.4 SPIN-TRANSFER-TORQUE SWITCHING

To program MTJ cells, one can apply external magnetic fields as illustrated in a resistance–magnetic field hysteresis loop (Figure 8.4a). The MTJ resistance is measured at a small read bias while sweeping static magnetic fields along the easy-axis. The MTJ cell has only two resistance states. Switching fields in both antiparallel-to-parallel (AP–P) and parallel-to-antiparallel (P–AP) directions are ~220 Oe, showing nearly zero offset field (H_{off}). Conventional MRAM has utilized magnetic fields for the write operation. Magnetic fields are generated by currents flowing through metal lines located near an MTJ cell to be programmed. The polarity of magnetic fields is controlled by changing the current direction. However, decreasing an MTJ cell size tends to increase switching fields, resulting in more write power consumption, hence, conventional MRAM does not provide good scalability.

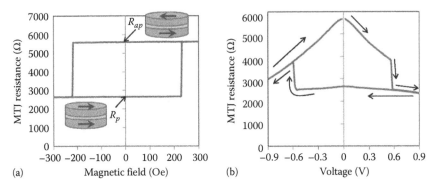

FIGURE 8.4 Hysteresis loop from (a) magnetic field switching and (b) spin-transfer-torque switching.

Another way to switch the free-layer magnetization is to apply spin-polarized currents directly through an MTJ cell without applying external magnetic fields. Current-induced magnetization switching was theoretically predicted by Slonczewski [12] and Berger [13] in 1996. The first experimental evidence was reported with metallic spin valves that have a nonmagnetic metallic spacer instead of the tunnel barrier [19]. Figure 8.4b shows a resistance–voltage hysteresis loop of an MTJ cell with no external magnetic fields applied. Positive voltages correspond to conduction electrons flowing from the fixed layer to the free layer, and vice versa for negative voltages. The initial MTJ state is R_{ap}, and the arrows indicate the direction of voltage sweeping. It is noteworthy that R_{ap} shows a strong bias dependence, whereas R_p is nearly independent of bias. AP–P switching occurs at about 0.6 V. Positive voltages produce parallelizing current, hence the MTJ state remains in the parallel configuration as the applied voltage increases further. Negative voltages produce antiparallelizing current, and P–AP switching occurs at about −0.6 V.

This phenomenon, called STT switching, is a result of interaction between spin-polarized conduction electrons and magnetization. In a simplified picture, STT switching can be understood using illustrations shown in Figure 8.5. First, assume that initial magnetization of the free layer is antiparallel to that of the fixed layer (Figure 8.5a). Free electrons in normal metals have equal populations of up-spin and down-spin. When the free electrons enter a ferromagnet, a substantial portion of

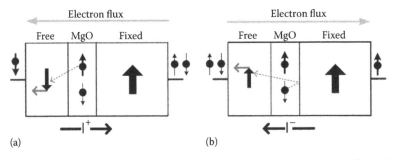

FIGURE 8.5 Simplified illustration of STT switching in an MTJ structure. (a) Parallelizing current and (b) antiparallelizing current.

conduction electrons are polarized to the magnetization direction of the ferromagnet, resulting in imbalanced spin populations. Hence, as conduction electrons pass through the fixed layer, the electrons are spin-polarized to the magnetization direction of the fixed layer. Substantial portion of these spin-polarized electrons, mostly residing in the 4s orbitals, tunnel through the insulating barrier without losing their polarization and exert torque on the free-layer magnetization. Since electrons in the 3d orbitals are responsible for magnetic moments observed in ferromagnetic materials, this interaction is essentially based on momentum exchange between electrons in the 3d orbitals and conduction electrons in the 4s orbitals. When sufficiently large currents are applied, the spin torque flips the free-layer magnetization. The threshold current is called critical switching current (I_c). On the other hand, substantial portion of the minority spin electrons are reflected at the barrier interfaces and exert torque on the fixed-layer magnetization. However, the fixed-layer magnetization is not switched because it takes a lot more energy to switch the SAF reference layers. For P–AP STT switching, conduction electrons are injected into the free layer first and polarized to the magnetization direction of the free layer (Figure 8.5b). Majority spin electrons tunnel through the barrier more easily because the free-layer magnetization is parallel to that of the fixed layer. Minority spin electrons, polarized to the opposite direction of the free-layer magnetization, are reflected at the barrier interfaces and then exert torque on the free-layer magnetization, eventually leading to P–AP switching.

According to the Slonczewski's model [12,20], intrinsic critical switching current (I_{c0}), defined as I_c at zero temperature, is described by the following equation:

$$I_{c0} = \frac{2e\alpha_{eff} M_s A t}{\hbar \eta} \left(H_{off} + H_{k\parallel} + 2\pi M_s \right) \tag{8.13}$$

where
e is the electron charge
α_{eff} is the effective damping constant
A is the MTJ area
t is the free-layer thickness
$\hbar$ is the reduced Planck's constant
$H_{k\parallel}$ is the uniaxial anisotropy field in the film plane
H_{off} is the magnetostatic offset field
η is the STT efficiency

With typical CoFeB-based free layers, α_{eff} is ~0.01 and M_s is ~1000 emu/cc. To a first-order approximation, η can be estimated by $(p/2)/(1+p^2\cos\theta)$, where p is the tunneling spin polarization of incident spin-polarized currents and θ is the angle between the free-layer and the fixed-layer moments. For symmetric MTJ junctions (e.g., CoFeB/MgO/CoFeB), p can be extracted from the Jullière formula, $TMR = 2p^2/(1 - p^2)$. $H_{k\parallel}$ is typically less than 0.5 kOe although it is affected by the aspect ratio (AR) and size of an MTJ cell, sidewall roughness, passivation, etc. H_{off} originates from interfacial roughness and dipolar coupling fields between the adjacent ferromagnetic layers;

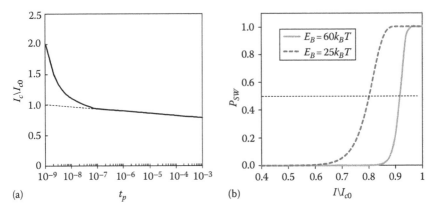

FIGURE 8.6 (a) Dependence on switching current on write pulse width (b) switching probability as a function of normalized switching current for MTJs with different E_B.

however, it is much smaller than $H_{k\parallel}$ and can be tuned to nearly zero by adjusting the thickness of reference layers. The $2\pi M_s$ term (typically larger than 6 kOe) originates from thin-film shape anisotropy and is a dominant factor that increases I_{c0}. This will be explained in detail in Section 7.4.1.

I_c measured as a function of current pulse width (t_p) typically shows two regimes of STT switching as shown in Figure 8.6a. When t_p is relatively long (>10 ~ 100 ns), thermal activation plays an important role in STT switching. In particular, Joule heating can increase the effective MTJ temperature. In the thermally activated regime, it has experimentally been shown that I_c increases linearly with exponentially decreasing t_p:

$$I_c = I_{c0}\left[1 - \frac{k_B T}{E_B}\ln\left(\frac{t_p}{\tau_0}\right)\right] \tag{8.14}$$

I_{c0} and E_B can be extracted from the y-intercept and the slope as indicated by the dotted extrapolated line in Figure 8.6a. E_B can also be obtained by analyzing the statistical nature of STT switching. At a finite temperature, switching probability (P_{sw}) of a single MTJ is given by

$$P_{sw} = 1 - \exp\left\{-\frac{t_p}{\tau_0}\exp\left[-\frac{E_B}{k_B T}\left(1 - \frac{I}{I_{c0}}\right)\right]\right\} \tag{8.15}$$

Figure 8.6b shows P_{sw} calculated for two different E_B values (solid line for 60 $k_B T$ and dotted line for 25 $k_B T$) when t_p is 100 ns. E_B can be estimated by fitting measured P_{sw} with Equation 8.15. However, it should be noted that the E_B values extracted using either Equation 8.14 or 8.15 do not represent the thermal barrier for standby data retention due to Joule heating and are much smaller, often by a factor of 2 or more, than those from Equation 8.12.

For sub-10 ns STT switching, I_c increases nonmonotonically. In this regime, magnetization reversal is dominated by precessional switching. The measured I_c typically follows the relationship:

$$I_c = I_{c0}\left(1 + \frac{\tau_{relax}}{t_p} \ln\left(\frac{\pi/2}{\theta_0}\right)\right)$$ (8.16)

where

τ_{relax} is the relaxation time
θ_0 is the root square average of the initial angle of the free-layer magnetization determined by thermal fluctuation

8.3   1T–1MTJ STT-MRAM BITCELL

Figure 8.7 shows a typical STT-MRAM array schematic with 1T–1MTJ bitcells. In 1T–1MTJ bitcell architecture, one MTJ is serially connected to an access transistor, which is an n-type metal oxide semiconductor device. To read a cell, the word line (WL) of the selected cell is turned on, and small read bias is applied to either the selected bit line (BL) or the source line (SL) while the other end of the cell is grounded. A sense amplifier (SA) determines the data of the cell by sensing the difference between the cell resistance and predefined reference resistance. The write operation is dependent on how an MTJ is connected to the access transistor in the 1T–1MTJ bitcell. In Figure 8.7, the fixed-layer side of an MTJ is connected to the access transistor. In this case, driving the BL with the SL grounded corresponds to AP–P switching (positive direction) and vice versa for driving the SL. In this section, the read/write margins and reliability of the 1T–1MTJ bitcell are discussed.

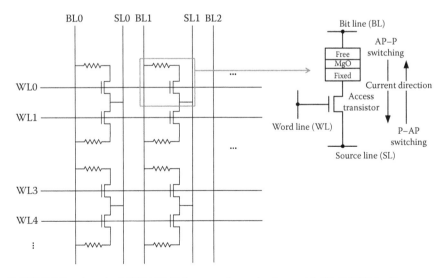

FIGURE 8.7   A typical STT-MRAM array schematic along with 1T–1MTJ bitcell.

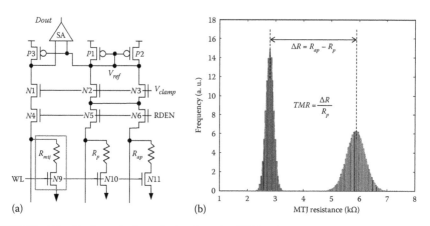

FIGURE 8.8 (a) A typical read circuitry for STT-MRAM. (b) Distribution of MTJ resistances.

8.3.1 READ MARGIN

Figure 8.8a shows a typical read circuitry for STT-MRAM. The reference cells are typically designed to average the currents through two MTJs, one in a parallel state and another in an antiparallel state. It is critical for these reference MTJs not to change their states in any circumstances. The SA detects the voltage difference (ΔV) between the voltage of the selected BL and the reference voltage, denoted as V_{ref} in Figure 8.8. High *TMR* is essential to develop ΔV in a very short time (~5 ns). However, this seemingly simple operation can be very challenging when one needs to ensure sufficient read margins across an entire memory array. Figure 8.8b shows typical distributions of R_p and R_{ap} that follow the normal distribution. Variations of MTJ size and uniformity of MgO barrier are primary factors that determine 1σ of these distributions. Covering 6σ from each distribution means that *TMR* must be larger than 12σ. With optimized MTJ patterning process, it has been shown that 1σ of R_p distribution can be as small as ~1%. Since scaling down MTJs tends to increase 1σ, MTJ patterning would be critical to achieve high-yielding STT-MRAM.

Usually, reference cells are inserted in a memory array, let's say, every 32 data bits, in order to track local variations of MTJ resistance. However, this means that reference cells themselves would exhibit some distributions. Ensuring enough separations among the distributions of R_p, R_{ap}, and reference resistance can be quite challenging, which may increase the *TMR* requirement. Decreasing the total number of reference cells or using a fixed number of predefined reference cells may help mitigate this problem. Toshiba recently showed that adopting a novel reference cell scheme can considerably relax the *TMR* requirement [10]. In addition to MTJ resistance distributions, transistor variations, particularly transistor mismatch in SA, can affect the read margin and need to be carefully examined. All the transistors connected in series with an MTJ as well as parasitic resistances from interconnect metals also decrease the effective *TMR* seen from SA. Considering all the factors listed earlier, it is not a simple task to analytically estimate the *TMR* requirement for 100% die yield because ΔV does not tend to follow the normal distribution.

Hence, it is recommended to run Monte Carlo circuit simulations with final read circuitry to ensure sufficient read margins at the worst-case operation corner.

TMR is dependent on temperature because R_{ap} decreases with increasing temperature while R_p does not nearly change. It has been reported that zero-bias TMR can decrease by more than 20% as temperature increases from 25°C to 125°C [21]. Such reduction in TMR can significantly decrease the read margin. As a consequence, the worst-case corner for read operations is positioned at the "hot" condition. However, a more challenging problem in read operations is read disturbance at elevated temperatures because MTJ becomes more susceptible to thermal disturbance. In particular, thermal reversal of reference cells could be detrimental, resulting in soft errors across WL. In the presence of read current (I_{read}), the effective E_B has been known to decrease by a factor of ($1 - I_{read}/I_{c0}$). At future technology nodes, I_{c0} is expected to continuously decrease due to reduced MTJ size and recent progress in material engineering for J_c reduction. However, I_{read} may not be reduced further because high-speed read operations require a certain amount of I_{read}. This means that preventing read disturbance may become very challenging at future technology nodes. One possible way to get around this problem is to optimize read circuitry and biasing conditions for ultrahigh-speed sensing operations. Since the switching probability is dependent on current pulse width, reducing read pulse width substantially lower than write pulse width will help prevent read disturbance, particularly when the effective read pulse width is shorter than 10 ns (boundary for precessional switching regime).

8.3.2 Write Margin

The maximum current available for MTJ switching is limited by the current-driving capability of the access transistor. For a given technology, the width of the access transistor must be carefully determined to provide sufficient current for MTJ switching at the worst-case corner. The output current from the access transistor becomes much lower when the transistor drives the MTJ at the source side, which is known as the source degeneration effect. In addition to this asymmetry of the transistor output current, it is typically more difficult to switch MTJ from a parallel state to an antiparallel state than vice versa. Additional parameter β, defined as $\beta = \left| I_c^{P-AP} / I_c^{AP-P} \right|$, can be introduced to describe the I_c asymmetry. The intrinsic I_c asymmetry (β_0) is similarly defined as $\beta_0 = \left| I_{c0}^{P-AP} / I_{c0}^{AP-P} \right|$. Prior work attributed the fundamental origin of β_0 to the asymmetric voltage dependence of the fixed-layer polarization factor (P_{fixed}) because the voltage-driven torque on the free-layer moment is proportional to P_{fixed} [20]. When electrons flow from the fixed layer to the free layer, P_{fixed} remains substantial with increasing voltages. However, P_{fixed} significantly decreases in the opposite case. This may lead to reduced STT efficiency for P–AP switching (η^P) in comparison to that for AP–P switching (η^{AP}). Assuming that thin-film shape anisotropy dominates in Equation 8.13, β_0 can be correlated to TMR by $\beta_0 \approx \eta^{AP}/\eta^P \approx 1 + TMR$. This indicates that MTJs with higher TMR may show stronger I_c asymmetry.

β is a critical parameter in designing an STT-MRAM bitcell. When the fixed-layer side of an MTJ is connected to the access transistor as shown in Figure 8.7, the transistor is subjected to the source degeneration effect when switching the MTJ

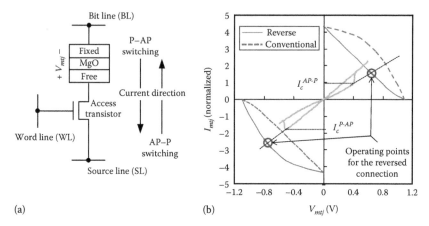

FIGURE 8.9   (a) Reversely connected 1T–1MTJ bitcell. (b) Loadline analysis of 1T–1MTJ.

from a parallel state to an antiparallel state. When this is coupled with $\beta > 1$, the P–AP switching is much more difficult to achieve than the AP–P switching. This problem can be mitigated by connecting the free-layer side of an MTJ to the transistor as shown in Figure 8.9a. In this case, driving the SL corresponds to AP–P switching, and vice versa for driving the BL. Loadline analysis (Figure 8.9b) clearly illustrates the benefit of reversing the connection between the MTJ and the transistor. Transistor output characteristics were transformed to a coordinate of MTJ current–voltage loop. The blue solid lines were obtained from the reversely connected 1T–1MTJ bitcell, and the dotted lines are from the conventional bitcell shown in Figure 8.7. The orange solid lines are the MTJ current–voltage loop, and I_c for AP–P and P–AP switching are indicated on the y-axis. While the conventional connection provides only a small margin in P–AP switching and a large margin in AP–P switching, the operating points of the reversely connected bitcell are well separated from the MTJ switching voltages in both AP–P and P–AP switching as indicated by solid red circles. The rule of thumb is to use the reversed connection when the output current asymmetry is larger than β.

Designing an STT-MRAM bitcell for robust write operations requires a bit more attention to parasitic resistances from interconnect metals and additional resistances from periphery transistors (write drivers, address decoders, etc.), particularly in case of a large-size memory array. Since STT-MRAM is a resistive memory, these additional resistors consume voltage headrooms, decreasing the transistor output current. Hence, when the write margin is tight, circuit designers are forced to increase the size of periphery transistors, which degrades array efficiency. In addition, PVT variations of the access transistor need to be considered to achieve a high-yielding memory array. Since thermal activation affects the STT switching process, I_c increases with decreasing temperature. The transistor output current also increases at low temperatures. Hence, the relative temperature sensitivity of I_c and transistor output current determines the worst-case temperature corner for write operations. The width of the access transistor must be determined to achieve a target write error rate (WER) at the worst-case PVT corner. For write pulse widths in the range of 20–100 ns,

WER is commonly fitted with the complementary error function, following a normal distribution. For relatively long write pulses, WER tends to deviate from the normal distribution and follow the Weibull distribution.

In addition to the switching margin, it is also important to secure a sufficient write reliability margin between switching voltages (V_c) and MgO breakdown voltage (V_{bd}). V_{bd} is also a function of pulse width and temperature. Longer pulse width and/or higher temperatures decrease V_{bd}. Since V_{bd} tends to increase faster than V_c with decreasing write pulse width [21], it is easier to secure the write reliability margin with shorter write pulses. With typical MgO thickness for STT-MRAM, the write reliability margin more than 13 $\sigma(V_c)$ has been demonstrated [22]; hence, securing this reliability margin is often considered less challenging than the read disturbance margin.

8.4   SCALABILITY OF EMBEDDED STT-MRAM

STT-MRAM provides good scalability in terms of write power. For a given J_c, scaling down MTJ size proportionally reduces switching energy. However, it is not trivial to obtain sufficiently large E_B from deeply scaled MTJs suitable for 20 nm CMOS technology node or beyond. Assuming that a patterned MTJ cell is a single-domain nanomagnet, Equation 8.11 predicts that the product of uniaxial anisotropy and free-layer thickness, $K_u - t = (M_s t)H_k /2$, should be increased to compensate for the reduced MTJ area and thereby maintain E_B. For example, to achieve E_B of 60 $k_B T$ from a circular MTJ cell with the diameter of 40 nm, the free-layer material should be engineered to provide $K_u - t$ of ~0.2 erg/cm^2. For a CoFeB-based free layer typically used in in-plane MTJ cells, it is reasonable to assume that $M_s t$ is ~0.2 memu/cm^2. This means that the H_k requirement is 2 kOe. As explained earlier, in-plane MTJs are patterned into elliptical shapes, and H_k is governed by shape anisotropy. Hence, increasing the AR of in-plane MTJ cells tends to enhance H_k. However, when the AR becomes larger than 3.0–3.5, H_k and E_B tend to saturate as the domain wall starts to appear and nucleation-driven switching dominates the magnetization reversal process. Therefore, it is very challenging to achieve $H_k > 1$ kOe for typical in-plane MTJs. In addition, as MTJ size scales down, it becomes more difficult to achieve good uniformity control.

For the 20 nm node, the MTJ size of 40 × 40 nm (corresponding to 25 × 62 nm for AR ~ 2.5) with a 40 nm pitch limits the minimum bitcell size to about 16 F^2. This is too large for STT-MRAM to compete with stand-alone DRAM (typically 8 F^2 or less), which means that the MTJ size must be decreased further to be cost-competitive. This, in turn, increases the H_k requirement. On the other hand, the bitcell size is not a dominant constraint for embedded STT-MRAM. This is partly because the SRAM bitcell size (in F^2) has been increasing as the technology node continues to shrink. It is expected to be more than 200 F^2 at the 20 nm node. Obviously, MTJ cells with a 40 nm or larger diameter can be used for embedded STT-MRAM as long as the access transistor can provide sufficiently large switching currents for reliable write operations. This essentially provides more rooms for scalability of embedded STT-MRAM technology in comparison to stand-alone applications. However, other factors such as read/write speeds and ultralow soft error rates are more important for embedded STT-MRAM. These factors are usually determined by system requirements.

All these challenges can be overcome by adopting perpendicular materials whose magnetization is perpendicular to the film plane. Various perpendicular magnetic materials such as L_{10}-ordered FePt, Co/Pd, or Co/Pt multilayers, rare-earth/transition metal alloys, etc., have been examined for optimizing MTJ performance metrics. The thermal barrier of perpendicular MTJs is provided by crystalline or surface magnetic anisotropy, not shape anisotropy. Hence, perpendicular MTJs can be patterned into a circular shape. Here we briefly introduce the basic concept of perpendicular magnetic anisotropy (PMA) and review challenges for high-performance STT-MRAM.

8.4.1 PERPENDICULAR MAGNETIC ANISOTROPY

Magnetization in nanomagnets with thin-film geometry normally lies in the film plane due to thin-film shape anisotropy. When an external field is applied perpendicular to the plane, magnetization linearly increases until it saturates to its maximum value, M_s. The saturation field (H_{sat}) represents the magnitude of thin-film shape anisotropy, and the uniaxial anisotropy energy (K_u) is given by $K_u = M_s H_{sat}/2$. Figure 8.10a illustrates out-of-plane magnetization curves of thin ferromagnetic films with and without PMA. In the absence of PMA (solid line), H_{sat} is identical to a demagnetizing field ($4\pi M_s$). For typical CoFeB-based free layers with an M_s of ~1000 emu/cc, H_{sat} is ~12 kOe. In this case, I_{c0} and E_B are given by

$$I_{c0} = \frac{2e\alpha_{eff}M_sV}{\hbar\eta}\left(H_{off} + H_{k\parallel} + 2\pi M_s\right) \sim \frac{\alpha_{eff}M_s^2V}{\eta} \quad (8.17)$$

$$E_B = \frac{M_s H_{k\parallel}V}{2} \sim (M_s t)^2 A \quad (8.18)$$

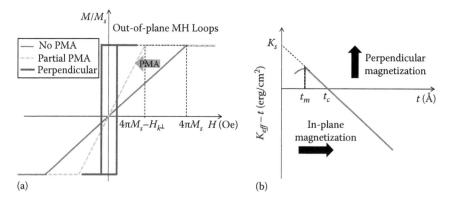

(a)   (b)

FIGURE 8.10 (a) Illustration of out-of-plane magnetization curves of thin ferromagnetic films with and without perpendicular magnetic anisotropy. (b) The product of effective anisotropy energy and film thickness vs. film thickness.

In Equation 8.17, the $2\pi M_s$ term originates from thin-film shape anisotropy, often called a demagnetizing field term. Since the spin-torque-driven switching process involves precessional oscillation of magnetization, thin-film shape anisotropy makes this oscillatory motion confined in the direction perpendicular to the film plane, resulting in an elliptical precession. Since $2\pi M_s$ is larger than $H_{k\parallel}$ by an order of magnitude, J_{c0} is nearly proportional to $\alpha_{eff} M_s^2 t/\eta$. This means that using low-M_s material for a free layer is the most efficient way for J_c reduction. However, $H_{k\parallel}$, in-plane shape anisotropy term resulting from the AR of an elliptically patterned MTJ cell, is proportional to $M_s t$, which leads to $E_B \sim (M_s t)^2$. Therefore, E_B is traded off as much as we gain from J_{c0} reduction when controlling M_s.

When PMA is introduced into the free layer of an in-plane MTJ with the free-layer magnetization remaining in the film plane, H_{sat} can be substantially reduced as shown in Figure 8.10a. The difference between $4\pi M_s$ and H_{sat} corresponds to the effective out-of-plane anisotropy field ($H_{k\perp}$), and the effective demagnetization field ($4\pi M_{eff}$) is defined as $4\pi M_s - H_{k\perp}$. In the presence of partial PMA, J_{c0} decreases without affecting E_B, particularly when $H_{k\perp}$ cancels a substantial portion of the demagnetization field ($4\pi M_s$):

$$I_{c0} = \frac{2e\alpha_{eff} M_s V}{\hbar\eta}\left(H_{off} + H_{k\parallel} + \frac{4\pi M_s - H_{k\perp}}{2}\right) \tag{8.19}$$

There are a number of experimental data that confirmed the existence of PMA in thin CoFeB films. Guan et al. showed $4\pi M_{eff} \sim 6$ kOe with $Co_{40}Fe_{40}B_{20}(2)/Ta/Ru$ (free layer/capping layers) MTJs, which corresponded to $\sim 50\%$ reduction in $4\pi M_s$ (thickness in nm) [23]. Yakata et al. investigated the effect of CoFeB compositions on $4\pi M_{eff}$ and J_{c0} with $(Co_xFe_{1-x})_{80}B_{20}(2)/Ta/Ru$ MTJs and found that significant reduction ($\sim 80\%$ with $Co_{20}Fe_{60}B_{20}$) in $4\pi M_{eff}$ was more pronounced with Fe-rich CoFeB while Co-rich CoFeB showed negligible $H_{k\perp}$ [24]. This trend in $4\pi M_{eff}$ was also well correlated with J_{c0}. All these results imply that J_{c0} can be considerably reduced by introducing PMA without trading off E_B. Although the physical origin of $H_{k\perp}$ is still ambiguous, Ikeda et al. claimed that surface anisotropy from the $MgO-Co_{20}Fe_{60}B_{20}$ interface is responsible for PMA [25].

In general, when PMA in thin magnetic films results from surface anisotropy, the effective magnetic anisotropy energy (K_{eff}) can be phenomenologically separated into the volume anisotropy (K_v) and the surface anisotropy (K_s) from the interfaces, obeying the following equation:

$$K_{eff} = K_v + \frac{K_s}{t} \tag{8.20}$$

where t is the film thickness. This relation is commonly used to extract K_v and K_s by plotting the product $K_{eff} - t$ vs. t. A typical illustration of this plot expected from CoFeB films with strong surface anisotropy is shown in Figure 8.10b. The positive portion of the $K_{eff} - t$ axis corresponds to perpendicular magnetization. K_v is obtained from the slope of the linear portion of the curve. For typical in-plane free layers,

K_v is $\sim 2\pi M_s^2$ as expected from thin-film shape anisotropy. K_s is extracted from the y-intercept of the line extrapolated from the linear portion of the curve. Positive K_s indicates the presence of PMA. Figure 8.10b shows that below a certain film thickness (t_c), the surface anisotropy exceeds the demagnetizing field (i.e., $H_{k\perp} > 4\pi M_s$), resulting in perpendicularly magnetized films. t_c values have been reported for a couple of different CoFeB films. For MgO/$Co_{20}Fe_{60}B_{20}$/Ta films, t_c was ~1.6 nm [25]. For Ta/$Co_{60}Fe_{20}B_{20}$/MgO films, t_c was ~1.1 nm [26]. Furthermore, perpendicular MTJs have been demonstrated utilizing perpendicularly magnetized CoFeB as a free layer [25,26]. Perpendicular MTJs do not rely on in-plane shape anisotropy to define magnetization states, hence can be patterned into a circular shape. For perpendicular MTJs with the effective perpendicular anisotropy field ($H_{k,eff}$) of $H_{k\perp} - 4\pi M_s$, I_{c0} and E_B are given by

$$I_{c0} = \frac{e\alpha_{eff}M_sV}{\hbar\eta}\left(H_{off} + H_{k,eff}\right) \sim \frac{\alpha_{eff}}{\eta}E_B \tag{8.21}$$

$$E_B = \frac{M_sH_{k,eff}V}{2} = (K_{eff}t)A \tag{8.22}$$

Note that I_{c0} is proportional to E_B. For a given target MTJ size, Equation 8.22 tells us that E_B is determined by the product $K_{eff} - t$. As shown in Figure 8.10b, the product $K_{eff} - t$ increases with decreasing film thickness until it reaches the maximum at a certain film thickness (t_m). Hence, the maximum E_B achievable is simply given by $(K_{eff} - t_m)A$. As the film thickness is decreased further, the product $K_{eff} - t$ starts to roll down. This deviation has been commonly observed in most of PMA systems and attributed to changes in magnetoelastic anisotropy, interdiffusion after annealing, and nonconformal films forming islands at a small film thickness. For example, for a circular MTJ with its diameter of 40 nm, E_B of 60 k_BT requires the product $K_{eff} - t$ of ~0.2 erg/cm^2, which can be achieved with thin CoFeB films. However, patterned MTJs often suffer from sidewall damages or edge roughness, particularly when the MTJ size is small. These can decrease E_B of patterned MTJs considerably. In addition, as the film thickness decreases, α_{eff} tends to increase. Hence, it is desirable to maximize K_s and achieve the target $K_{eff} - t$ value with the film thickness below t_m. With the top and bottom interfaces of a free layer carefully engineered, it seems feasible to increase K_s more than what has been reported to date.

8.4.2 MTJ Material Engineering for Write Power Reduction

To compete with conventional embedded memory at the advanced technology nodes, it is critical to reduce write energy per bit. In this sense, capturing product opportunities to replace state-of-the-art embedded memory would require more aggressive J_c reduction. Considering that the performance requirements for conventional embedded memory have been ever increasing, it is desirable to reduce the operating frequency of embedded STT-MRAM at least below 100 MHz, which requires reliable

STT switching in the precessional switching regime. J_c reduction would allow faster write speed. Also, for a given target write speed, decreasing J_c leads to reduced bit-cell size, and thereby less cost.

Regardless of PMA in the film structures, it is always beneficial to optimize α_{eff} and η because J_{c0} can be reduced without trading off E_B. For bulk-type materials, damping is an intrinsic material property; however, damping in thin ($\sim$nm) magnetic films is also affected by adjacent nonmagnetic layers. For CoFeB-based MTJs, α_{eff} can be tuned by optimizing material compositions of the free layer (intrinsic contribution) or the capping layers adjacent to the free layer (extrinsic contribution). For the intrinsic contribution, it has been reported that Fe-rich $Co_{20}Fe_{60}B_{20}$ exhibits lower damping constant in comparison to $Co_{40}Fe_{40}B_{20}$, contributing to J_{c0} reduction [27]. The thickness and crystalline states of a free layer are also known to affect α_{eff} [28]. For the extrinsic contribution, α_{eff} of a thin ferromagnetic layer adjacent to a normal metal layer can be greatly increased, particularly when the normal metal layer has a short spin relaxation time (e.g., Pt), which is known as the spin pumping effect. This implies that optimization of the capping layer may reduce α_{eff} when the spin pumping effect is suppressed. Insulating capping layers may be suitable for this purpose. However, the capping-layer optimization is often not trivial because it also affects the crystallization process of the free layer underneath and modifies other MTJ properties such as *TMR*. In addition, it is not clear whether there is room for reducing α_{eff} further. α_{eff} value reported from $Co_{20}Fe_{60}B_{20}$ is $\sim$0.007, which is close to its intrinsic damping of CoFeB.

η represents the efficiency of spin-polarized current driving magnetization reversal. To the first-order approximation, η is a function of spin polarization factor and can be increased by enhancing *TMR*. However, for MTJs with sufficiently high *TMR* ($\sim$150%), it has been predicted that enhancing *TMR* further would not result in considerable J_{c0} reduction because increased spin polarization would not lead to significantly stronger spin torques [29]. Recently, it has been suggested that η can also be increased by enhancing out-of-plane spin torques generated by a spatially nonuniform spin current within a tapered nanopillar spin valve [30]. Adding a perpendicular polarizer in an MTJ film stack may introduce additional out-of-plane spin torques and enhance η [31]. While the out-of-plane spin torques are negligible in metallic spin valves, it has recently been found that the out-of-plane spin torques play a significant role in MTJs. In general, the spin torque (Γ) has both in-plane and perpendicular components and can be written as

$$\Gamma = a_J \frac{M}{M} \times (M \times n_s) + b_J M \times n_s \qquad (8.23)$$

where a_J and b_J represent the in-plane spin torque and the perpendicular spin torque, respectively. Note that the second term acts as a field-like torque and needs to be added to Equation 8.9 to consider the effect of b_J on magnetization dynamics. The magnitude, polarity, and voltage dependence of b_J have been investigated using various measurement techniques [29,32–35]. However, clear experimental evidence for the benefits of enhancing b_J for J_c reduction is yet to be explored.

8.4.3 BIT ERROR RATES OF STT-MRAM

Since STT switching is inherently probabilistic, it is imperative to secure sufficient read/write margins to meet bit error rate requirements that are typically determined by a system specification. There are three types of bit error rates that need to be considered for STT-MRAM: thermal disturb rate (TDR), read disturb rate (RDR), and WER. For an MTJ array, the TDR is determined by the population of tail bits with relatively small E_B. Hence, to minimize the TDR of an STT-MRAM chip, it is critical to tighten the distributions of MTJ size and H_k while ensuring the typical MTJ cell to have sufficiently large E_B.

Figure 8.11 illustrates the typical RDR and WER characteristics of an MTJ device. For typical read voltage (V_{read}), RDR can be estimated by performing a Taylor expansion of Equation 8.15 with current replaced by voltage [36]:

$$\ln(\text{RDR}) = \ln\left(\frac{t_p}{\tau_0}\right) - \frac{E_B}{k_B T}\left(1 - \frac{V_{read}}{V_{c0}}\right) \tag{8.24}$$

where

V_{c0} is the intrinsic critical switching voltage
V_{read} is assumed to be much smaller than V_{c0}

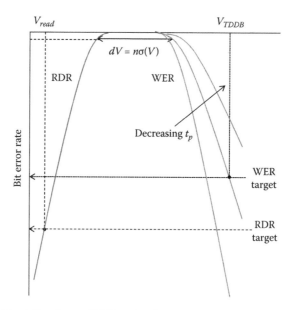

FIGURE 8.11   Read disturb rate (RDR) and write error rate (WER) of an MTJ device. Read voltage (V_{read}) and time-dependent dielectric breakdown voltage (V_{TDDB}) are usually predetermined. V_{read} and V_{TDDB} set the limits for WER and RDR targets. Steep slopes in RDR and WER characteristics are critical for meeting ultralow bit error rates. This is particularly important for high-speed write operations because WER slopes degrade with decreasing write pulse width.

RDR decreases exponentially with decreasing V_{read}, and the RDR slope is determined by E_B. As MTJ scales down, I_c proportionally decreases. For example, assuming J_c of 2 MA/cm^2, I_c of a perpendicular MTJ device with the diameter of 20 nm would be only 15 µA. However, read current cannot be reduced below a certain threshold because faster read operations typically require more current. Hence, the read disturb margin is squeezed with deeply scaled MTJ. Hence, for a given V_{read}, it is desirable to increase E_B to meet the RDR target at advanced technology nodes. However, from the energy-scaling perspective, the fundamental problem is that read energy does not scale down with conventional SA schemes while write energy continues to decrease with MTJ size shrinking. This implies that fully enabling high-speed STT-MRAM may require innovations in read circuitry to sense resistance differences with lower energy and faster speed.

For in-plane MTJs, the WER tends to follow a Gaussian CDF for t_p of 20–200 ns [37]. Above 500 ns, the WER tends to fall off faster and is often described by a Weibull CDF. Since the maximum write voltage is limited by time-dependent dielectric breakdown of the MgO barrier, it is imperative to obtain steep WER slopes to meet an ultralow WER target. However, as illustrated in Figure 8.11, WER slopes tend to degrade with decreasing t_p. This means that for faster write operations, one needs to increase write voltages to meet the WER target. This is particularly challenging for in-plane MTJs because of their inefficient magnetization reversal dynamics that involves preswitching oscillations prior to magnetization reversal [38,39]. At the moment, it doesn't seem feasible to achieve an ultralow WER target ($<10^{-12}$) below 20 ns using in-plane MTJs. This challenge can be substantially mitigated by utilizing perpendicular MTJs. Recently, significantly improved WER characteristics of perpendicular MTJs (at 10 ns) have been reported, demonstrating error-free write operations (WER $< 10^{-11}$) at 10 ns below 0.5 V [40]. Therefore, it is inevitable to enable perpendicular MTJ technology in order to realize reliable high-speed STT-MRAM at advanced technology nodes.

REFERENCES

1. R. Nebashi, N. Sakimura, H. Honjo, S. Saito, Y. Ito, S. Miura, Y. Kato et al., A 90 nm 12ns 32 Mb 2T1MTJ MRAM, *ISSCC Tech. Dig.*, 462–463, February 2009.
2. M. Hosomi, H. Yamagishi, T. Yamamoto, K. Bessho, Y. Higo, K. Yamane, H. Yamada et al., A novel nonvolatile memory with spin torque transfer magnetization switching: Spin-RAM, *IEDM Tech. Dig.*, 459–462, December 2005.
3. T. Kawahara, R. Takemura, K. Miura, J. Hayakawa, S. Ikeda, Y. Lee, R. Sasaki et al., 2 Mb Spin-Transfer Torque RAM (SPRAM) with bit-by-bit bidirectional current write and parallelizing-direction current read, *ISSCC Tech. Dig.*, 480–481, February 2007.
4. R. Beach, T. Min, C. Horng, Q. Chen, P. Sherman, S. Le, S. Young et al., A statistical study of magnetic tunnel junctions for high-density spin torque transfer-MRAM (STT-MRAM), *IEDM Tech. Dig.*, 1–4, December 2008.
5. C. J. Lin, S. H. Kang, Y. J. Wang, K. Lee, X. Zhu, W. C. Chen, X. Li et al., 45 nm low power CMOS logic compatible embedded STT MRAM utilizing a reverse-connection 1T/1MTJ cell, *IEDM Tech. Dig.*, 1–4, December 2009.
6. A. Driskill-Smith, S. Watts, V. Nikitin, D. Apalkov, D. Druist, R. Kawakami, X. Tang et al., Non-volatile Spin-transfer Torque RAM (STT-RAM): Data, analysis and design requirements for thermal stability, *Symp. VLSI Tech. Dig.*, 51–52, June 2010.

7. Y. M. Lee, C. Yoshida, K. Tsunoda, S. Umehara, M. Aoki, and T. Sugii, Highly scalable STT-MRAM with MTJs of top-pinned structure in 1T/1MTJ cell, *Symp. VLSI Tech. Dig.*, 49–50, June 2010.

8. S. C. Oh, J. H. Jeong, W. C. Lim, W. J. Kim, Y. H. Kim, H. J. Shin, J. E. Lee et al., On-axis scheme and Novel MTJ structure for sub-30 nm Gb density STT-MRAM, *IEDM Tech. Dig.*, 12.6.1, December 2010.

9. S. Chung, K.-M. Rho, S.-D Kim, H.-J. Suh, D.-J. Kim, H.-J. Kim, S.-H. Lee et al., Fully integrated 54 nm STT-MRAM with the smallest bit cell dimension for high density memory application, *IEDM Tech. Dig.*, 12.7.1, December 2010.

10. K. Tsuchida, T. Inaba, K. Fujita, Y. Ueda, T. Shimizu, Y. Asao, T. Kajiyama et al., A 64 Mb MRAM with clamped-reference and adequate-reference schemes, *ISSCC Tech. Dig.*, 258–259, February 2010.

11. D. C. Worledge, G. Hu, P. L. Trouilloud, D. W. Abraham, S. Brown, M. C. Gaidis, J. Nowak et al., Switching distributions and write reliability of perpendicular spin torque MRAM, *IEDM Tech. Dig.*, 12.5.1, December 2010.

12. J. C. Slonczewski, Current-driven excitation of magnetic multilayers, *J. Magn. Magn. Mater.*, 159, L1–L7, June 1996.

13. L. Berger, Emission of spin waves by a magnetic multilayer traversed b a current, *Phys. Rev. B*, 54, 9353–9358, October 1996.

14. M. Jullière, Tunneling between ferromagnetic films, *Phys. Lett. A*, 54, 225–226, September 1975.

15. W. H. Butler, X.-G. Zhang, T. C. Schulthess, and J. M. MacLaren, Spin-dependent tunneling conductance of Fe/MgO/Fe sandwiches, *Phys. Rev. B*, 63, 054416, January 2001.

16. J. Mathon and A. Umerski, Theory of tunneling magnetoresistance of an epitaxial Fe/MgO/Fe (001) junction, *Phys. Rev. B*, 63, 220403, May 2001.

17. S. Ikeda, J. Hayakawa, Y. Ashizawa, Y. M. Lee, K. Miura, H. Hasegawa, M. Tsunoda, F. Matsukura, and H. Ohno, Tunnel magnetoresistance of 604% at 300 K by suppression of Ta diffusion in CoFeB/MgO/CoFeB pseudo-spin-valves annealed at high temperature, *Appl. Phys. Lett.*, 93, 082508, August 2008.

18. S. Ikeda, J. Hayakawa, Y. M. Lee, F. Matsukura, Y. Ohno, T. Hanyu, and H. Ohno, Magnetic tunnel junctions for spintronic memories and beyond, *IEEE Trans. Electron Dev.*, 54, 991–1002, May 2007.

19. J. A Katine, F. J. Albert, R. A. Buhrman, E. B. Myers, and D. C. Ralph, Current-driven magnetization reversal and spin-wave excitations in Co/Cu/Co pillars, *Phys. Rev. Lett.*, 84, 3149–3152, April 2000.

20. J. C. Slonczewski, Currents, torques, and polarization factors in magnetic tunnel junctions, *Phys. Rev. B*, 71, 024411, January 2005.

21. K. Lee and S. H. Kang, Development of embedded STT-MRAM for mobile system-on-chips, *IEEE Trans. Magn.*, 47, 131–136, January 2011.

22. Q. Chen, T. Min, T. Torng, C. Horng, D. Tang, and P. Wang, Study of dielectric breakdown distributions in magnetic tunneling junction with MgO barrier, *J. Appl. Phys.*, 105, 07C931, March 2009.

23. Y. Guan, J. Z. Sun, X. Jiang, R. Moriya, L. Gao, and S. S. Parkin, Thermal-magnetic noise measurement of spin-torque effects on ferromagnetic resonance in MgO-based magnetic tunnel junctions, *Appl. Phys. Lett.*, 95, 082506, August 2009.

24. S. Yakata, H. Kubota, Y. Suzuki, K. Yakushiji, A. Fukushima, and S. Yuasa, Influence of perpendicular magnetic anisotropy on spin-transfer switching current in CoFeB/MgO/CoFeB magnetic tunnel junctions, *J. Appl. Phys.*, 105, 07D131, April 2009.

25. S. Ikeda, K. Miura, H. Yamamoto, K. Mizunuma, H. D. Gan, M. Endo, S. Kanai et al., A perpendicular-anisotropy CoFeB-MgO magnetic tunnel junction, *Nat. Mater.*, 9, 721–724, July 2010.

26. D. C. Worledge, G. Hu, D. W. Abraham, J. Z. Sun, P. L. Trouilloud, J. Nowak, S. Brown et al., Spin torque switching of perpendicular Ta|CoFeB|MgO-based magnetic tunnel junctions, *Appl. Phys. Lett.*, 98, 022501, January 2011.

27. J. Hayakawa, S. Ikeda, K. Miura, M. Yamanouchi, Y. M. Lee, R. Sasaki, M. Ichimura et al., Current-induced magnetization switching in MgO barrier magnetic tunnel junctions with CoFeB-based synthetic ferrimagnetic free layers, *IEEE. Trans. Magn.*, 44, 1962–1967, July 2008.

28. C. Bilzer, T. Devolder, J.-V. Kim, G. Counil, C. Chappert, S. Cardoso, and P. P. Freitas, Study of the dynamic magnetic properties of soft CoFeB films, *J. Appl. Phys.*, 100, 053903, September 2006.

29. J. C. Sankey, Y.-T. Cui, J. Z. Sun, J. C. Slonczewski, R. A. Buhrman, and D. C. Ralph, Measurement of the spin-transfer-torque vector in magnetic tunnel junctions, *Nat. Phys.*, 4, 67–71, January 2008.

30. P. M. Braganca, O. Ozatay, A. G. F. Garcia, O. J. Lee, D. C. Ralph, and R. A. Buhrman, Enhancement in spin-torque efficiency by nonuniform spin current generated within a tapered nanopillar spin valve, *Phys. Rev. B*, 77, 144423, April 2008.

31. H. Liu, D. Bedau, D. Backes, J. A. Katine, J. Langer, and A. D. Kent, Ultrafast switching in magnetic tunnel junction based orthogonal spin transfer devices, *Appl. Phys. Lett.*, 97, 242510, December 2010.

32. S. Petit, C. Baraduc, C. Thirion, U. Ebels, Y. Liu, M. Li, P. Wang, and B. Dieny, Spin-torque influence on the high-frequency magnetization fluctuations in magnetic tunnel junctions, *Phys. Rev. Lett.*, 98, 077203, February 2007.

33. H. Kubota, A. Fukushima, K. Yakushiji, T. Nagahama, S. Yuasa, K. Ando, H. Maehara et al., Quantitative measurement of voltage dependence of spin-transfer torque in MgO-based magnetic tunnel junctions, *Nat. Phys.*, 4, 37–41, January 2008.

34. Z. Li, S. Zhang, Z. Diao, Y. Ding, X. Tang, D. M. Apalkov, Z. Yang, K. Kawabata, and Y. Huai, Perpendicular spin torques in magnetic tunnel junctions, *Phys. Rev. Lett.*, 100, 246602, June 2008.

35. S. C. Oh, S. Y. Park, A. Manchon, M. Chshiev, J. H. Han, H. W. Lee, J. E. Lee et al., Bias-voltage dependence of perpendicular spin-transfer torque in asymmetric MgO-based magnetic tunnel junctions, *Nat. Phys.*, 6, 898–902, October 2009.

36. R. Heindl, W. H. Rippard, S. E. Russek, M. R. Pufall, and A. B. Kos, Validity of the thermal activation model for spin-transfer torque switching in magnetic tunnel junctions, *J. Appl. Phys.*, 109, 073910, 2011.

37. D. D. Tang and Y. J. Lee, *Magnetic Memory: Fundamentals and Technology*, 1st edn., Cambridge University Press, New York, pp. 136–138, 2010.

38. M. D. Stiles and J. Miltat, *Spin-Transfer Torque and Dynamics. Spin Dynamics in Confined Magnetic Structure III Springer*, Berlin, Germany, p. 255, 2006.

39. T. Devolder, J. Hayakawa, K. Ito, H. Takahashi, S. Ikeda, P. Crozat, N. Zerounian et al., Single-shot time-resolved measurements of nanosecond-scale spin-transfer induced switching: stochastic versus deterministic aspects, *Phys. Rev. Lett.*, 100, 057206, 2008.

40. J. J. Nowak, R. P. Robertazzi, J. Z. Sun, G. Hu, D. W. Abraham, P. L. Trouilloud, S. Brown et al., Demonstration of ultralow bit error rates for spin-torque magnetic random-access memory with perpendicular magnetic anisotropy, *IEEE Magn. Lett.*, 2, 3000204, 2011.

9 Magnetic Domain Wall "Racetrack" Memory

Michael C. Gaidis and Luc Thomas

CONTENTS

9.1   ELEMENTS OF RACETRACK MEMORY

9.1.1   GENERAL DESCRIPTION OF RACETRACK MEMORY

The concept of racetrack memory (RTM) was invented and patented by Dr. Stuart Parkin in 2004 (Parkin 2004). It has been described in several review articles (Parkin et al. 2008) and book chapters (Thomas and Parkin 2007, Parkin et al. 2011) in the past few years. The following sections provide an updated overview of progress made in materials and device fabrication for RTM.

The typical RTM cell comprises three elements: (1) a magnetic racetrack nanowire (RTNW) in which data bits are stored in the form of magnetic domains or domain walls (DWs); (2) a write element that is used to inject one or two DWs in the RTNW, for example, by reversing the magnetization direction locally; and (3) a readout element (typically magnetoresistive), which can transduce the local orientation of the magnetization in the racetrack into an electrical signal. A schematic view can be found in Figure 9.1. RTM can be designed with RTNWs

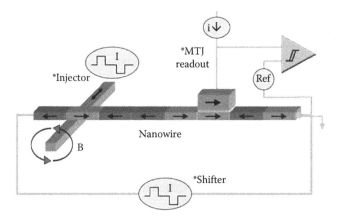

FIGURE 9.1   The components involved in an RTM cell element. The magnetic nanowire stores the bits as multiple magnetic domains in the wire. An injection element such as a field-producing wire is used to set bits in a desired state at a particular location in the nanowire. The shift element provides current needed to move the domain walls along the nanowire. Finally, an MTJ readout mechanism is placed on or near the nanowire, and a small current through the MTJ can generate a voltage for comparison against a reference element. The asterisked elements each require a select transistor for the isolation of the memory cell from others in an array, although the injector can be shared to some extent. The key driver for RTM as a high-density memory alternative is the use of only a few (two or three) transistors to control a hundred or more bits in the magnetic nanowire.

fabricated parallel to the plane of the silicon wafer or perpendicular to the wafer. These architectures are called horizontal RTM (HRTM) and vertical RTM (VRTM), respectively (Parkin et al. 2008). Note that these terms refer to the physical layout of the RTNW and not to the direction of the magnetization within the nanowire.

9.1.2 DOMAIN WALL STORAGE IN A NANOWIRE

9.1.2.1 Material Considerations: Perpendicular versus In-Plane Magnetic Anisotropies

Whether bits of data are coded using magnetic domains or DWs, data manipulation in the RTM is based on the motion of DWs along the racetracks. Both the static and dynamical properties of the DWs depend strongly on the class of material used to fabricate the racetrack (see, e.g., Hubert and Schäfer (1998) for a review of the physics of magnetic domains and DWs). There are two main classes of materials suitable for RTM. The first class comprises materials with small intrinsic anisotropy, which are usually referred to as magnetically "soft" materials. The most widely used material is permalloy, a metallic alloy made of 20 at% Fe and 80 at% Ni. Current-driven DW motion has also been studied in other soft materials, but the results are usually inferior to that of permalloy devices. In such soft materials, the magnetostatic energy plays a crucial role. Magnetostatic interactions lead to the well-known shape anisotropy, which causes the magnetization to be parallel to the longest dimensions of the device. For a thin magnetic film, the magnetization lies in the plane of the film, leading to an effective in-plane magnetic anisotropy (IMA). In the case of a racetrack shaped as an elongated wire or a strip, the magnetization is aligned along the length of the racetrack. As we will see later, because of this shape anisotropy, many of the DW properties depend on the physical dimensions of the racetrack.

On the contrary, the second class of materials has a very large anisotropy, which is perpendicular to the plane of the wafer. If it is strong enough, this so-called perpendicular magnetic anisotropy (PMA) causes the magnetization to be aligned in the direction perpendicular to the plane of the wafer, despite the increase in magnetostatic energy thus caused. The PMA in materials relevant for RTM is usually caused by surface anisotropy arising from the interfaces between the different layers of a multilayered structure (Johnson et al. 1996), although the magnetocrystalline anisotropy in a single crystal or an epitaxial film can also be strong enough. One of the first examples of PMA caused by interface anisotropy was found in Co/Ni multilayers. Even stronger PMA has been found in Pt/Co or Pd/Co interfaces. Most of the work on DW motion in PMA materials has been done in multilayered structures including Pt, Co, and Ni layers. Other very promising structures are very thin CoFeB/MgO bilayers, which are widely used in perpendicular spin-transfer-torque (STT)-MRAM cells (Ikeda et al. 2010, Worledge et al. 2011). Contrary to IMA materials, the structure and properties of DWs in PMA materials are mostly determined by material properties and are thus much less dependent on the physical dimensions of the device.

9.1.2.2 Domain Wall Structure and Width

The DW structure is very different for IMA and PMA materials. For IMA materials, since the shape anisotropy forces the magnetic domains to be aligned along the racetrack length, these domains are separated by head-to-head or tail-to-tail DWs. The competition between exchange energy and shape anisotropy can lead to complex wall structures. For a racetrack shaped as a strip, that is, having a width w much larger than its thickness t, there are two stable DW structures called vortex (V) and transverse (T) DWs (Thiaville and Nakatani 2006) (Figure 9.2a). Which of these two structures has the lowest energy depends on w and t. The complete phase diagram in the (w,t) space has been calculated by micromagnetic simulations. The boundary between separating these two states is given by $t \times w = C\delta^2$, where C is a numerical constant (C $\sim$ 128) and δ is the exchange length determined by the material parameters A (exchange stiffness) and M_S (saturation magnetization): $\delta = \sqrt{(A/\mu_0 M_S^2)}$. Note that both structures can coexist for dimensions close to the boundary. For very large devices, more complex DW structures including several vortices can be stable. However, these structures are usually not stable for dimensions suitable for device applications (less than a few 100 s of nm in width). As shown in Figure 9.2a, V and T walls are two-dimensional structures for which it is difficult to define an actual width. However, an effective width can be estimated by fitting the two-dimensional profile to a simpler one-dimensional Bloch DW. The width parameter Δ is proportional to the racetrack width w, and the scaling factor is different for V and T walls: $\Delta \sim w/\pi$ for T-DWs, and $\Delta \sim 3w/4$ for V-DWs. Note that the physical width of the DW over which most of the magnetization change takes place is given by $\pi\Delta$.

In the case of PMA materials, DWs have an almost ideal Bloch (B) or Néel (N) profile (Malozemoff and Slonczekwski 1979). The magnetization within the DW can rotate either in the plane (B type) or perpendicular to the plane (N type) (Figure 9.2b). Since the difference in energy between these two states is due to magnetostatic interactions, which of these configurations has the minimum energy also depends on the

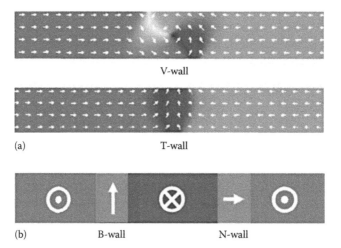

(a)   V-wall

                                       T-wall

(b)   B-wall   N-wall

FIGURE 9.2 (a) Micromagnetic simulations of vortex and transverse DWs in IMA RTNWs. (b) Cartoon of the magnetization configuration for Bloch and Néel DWs in a PMA RTNW.

width and thickness of the RTNW. For the B-DW, the magnetostatic energy arises from the edges of the nanowire, where the magnetization in the DW is perpendicular to the surfaces of the nanowire (creating what is often called "surface magnetostatic charges"). On the contrary, for the N-DW, the magnetostatic energy comes from the nonzero divergence of the magnetization within the DW (called "volume magnetostatic charges"). A phase diagram in terms of nanowire width and thickness can be calculated numerically, but a simple rule of thumb to estimate the boundary between the two configurations is to equate the physical width of the DW $\pi\Delta$ with the RTNW width w. This rule holds except for very thin and wide RTNWs. Contrary to the IMA case, the magnetostatic contribution to the DW energy is weak for PMA. To first order, N and B walls have the same width, given by $\Delta = \sqrt{(A/K_{eff})}$. The effective anisotropy energy is equal to the PMA constant K reduced by the demagnetizing energy term $\mu_0 M_s^2/2$. Note that the demagnetizing energy is slightly different for B and N walls, thus leading to slightly different widths. In most cases, this is a very small effect that can be neglected.

9.1.2.3 Scaling and Storage Density Considerations

The fundamental difference between IMA and PMA materials leads to different scaling properties. Scaling considerations differ also greatly for horizontal or vertical RTM designs. Let us first discuss the case of HRTM, which is more challenging in terms of storage density. For IMA RTNWs, the DW width is directly proportional and always larger than the RTNW width w. This limits the use of IMA materials in RTM and makes V-DWs unsuitable for high-density application. Indeed, the physical width of V-DWs is about 2.4w. If we assume that the minimum distance between adjacent DWs is the same as the DW physical width (an aggressive assumption as we will see later), the minimum cell size along the RTNW is about 5w. If we further assume that w is the minimum lithographic feature at the current technology node F, and that two adjacent RTNWs are also separated by a distance F, the cell size for a V-DW is $10F^2$, which is very large for storage applications. Another key limitation of V-DWs is that they are only stable for relatively wide RTNWs and would not scale down to small dimensions. For permalloy, V-DWs are stable only for RTNWs wider than ~100 nm. However, it is possible that material engineering would allow using V-DWs for much smaller devices, for example, by reducing the exchange length. Density and scaling are much more favorable for T-DWs. First, the physical width of a T-DW is w, leading to a minimum cell size of only $4F^2$. Second, T-DWs are stable down to arbitrarily small values of F.

In the case of PMA materials, the DW width is independent of F, but depends only on the magnetic properties of the RTNW material. Widths as small as ~10 nm have been reported in high-anisotropy materials (remember that the relevant dimension is the physical width $\pi\Delta$ and not the width parameter Δ). Thus, PMA materials allow for a minimum cell size smaller than $4F^2$ down to the 20 nm node.

The density can of course be increased by stacking RTNWs on top of one another in the HRTM design, in an approach similar to that used, for example, for FLASH memory devices. However, RTM reaches its full potential only when the RTNWs are oriented perpendicular to the plane of the wafer, in what we call the VRTM design. In the VRTM design, the density is essentially given by the diameter of the RTNW divided by the number of DWs in the RTNW. Since the vertical dimension is essentially free, the only limit in the density comes from circuit considerations rather than limitations in the magnetic properties. Assuming a square network of RTNWs of diameter F, each of which contains N DWs, the density is simply $4F^2/N$.

9.1.3 DOMAIN WALL WRITING

Writing data bits in RTM requires injecting one or two DWs in the RTNW. This can be done either by reversing the magnetization direction in a small portion of the RTNW or by moving a DW located in a reservoir into the RTNW. The first method is very similar to the writing of an MRAM cell. It requires overcoming the energy barrier associated with the reversal of magnetization of a small volume of the RTNW. This can be done in several ways, for example, by applying a local magnetic field or by using the STT from a reference layer. The second method is related to the propagation of a DW, and in many aspects, the energentics are very similar to that of DW shifting, which will be discussed later in more detail.

The easiest method to implement to inject DWs in the RTNW is to use the Oersted field from the current passing in a metallic line perpendicular to the RTNW. Since the Oersted field has both in-plane and out-of-plane components, this method can be used for both IMA and PMA racetracks. However, this method requires large currents, and the magnetic field created is not only at the write site but also all along the injection line. Moreover, even though the magnetic field decreases rapidly away from the injection line, the tails extend over long distances. This can be very significant if DWs are weakly pinned in the RTNW, as is the case for permalloy RTNWs. For example, we have shown that the field from the injection line can reach several gauss micrometers away from the injection and affect the motion of V-DWs in a permalloy RTNW (Hayashi et al. 2008).

DWs can also be written by the STT from a magnetic tunnel junction (MTJ) fabricated above or below the RTNW. Although this method has not been demonstrated experimentally, it is conceptually similar to the writing of an MRAM cell and thus could take advantage of the rapid development of this technology.

9.1.4 READOUT

9.1.4.1 Methods of Electrical Domain Wall Detection

Reading data in RTM is done electrically by taking advantage of the magnetoresistance of the device. In a lab environment, several magnetoresistive effects can be harnessed to probe the presence and the position of DWs in an RTNW. Even though many of these effects are too small to be useful in an actual device, it is useful to understand their advantages and limitations.

The first phenomenon is the anisotropic magnetoresistance (AMR), which exists in many ferromagnets such as permalloy, Co, Ni,... AMR is a bulk effect arising from the spin–orbit interaction, whereby the resistance of the material is slightly dependent on the angle between the current and the magnetization (McGuire and Potter 1975). The AMR ratio $(R_{//} - R_{\perp})/R$ can reach a few% at room temperature.

Note that AMR is only sensitive to the direction of the magnetization and not to its orientation. AMR detection is particularly suited to IMA materials (Thomas and Parkin 2007). Indeed, since the magnetization is constrained along the RTNW in the domains, the current is parallel to the magnetization everywhere but in the DWs. Thus, AMR is extremely sensitive to the volume of the DW, which itself is directly related to its structure (Hayashi et al. 2006). It is easy to show that the AMR contribution of a DW is directly proportional to its width. If several DWs are present in the RTNW, the total AMR signal is simply the sum of the signals from the individual DWs. However, AMR is not sensitive to the position of the DWs along the RTNW because it does not depend on the orientation. The typical AMR signal of V- and T-DWs in permalloy is ~ 0.1–0.5 Ω. Note that the RTNW resistance is usually reduced in the presence of DWs in the IMA case. AMR can also be used to probe N-DWs in PMA materials. The signal is reversed compared to the IMA case because the magnetization is perpendicular to the current in the magnetic domains and parallel to the current in the DW. In the case of B-DWs, the magnetization remains perpendicular to the current in the DW, and no AMR signal is expected. Indeed, the onset of the AMR from B-DW to N-DW has been used to probe the transition between these two configurations as a function of the wire width in Co/Ni multilayers (Koyama et al. 2011a). The signal due to N-DWs in PMA materials is also of the order 0.1 Ω. Note that additional contributions due to spin-dependent scattering at the DW can also contribute to this signal.

The giant (GMR) and tunnel (TMR) magnetoresistance can also be used to probe DW dynamics in RTNWs. GMR and TMR are observed when two magnetic layers are separated by a nonmagnetic metallic layer such as Cu (for GMR) or a tunnel barrier such as AlOx or MgO (for TMR) (there is an extensive literature on these topics, see, e.g., Tsymbal (2011) for recent reviews). Even though the microscopic origin of TMR and GMR is different, they are very similar in terms of DW detection. Both GMR and TMR are proportional to the relative orientation (parallel or antiparallel) of the two magnetic layers. The GMR and TMR ratios are defined as $(R_{AP} - R_P)/R_P$. Both GMR and TMR require that the RTNW also include not only a layer in which DWs are located (the free layer) but also a fully magnetized reference layer. When a DW is moving along the free layer, the device resistance varies monotonically and is simply given by the difference between portions of the RTNW parallel or antiparallel to the reference layer. Thus, contrary to AMR, GMR and TMR are sensitive to the position of the DWs if the RTNW contains an odd number of DWs. No signal is expected for an even number of DW move in lockstep. Note that GMR and TMR can be measured with the current being applied either within or perpendicular to the plane of the layers, in the so-called current in-plane or current perpendicular to the plane (CPP) geometries. The signal from GMR or TMR can reach tens of Ω (or even hundreds of Ω in the case of CPP-TMR). It is large enough to enable single-shot measurement of the DW dynamics with a subnanosecond time resolution.

All the aforementioned methods are global detection techniques, which are sensitive to the full extent of the RTNW. In the case of PMA RTNWs, the anomalous Hall effect (AHE) gives an easy way of probing the local magnetization direction. The Hall voltage created across the width of the RTNW can be measured readily either attaching contacts to the sides of the racetrack or more simply by depositing a resistive line directly in contact and across the RTNW. The second method leads to slightly smaller signals due to shunting but is much simpler in terms of nanofabrication. In most materials, the Hall resistance is much smaller than 1 Ω, and the AHE voltage is of the order of a few μV for small read currents (note that extremely large Hall resistance of ~ 60 Ω has been recently reported in Ta/CoFeB/MgO devices (Kim et al. 2013)). Moreover, Hall voltage measurements at high speed are extremely challenging.

9.1.4.2 Readout with a Magnetic Tunnel Junction

The most promising readout method for RTM relies on a local TMR measurement. In this approach, the RTNW is the free layer of the MTJ, and the tunnel barrier and reference layer are patterned in contact with the RTNW. This approach allows for a large readout signal because the TMR in the CPP geometry can be very large (>100%). Moreover, lateral resolution is achieved by patterning the reference layer down to a size comparable to the bit length (typically the DW width). This method of readout has been demonstrated by NEC Corp (Fukami et al. 2009). The device described in this paper is a three-terminal MRAM cell in which the free layer is written by using the current-driven motion of a DW. It is essentially a single DW RTM.

9.1.5 DOMAIN WALL SHIFTING

The unique characteristics of RTM rely on the ability of shifting series of DWs in lockstep to and fro along the RTNWs. Indeed, while both writing and reading operations share many similarities with STT-MRAMs, the concept of shifting bits of information is specific to RTM. While it is possible to shift series of DWs with magnetic fields, this requires the use of complex sequences of space and/or time-varying fields. Nonetheless, device concepts based on the motion of DWs by varying magnetic fields have been proposed in the past, the most prominent example being the bubble memory (Malozemoff and Slonczewski 1979). Similar devices have also been discussed very recently (Allwood et al. 2005, Franken et al. 2012). Besides the complexity of design inherent to the use of field gradients, scaling is also a major issue of this type of devices. Indeed, as lateral dimensions shrink, it becomes increasingly difficult to address single devices with localized magnetic field, and the energy efficiency worsens significantly.

For all these reasons, the motion of DWs in RTM is not based on magnetic fields but rather on electrical current passing directly though the RTNWs.

9.1.5.1 Spin-Polarized Current and Spin-Transfer Torque

The STT (Berger 1996, Slonczewski 1996) responsible for current-driven DW motion arises for the spin polarization of the electrical current that flows within ferromagnetic metals. Indeed, when the current flows through a ferromagnet, scattering of the conduction electrons carrying the current is spin dependent. This means that electrical conductivity is different for electrons having their spin parallel or antiparallel to the magnetization. As a result, the net current becomes spin-polarized, that is, it is carried primarily by electrons having one spin orientation. This two-spin channel picture of electrical transport is at the center of spintronic phenomena of metal, for example, GMR and TMR. However, the phenomenon relevant to current-driven DW motion discussed in this chapter is the spin polarization in the bulk of the ferromagnet, rather than the interfacial polarization across the tunnel barrier, which is key to the properties of MTJs used in STT-MRAMS. The spin polarization P varies from 0 in a nonmagnetic metal (e.g., Cu) to 1 in a half metal, for which one spin channel is metallic and the opposite spin channel is insulating. Typical values for materials in which current-driven DW motion has been reported are typically in the range 0.2–0.6.

When current flows across a DW, its spin polarization rotates to follow the change of direction of the local magnetization. In most cases relevant to applications, the DW width is large enough (>10 nm) so that the polarization remains almost parallel to the local magnetization across the DW. This regime is called the adiabatic limit.

The spin-polarized electrons carry a spin-angular momentum, which must be conserved in a closed system. Thus, the momentum lost when the polarization direction reverses from one magnetic domain to the next is transferred to the local magnetization. As a result, there is a torque on the local magnetization. STT is proportional to the current density, and it can lead to DW motion when the current density J is large enough (Berger 1986, 1988, Li and Zhang 2004, Tatara and Kohno 2004, Zhang and Li 2004, Barnes and Maekawa 2005, Thiaville et al. 2005).

9.1.5.2 Adiabatic and Nonadiabatic Spin Transfer Torques

Since STT derives directly from an argument of conservation of spin-angular momentum, its magnitude can be quantified easily by equating the amount of spin-angular momentum carried by the current per unit time and per unit cross-sectional area, $P \cdot (J/e) \cdot g\mu_B$ (μ_B is the Bohr magneton, g the Lande factor, and e the electron charge), to the change in magnetization due to DW motion at velocity u, $2 \cdot M_S \cdot u$ (M_S is the saturation magnetization). In the context of DW motion, the STT is quantified by this quantity $u = g\mu_B PJ/(2eM_S)$.

Both experimental and theoretical studies over the last decade have revealed that the STT has two components, which are orthogonal to each other. These two components are both parameterized by u and both play a key role in current-driven DW motion. These two torques are often referred to as adiabatic and nonadiabatic STTs in the literature, even though they both coexist even in the adiabatic limit (i.e., for wide DWs). The nonadiabatic (or field-like) torque is proportional to βu, where β is a dimensionless coefficient, which is of the order of the Gilbert damping constant α. In most cases, $\beta \ll 1$, which means that the nonadiabatic torque is much smaller than its adiabatic counterpart. Nonetheless, this torque plays a very important role in current-driven DW motion.

9.1.5.3 Critical Current Density for Depinning:
Intrinsic versus Extrinsic Pinning

DW motion only takes place when the current density exceeds a critical value J_C. Theoretically, the origin of a nonzero value of J_C depends on the details of the STT. Under the sole effect of the adiabatic STT, J_C takes a finite value even for an ideal RTNW without any defect. This is quite different from the effect of a magnetic field.

Indeed, for an ideal wire, DWs would move under any nonzero applied field, albeit at very low velocity. The existence of this so-called intrinsic pining has been widely debated in the last decade. When the nonadiabatic STT is present, J_C depends on pinning defects along the RTNW, much like the propagation field needed to move DWs.

Experimentally, values of J_C reported in the literature are typically in the range 1–2 10^8 A/cm^2 for IMA materials (Parkin et al. 2008), and between 0.1 and 2 10^8 A/cm^2 for PMA materials (Koyama et al. 2011a). In the latter case, much smaller values ($\sim 10^5$–10^6 A/cm^2) have also been reported in the creep regime, in which DWs move under the combined action of current and thermal activation at extreme low speed.

9.1.5.4 Spin–Orbit Torques

Recent experimental studies of ultrathin PMA multilayers show very promising results for RTM (Miron et al. 2011, Ryu et al. 2012). In these structures, the RTNW comprises atomically thin magnetic layers (typically Co or Co/Ni/Co multilayers) in contact with nonmagnetic layers made of heavy metals such as Pt. Extremely efficient DW motion has been reported in such systems, with very high velocity (>400 m/s) and low critical current density (as low as a few 10^5 A/cm^2 for slow motion). These results are inconsistent with conventional STT. Moreover, in many reports, the motion direction is along the current flow, opposite to that expected for conventional STT. Although the detailed origin of this remarkable behavior is still unknown, it appears likely that it originates from the strong spin–orbit interaction in the heavy metal or at the interface between this metal and the magnetic layer. The spin–orbit coupling gives rise to additional torques on the magnetization due to Rashba or spin Hall effects.

9.1.5.5 Current-Driven Domain Wall Velocity and Clock Speed

The velocity of DW driven by current is an important quantity for RTM because it is a key component of the RTM clock speed. The typical RTM clock cycle includes writing, shifting, and reading operations. Writing and shifting are the most time consuming. The writing time depends on the method used to inject DWs. If writing is done by magnetic field or by STT using an MTJ, writing times of the order of 1–10 ns can be estimated. If writing is done by moving DWs from a reservoir, then both the writing and shifting times are directly related to the DW velocity.

A lower bound of the time needed to shift DWs along the racetrack is given by the shifting distance (i.e., the sum of the DW width and the spacing between DWs) divided by the DW velocity v. Thus, for a given clock speed, the requirements on the velocity are more or less stringent depending on the storage density. If, for example, the shifting distance is ~100 nm, a velocity of 100 m/s is sufficient to complete the operation in ~1 ns. Of course, this time increases linearly with the shifting distance.

Theoretically, the velocity of DW driven by STT is given by $v_{DW} = (\beta/\alpha)u$ for moderate current density (see Section 9.2.5.2 for the definition of the parameters). In other words, v_{DW} is directly proportional to the current density J and to the strength of the nonadiabatic STT. The linear dependence in J is seldom observed experimentally because of the limited range of current densities available between the threshold for motion due to pinning (at low J) and the onset of Joule heating and/or precessional motion, which leads to a saturation of v_{DW} (at high J).

Experimental values reported in the literatures are of the order of 100–150 m/s (Hayashi 2007, Thomas and Parkin 2010) for IMA materials (V-DWs in permalloy) and 40–80 m/s (Koyama et al. 2011b, Thomas et al. 2011) for thick PMA materials (several repetitions of Co/Ni). The difference between the two cases is attributed to smaller values of β/α in the latter case. As mentioned earlier, much faster velocities (up to 400 m/s) are observed in thin PMA layers fabricated on a Pt seed layer because of additional torques at the Pt interface (Miron et al. 2011, Ryu et al. 2012).

9.1.5.6 Domain Wall Positioning

For RTM to operate successfully, DWs must be moved by a controlled distance for each shifting operation. This is in principle nontrivial because the DW velocity is in general not constant throughout the application of the current pulse. Indeed, DWs have a mass and exhibit inertia. However, we have shown that the net distance traveled by DWs in response to a current pulse in IMA materials is directly proportional to the length of the pulse. Indeed, the lag due to the DW acceleration just after the pulse onset is compensated by the relaxation that takes place after the end of the current pulse (Thomas et al. 2010). Similar results have also been reported in PMA materials (Vogel et al. 2012). This allows for the precise positioning of DWs along the RTNW.

9.1.5.7 Motion of a Series of Domain Walls by Current Pulses

The key feature of current-driven DW motion is that the motion direction of all DWs in the RTNW is solely determined by the current polarity, and not by the magnetic charge of the DWs (e.g., head-to-head or tail-to-tail for IMA, or up/down and down/up for PMA). The reason for this behavior is that the current is repolarized in each domain separating consecutive DWs. By contrast, if the STT was generated from a fixed polarizer, such as the reference layer of a spin valve of an MTJ, DWs would move in opposite directions. The experimental demonstration of the motion of series of DWs by current pulses has been the key breakthrough needed to validate the concept of RTM. The first demonstration was given in 2008 for two and three DWs in permalloy nanowires (Hayashi et al. 2008). More recently, successful shifting of up to five DWs has been demonstrated in both IMA and PMA RTNWs (Thomas et al. 2011).

An experimental demonstration of a 6-bit (5 DWs) unidirectional first-in first-out shift register is shown in Figure 9.3. This test device comprises a single permalloy RTNW made by electron beam lithography. Sensing is done by measuring the resistance of the RTNW, which is reduced by a fixed amount for every additional DW located within the RTNW. The sequence of injection and shift pulses used in the experiments together with the device resistance is shown in the figure. The shift pulses used here are 34 ns long. During the first five iterations, one additional DW is injected and moved along the RTNW until five DWs are located in the RTNW. For subsequent iterations, one DW exits the left end of the RTNW at each shift pulse, and one DW is written at the right end of the RTNW at each injection pulse, such that the total number of DWs in the RTNW oscillates between four and five. Cartoons of the magnetic configuration inferred from the resistance data for the first seven steps of the experiment are shown using a color code overlayed on a scanning electron micrograph image of the device.

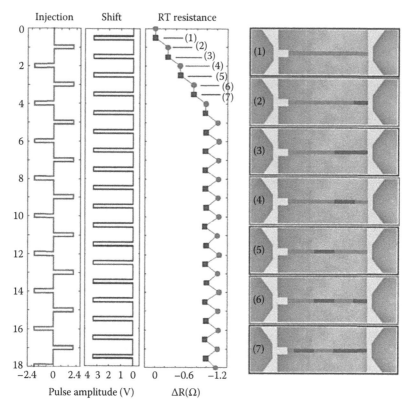

FIGURE 9.3 Experimental demonstration of a 5 DW shift register in a horizontal RTM device. The RTNW is a 20 μm long, 200 nm wide permalloy nanowire. Writing is done by the Oersted field from a metallic contact line. Reading is done by measuring changes in the RTNW resistance. The polarity of the shift pulse is such that DWs propagate from right to left along the electron flow. The current density is ~1.2×10^8 A/cm^2.

9.2 MANUFACTURING CONSIDERATIONS

9.2.1 EVOLUTIONARY VERSUS REVOLUTIONARY

To bring the RTM out of the research laboratory and into working products requires tradeoffs in performance vs. manufacturability, and additional integration complexity for reliable operation and longevity. With NAND flash memory as the gold standard for high-density memory, an equivalent area of ~$4F^2$ per bit must be targeted to be competitive. One can argue the benefits of speed, nonvolatility, endurance, etc., but without competitive density, a memory will be unlikely to capture much market share and will be just as unlikely to play a prominent role in "big data" applications.

Conventional semiconductor memories use a randomly accessible matrix configuration with a bit line and a word line intersecting at a given memory cell. The area of the memory cell is set by the size of the memory element and the additional space needed for connecting wires and drive components. A square memory element with minimum feature size F equal to the lateral dimension generally requires isolation from neighboring elements by at least the minimum feature size as well. This gives a $2F \times 2F$, or $4F^2$ cell size for the smallest conventional memory cell. Additional components, such as wiring connections for a three-terminal transistor, will rapidly grow the cell size to the order of 10 F^2 for DRAM or 100 F^2 for six-transistor SRAM. The increasing popularity of two-terminal memory elements such as phase change memory or STT-MRAM comes from the ability to create a cell with a reduced number of additional, space-consuming access connections.

Evolutionary improvements in the conventional structures include sublithographic patterning to create elements of lateral size below F, or the creation of multistate memory elements that can store more than 1 bit per memory cell. Other recent work utilizes package stacking, die stacking with multilevel wire bonding, and now stacking of chips with through-silicon vias (TSVs) to connect one planar layer of memory to another—for perhaps a factor of 4–16 improvement in effective memory density (Knickerbocker et al. 2006). Also, through the use of multilayer polysilicon or Si/SiGe epitaxial silicon growth techniques, NAND flash has been stacked at the device level within a silicon chip (Kim et al. 2012). These modest improvements to memory density come at great expense, either in chip manufacturing complexity or in test and assembly.

The attraction of RTM is in its ability to increase the number of stored bits per transistor used to access the bits. As described in the previous section, each RTM nanowire requires a controller to send shifting current through the nanowire to move the bits, a controller to multiplex the readout of several RTM array elements into a single sense amplifier, and a controller to manage the insertion of DWs when writing to the RTM. This third controller can be amortized over several RTM array elements, such that we can consider RTM as requiring order of two control units (transistors) per RTNW (potentially > 100 bits). Comparing this to the 2 to 16 times improvements in memory density with evolutionary techniques, the potential for 50 times with RTM is revolutionary.

9.2.2 Horizontal (In-Plane) versus Vertical (3-D) Structures

The cost of two-dimensional (planar) memories generally revolves around the areal cell size per bit. Figure 9.4 shows a perspective view of the type of HRTM manufactured in IBM's T.J. Watson Research Center (Annunziata et al. 2011). With reference to Figure 9.1, one can see that in the simplest case, the silicon beneath the memory is poorly utilized. The RTM's revolutionary improvement in density is largely lost because the area to house the RTNWs is substantially larger than the area required of the silicon. One can improve things perhaps a factor of 2 by improving array efficiency—placing sense amplifiers, drivers, multiplexers, etc., underneath the arrays of nanowires. Shown in Figure 9.5, another factor of two or more can be realized by stacking horizontal RTNW arrays atop one another. Since the RTM elements are purely "back-end" devices, the number of stacked layers is limited only by fabrication cost and the ability to connect each RTNW to the control transistors beneath. These realizations of RTM can be expected to be competitive with flash memory in terms of the effective area per bit, but are not a substantial improvement in density.

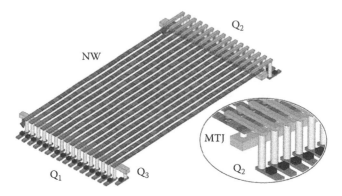

FIGURE 9.4 Drawing of a horizontal RTM array structure, highlighting the large number of bits (light vs. dark contrast) per nanowire, and the relatively few driving transistors (approx. 2× number of RTNWs). Q_1 transistors gate the shifting current that moves DWs. Q_2 is used to direct the desired MTJ readout element to the sense amplifier. Q_3 is used to switch the current for injecting a domain wall into each of the nanowires and, in this configuration, can be amortized over the array.

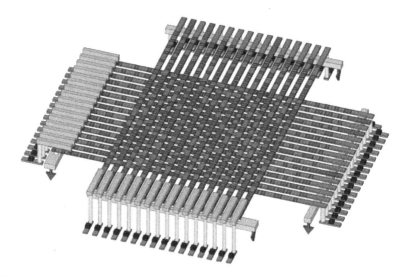

FIGURE 9.5 Drawing of a two-layer stacked horizontal RTM array. (The crossing nanowires do not touch as they are separated spatially by a suitable insulating film.)

They may still be desirable if one makes good use of the speed, endurance, reliability, and nonvolatility of the RTM as compared to flash or other novel memories.

A more revolutionary structure is conceptually shown in Figure 9.6. This describes the vertical RTM in one potential embodiment. Others can be found in earlier literature, including direct-coupled MTJ elements and curved RTNW elements that loop into a racetrack-like oval shape—giving the name to this particular type of memory (Parkin 2004, Parkin et al. 2011). Although not considered impossible, the fabrication of such a structure will require substantial breakthroughs to achieve the ultimate

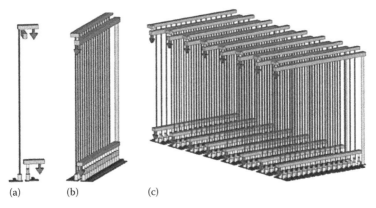

(a)   (b)   (c)

FIGURE 9.6 One concept for vertical RTM structure. (a) Individual RTM element with vertical RTNW. A field-based DW injector is represented as a detached rectangular wire near the top of the structure. The readout element in this device is shown as a non-contacting MTJ. Although the signal of a field-coupled MTJ will be smaller than a directly contacted MTJ, such a configuration can make fabrication simpler. (b) A row of identical vertical RTM elements, sharing a common DW injection element. (c) A full array of identical vertical RTM elements for memory density approaching that of hard disk drives.

density predicted for vertical RTM. Of particular concern are the aspect ratio of each RTNW (>1 μm tall and <50 nm diameter) and the difficulty in growing an MTJ readout element in contact with the vertical wire, or nearby.

9.2.3 OVERVIEW OF IN-PLANE FABRICATION

The fabrication process for a successfully fabricated HRTM array like that shown in Figure 9.4 is explained here. Referring to a single element for simplicity in explanation (Figure 9.7), the RTM-specific elements are all fabricated in the back-end of line at a level above the underlying silicon-based CMOS driving circuitry. The gray-colored shapes in Figure 9.7 represent industry-standard damascene copper wiring beneath the RTM NW. The elements with herringbone-patterned faces are shallow vias that connect the copper to the RTM NW, which is represented in black. The MTJ readout element is shown as a white box and resides atop the RTM NW near one edge.

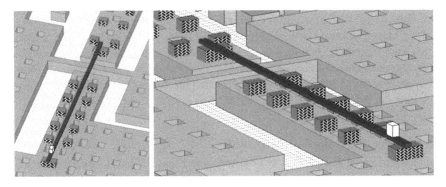

FIGURE 9.7 Drawings of an isolated RTNW to illustrate the fabrication process.

9.2.3.1   Base Copper Wiring

As illustrated in (Figure 9.7), the damascene copper wiring beneath the RTNW is "cheesed" in wide copper regions to assist with across-feature uniformity during the damascene polish. The "cheese" refers to holes placed at various locations within the copper so that the chemical–mechanical planarization (CMP) will see a roughly uniform background combination of copper interspersed with dielectric. Similarly, in regions with little copper, dummy "fill" structures are placed so the polish is not presented with large regions of one material or another, but instead with a near-uniform pattern of dielectric and copper.

The majority of the RTNW resides atop a large block of copper, which serves the purpose of a heat sink when high currents are used to shift the domain walls in the RTNW. Such a heat sink can improve device resistance to burn out by as much as an order of magnitude in RTNW shift current density. At the end of the RTNW opposite the MTJ readout, there is a narrow uncheesed copper wire perpendicular to the RTNW. This is the simple injection wire configuration for IMA RTM, where a current passed through the wire will generate a field strong enough to nucleate a domain wall within the RTNW and localized above the injection wire.

9.2.3.2   Shallow via Nanowire Connection

The gray structures in (Figure 9.7) represent relatively short (~50 nm tall) conductive vias that connect the underlying copper wiring to the RTNW while allowing certain copper features (like the injection wire) to cross beneath the RTNW without contact to the RTNW. It is desirable for these vias to be of minimal height so as to maximize field coupling from the injection wire for lowest-power operation. In addition, proximity of the RTNW to the underlying heat sink provides greater latitude in terms of current density used to shift the domain walls. The vias are best formed with a damascene polish process for compatibility with existing CMOS tooling options and for optimal RTM operation.

It is seen particularly in IMA RTNWs that domain-wall pinning effects can dramatically increase the necessary current for shifting DWs along the RTNW. For sufficiently large pinning, the operation of the RTM is impractical, as the large currents needed to overcome the pinning can result in unpredictable formation and annihilation of domains and can even cause fusing of the RTNW. It is therefore quite important to create a smooth surface on which to grow the nanowire so that undesired imperfections in the surface do not serve as localized pinning sites. Conventional CMOS CMP techniques can be slightly modified to provide surfaces with better than 0.1 nm rms roughness on which to grow the RTNW films. The astute reader will notice also the additional gray-colored via features that do not connect to the RTNW. These are dummy "fill" elements that are placed throughout the layout to provide a more uniform pattern for the polish operation, preventing overpolish or underpolish in various regions.

9.2.3.3   Nanowire and MTJ Formation

The blue and green RTNW and MTJ elements in Figure 9.7 are formed in one preferred method as described here. After shallow-via polish surface smoothing, a magnetic film stack is deposited in blanket form over the entire wafer. The magnetic film stack includes (from the bottom to the top) a thin seed layer; the RTNW magnetic film; an MgO tunnel barrier; an antiferromagnetically pinned, flux-closed, synthetic antiferromagnet (SAF) reference layer; and finally an etch stop/cap layer and a hard mask layer. These films are deposited in situ so as to eliminate undesirable inter-layer interface contamination. Patterning of the structures is more difficult with the full-stack deposition technique, but device performance is substantially better than that with ex situ multistep deposition and patterning methods. Figure 9.8 gives an example of the thin films one would deposit for IMA RTM. The NiFe permalloy layer underneath the MgO tunnel barrier will be patterned in nanowire form to house the magnetic domains. The layers above the MgO tunnel barrier are patterned into a relatively small MTJ top electrode that is used to sense the domain walls being shifted along the permalloy by RTNW currents. These top electrode layers are con-figured in a SAF configuration to minimize fringing fields that could affect DW motion. A top PtMn antiferromagnet layer is exchange-coupled to the SAF to pin the SAF magnetization for use as an unchanging reference layer. Finally, a conductive hard mask is employed for self-aligned contact to the reference layer films.

The patterning of the RTNW and MTJ elements in a conventional sense is done by first lithographically defining the reference layer of the MTJ and then etching the hard mask and reference layers with a reactive ion etch (RIE) or ion beam etch (IBE) technique. Using etch endpoint detection or good etch selectivity, one can realize a clean stop on the MgO tunnel barrier. Due to the need for high-density memory at

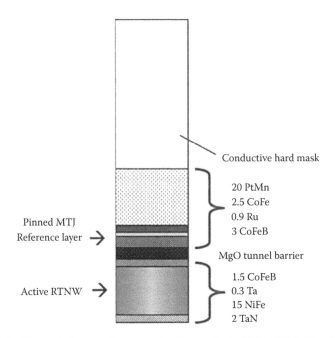

FIGURE 9.8 Example layers of a magnetic film stack for IMA RTM. The full stack is deposited in situ over the entire wafer after shallow via formation. Numbers refer to thick-nesses in nm. Exact film stoichiometries are chosen for high magnetoresistance, desirable DW characteristics, and manufacturability.

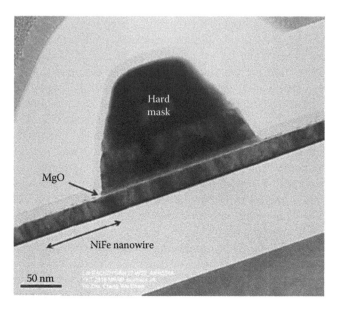

FIGURE 9.9 TEM cross-sectional image showing formation of the MTJ reference layer using an etch in conjunction with a conductive hard mask. The etch stops on the MgO tunnel barrier. At this stage, the NiFe "nanowire" is a blanket film across the entire wafer. The reference layer has been patterned into submicrometer-sized pillars.

very fine pitch, it is impractical to consider a via contact to the top of the MTJ, and one benefits from utilizing a self-aligned connection to the top electrode of the MTJ. This is most easily realized by using a tall conducting hard mask at the MTJ patterning step. After patterning the MTJ, the protruding hard mask can be contacted simply with a wiring layer above the MTJ. Figure 9.9 shows an SEM cross section of a magnetic stack like that in Figure 9.8, after etching of the reference layer films to form a submicron MTJ.

A second lithography step then defines the RTNW, masking the previously patterned MTJ. RIE and/or IBE is used to pattern the NW by removing unwanted metal, down to the insulating shallow-via dielectric. Figure 9.10 shows representative images of the MTJ/RTNW patterning as performed on IMA HRTM. Modified patterning techniques requiring substantially greater process control can create a self-aligned MTJ/RTNW combination that shows reduced DW pinning as DWs are shifted beneath the MTJ readout. Figure 9.11 shows a representative image of the self-aligned MTJ after MTJ/RTNW patterning. Such patterning is considered essential for narrow RTNWs that are to be used in realizing RTM's greatest potential for densely packed memory.

Figure 9.12 shows lower-magnification views of an isolated RTNW structure after MTJ and RTNW patterning, corresponding to the drawings of Figure 9.4. Underlying copper metal can be seen through the dielectric of the shallow vias, as relatively large SEM acceleration voltage was used. This particular device makes use of a flared wider RTNW section near the injection wire to reduce the magnetic field

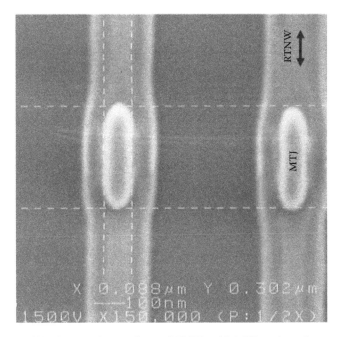

FIGURE 9.10 SEM image of two adjacent RTNWs with MTJ readout elements (elliptical shapes) atop the RTNWs. The RTNW is roughly 175 nm wide, and the MTJ is roughly 90 × 300 nm in area. Note the bulge in the RTNW near the MTJ. With the aforementioned process technique, etching of the RTNW is shadowed by the topography of the MTJ and creates a nonlinear edge to the RTNW.

needed to inject a DW. The self-aligned MTJ readout element is faintly visible in the top image, just to the right of the leftmost RTNW-to-shallow via contact.

9.2.3.4 Encapsulation and Top Contact

After patterning the metals and MTJ barrier, a dielectric encapsulant is deposited to protect the device during further processing. Essential characteristics of the dielectric include good adhesion, low current leakage, negligible effects on magnetic or tunnel barrier properties, and resistance to diffusion of impurities through the dielectric or along the dielectric interfaces. Industry-standard copper cap films generally provide the desirable properties. SiN, SiON, SiCN, etc., are readily available in fab facilities and can be deposited at relatively low temperatures (<300°C) to minimize the redistribution of magnetic stack elements during deposition.

Before the formation of the top contact, the surface of the wafer should be planarized for best results in lithographic steps that follow. The encapsulating film can itself be used for planarization of the top surface of the structure (e.g., with a CMP operation), or additional dielectric layers can be added for desired properties. For example, used with chemical vapor deposition (CVD), tetra-ethyl-ortho-silicate precursor silicon oxide films exhibit very good gap-fill properties and prevent void or seam formation near the topography of the NW and MTJ. CVD of silicon oxide films with a high-density plasma silane precursor can result in nonconformal topography

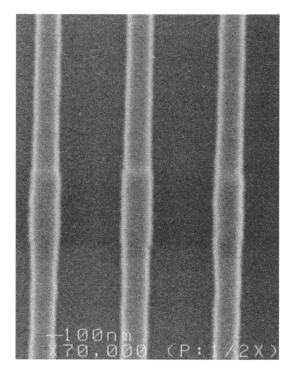

FIGURE 9.11 SEM image of three adjacent RTNWs with self-aligned MTJ readout elements faintly visible as rectangular bulges atop the RTNWs. The RTNW is roughly 125 nm wide, and the MTJ is roughly 125 × 300 nm in area.

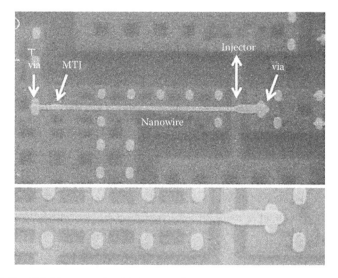

FIGURE 9.12 SEM images of a fabricated RTM NW corresponding to the drawings in Figure 9.7. Low mag (top) and higher mag (bottom). NW width is approximately 150 nm for this device.

above the patterned device such that a very brief CMP operation can be used for planarization. The shorter CMP process implies improved across-wafer uniformity in terms of absolute dielectric thickness. Modern spin-on glasses such as methyl silsesquioxane have robust mechanical properties and offer the benefit of planarization without a costly CMP operation. The spin-on glasses also provide extremely uniform coating thickness, which can be essential in the following contact formation steps where an etch-stop layer may not be practical.

The wiring above the RTM element in a preferred embodiment uses dual-damascene copper processing to create a via structure for contact to low-lying conductors underneath, and a trench structure for contact to the elevated MTJ top electrode (e.g., conductive hard mask). There is a tradeoff in the optimal height of the conductive hard mask, as thinner masks will ease the stop-on-barrier etching for MTJ patterning, and further integration with aligned NWs. Thicker masks ease the top-contact formation procedure by affording more latitude to the depth of the trench being etched in the planarizing dielectric to form the wire that contacts the MTJ top electrode. If the planarizing dielectric exhibits poor across-wafer uniformity, or if the trench etch is nonideal in a similar sense, there may be only a very small process window between trench etch exposure of the tops of all MTJ hard masks and trench etch exposure of the MTJ barrier and underlying nanowire. These concerns can be better understood by studying the TEM cross-sectional image in Figure 9.13.

An alternative technique to contact the top electrode of the MTJ involves using a polish-resistant hard mask for the MTJ. In this situation, after deposition of the

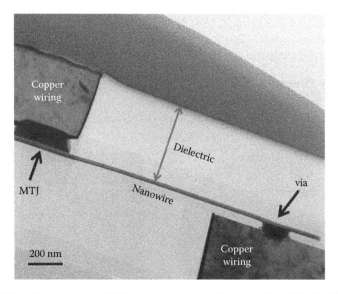

FIGURE 9.13   Cross-sectional TEM image of the readout end of an RTM NW. The self-aligned conductive hard mask makes electrical contact between the MTJ reference layer and the copper wiring layer above. The dielectric encapsulation serves to protect the magnetics as well as planarize for damascene copper wire formation. A shallow via makes contact to copper wiring beneath and carries the relatively small read current from the MTJ to ground. The via also sinks the much larger DW shifting current from the NW.

planarizing dielectric, the CMP operation proceeds past simple planarization and removes enough dielectric to expose the top of the MTJ hard mask. Relatively high selectivity between polish rates of the dielectric and the hard mask can be achieved through the use of customized CMP slurries (e.g., ceria-based) or with polish-resistant MTJ hard masks such as diamond-like carbon. After the polish exposes the hard mask and any necessary steps are taken to remove undesired insulating films, a blanket metal layer can be deposited and patterned with a subtractive etch technique (the inverse of the damascene process technique). Dielectric fill and polish can then planarize above this so that additional fine-pitch wiring layers can be created atop the RTM devices.

9.2.4 Integration with CMOS

As a demonstration of the potential of RTM, the group at IBM has integrated RTM arrays with CMOS driving circuitry in a product-like configuration. The CMOS circuitry was fabricated with 90 nm ground rules on 200 mm wafers, and the RTM layers described in Section 9.3.3 were integrated between the fifth and sixth copper metal layers. The additional patterning to create the RTM elements required three nonstandard photomasks: one for the shallow via, one for the MTJ, and one for the RTNW. The RTNWs on each die were designed to cover a range of widths from 60 to 240 nm, and with lengths varying from 6 to 12 µm. Each array had 256 active racetrack cells with a line-to-line pitch of 560 nm, as shown in Figures 9.14 and 9.15. The MTJ read-out elements were designed to cover the size range 80 × 160 nm to 240 × 400 nm. Successful operation of the RTM is described in detail in (Annunziata et al. 2011).

9.2.5 Practical Pitfalls

9.2.5.1 DW Pinning

IMA HRTM is plagued particularly by the sensitivity of the DWs in the nanowire to structural or magnetic defects. Such defects can pin the DWs, effectively

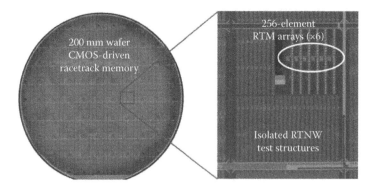

FIGURE 9.14 Wafer image showing CMOS foundry process compatibility for RTM fabrication. The die within the wafer contains various experimental test structures in addition to the CMOS-driven arrays.

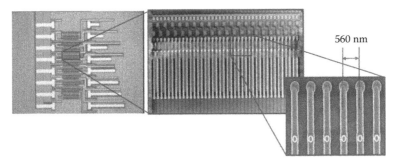

560 nm

FIGURE 9.15 Top-down optical microscope view of fully fabricated CMOS-driven RTM array (at left) and top-down SEM images detailing portions of the array as seen after RTNW and MTJ patterning (at right, outlined by the small box). The full array consists of 8 rows of RTNWs, with each row containing 32 RTNWs. There are "dummy" devices surrounding the active 8 × 32 array to ensure uniformity of critical areas during processing.

increasing the necessary shift current by as much as a factor of 2 or more. In extreme cases, the depinning current exceeds the threshold for spontaneous DW formation in the RTNW or even exceeds the current-carrying capacity of the RTNW. Much effort therefore goes into the fabrication of smooth-sided RTNWs and integration of MTJ readout elements with minimal perturbation to the RTNW. Misalignment between the MTJ element and the RTNW is a troublesome source of RTNW shape nonideality, and testing shows pinning of DWs at the site of the MTJ. Refined alignment schemes and masking methods for smooth-sidewall etching are effective ways to reduce the pinning problem. Etching the magnetic materials for smooth sidewalls is not a trivial procedure, as the RIE by-products created in reactions with the magnetic stack elements are typically nonvolatile and require a substantial ion-bombardment component to dislodge the atoms. Redeposition of these "nonvolatile" elements on the sidewalls of device topography is common and can result in veil formation that masks the etch in a nonuniform manner. Figure 9.16 compares an older process flow with a more advanced flow that results in reduced DW pinning as seen at final wafer test as a reduction in required shift current.

In idealized device concepts, the RTNW is well behaved and even the smallest of applied shift currents or thermal energy can cause DW creep. These ideal devices would actually benefit from artificial pinning sites, regularly spaced at desired DW-to-DW pitch along the RTNW. Applied shift current would move a DW from one pinning site to the adjacent site. The local energy minimum at the pinning site keeps the DW from undesirable motion along the RTNW, but is not so low an energy as to require excessive shift currents. Examples can be found in the patent literature (e.g., Gaidis et al. 2011). PMA nanowires are expected to behave better with regard to pinning in that lower currents are required to overcome the pinning strength. This, along with the promise of more densely packed DWs, makes PMA magnetics arguably the leading candidate for future high-density RTM.

Self-aligned process

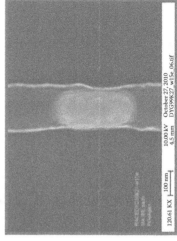

RTNW-aligned to MTJ process

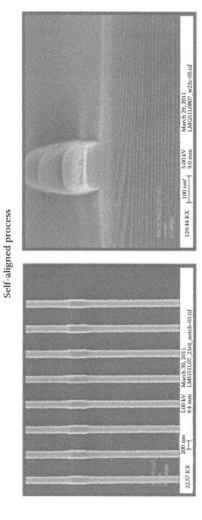

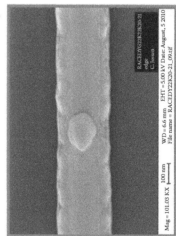

FIGURE 9.16 SEM images of RTNWs with MTJs patterned above, using different methods for processing. The self-aligned process shows less perturbation of the RTNW shape by the MTJ, and improved masking techniques result in smoother sidewalls on the RTNW.

9.2.5.2 Vertical Racetrack Memory

The ease of planar device fabrication has directed research efforts toward the in-plane structures referred to as HRTM. However, the ultimate in memory density will come when the RTNWs can be arranged vertically atop drive transistors, and most efficient use of silicon area is realized. VRTM development is difficult in particular because one must adapt conventional ("planar") processes and tooling to a structure that can be an order of magnitude taller than conventional multilayer planar circuits. Integration ideas involving self-assembly or superfill plating can potentially create vertical nanowires with somewhat desirable magnetic properties. The challenge then becomes finding a way to couple a low-noise readout device to the RTNW. Parkin's original patent (Parkin 2004) suggests an elegant design where the RTNW is gently curved to present a planar region for readout and DW injection, but the majority of the RTNW is oriented vertically for minimal wafer area usage. DW motion along curved wire trajectories has been demonstrated with in-plane wires, but it will be more challenging to create a smoothly curved RTNW out of plane. It is impractical at this time to consider coupling an MTJ along a vertically oriented RTNW, as the MTJ performance is extremely sensitive to the quality and uniformity of the tunnel barrier and other magnetic film thicknesses. We do not yet have a good way to create such a perfect structure on a vertically oriented topographical structure.

An additional consideration with VRTM is the potential use of PMA magnetics for tighter DW packing density. Conventional thin-film PMA magnetics suitable for RTM generally would involve strained multilayers (e.g., Co/Ni) to induce the PMA character. Such multilayers would likely require atomic-level manipulation of the elements during deposition. Atomic layer deposition is a possible solution to this quandary, but suitable precursors for the magnetic films must be developed first. Such techniques can also be envisioned as suitable for forming near-perfect interfaces for interface-mediated, current-controlled DW motion. There is great promise in the use of interface effects for lower-current DW shifting.

9.3 CONCLUSION

This chapter has provided an updated overview of the physics and engineering behind magnetic domain wall RTM. RTM is a "real" memory in the sense that it has been demonstrated in integration with CMOS driving circuitry, with fab-friendly materials and processes. However, the promise of ultrahigh density memory remains in the not-so-near future, as substantial material development is required to demonstrate, for example, PMA HRTM with low shift current, adequate thermal stability/retention, high-speed DW motion for high-speed device operation, and robust repeatable shifting of DWs in lockstep with others on the same RTNW. Vertical RTM for the ultimate in memory density requires further development of processing techniques unimaginable a decade ago, but perhaps now in the realm of possibility. However, even the simpler HRTM promises very high density memory with speed, reliability, nonvolatility, and low-power operation. Array efficiency can be quite high as the peripheral drive circuitry can be moved beneath the RTM arrays.

We have taken significant steps toward the realization of this new type of magnetic memory. Future materials and process improvements are on the horizon: improvements that should make HRTM practical for common use and eventually VRTM for the most challenging of "big data" applications.

ACKNOWLEDGMENTS

We gratefully acknowledge the contribution of team leaders S.S.P. Parkin and W.J. Gallagher, along with the dedication and valuable contributions of A. Annunziata. It is also a pleasure to acknowledge our many co-workers at the IBM Almaden Research Center, in particular C. Rettner, B. Hughes, S.-Y. Yang, M. Hayashi, R. Moriya, K.-S. Ryu and T. Phung, whose many contributions over the past few years have been instrumental in advancing our understanding of the physics of RTM. The myriad contributions of process integrators and test engineers made the work possible. In particular, C.C. Hung, C.W. Chien, J.P. Hummel, E. O'Sullivan, M. Rothwell, C. Tyberg, E.A. Joseph, and M. Lofaro were responsible for successfully creating the RTM structures that seemed so challenging less than a decade ago. P. Chevalier and P. Trouilloud donated valuable time and advice for the magnetic testing described earlier. TEM imaging was done by the capable hands of P. Rice, T. Topuria, E. Delenia, Y. Zhu, and C. Breslin. Significant additional contributions were provided by S.L. Brown, G. Hu, E. Galligan, and C. Jessen.

Finally, we also thank IBM's Microelectronics Research Laboratory (MRL) for process development and fabrication. A research environment that can take a device concept from paper to reality is invaluable. The MRL is a prime example of such a place, and without it this work would not have been successful. Thank-you to J. Silverman and the MRL team for smooth and fast-paced wafer processing.

REFERENCES

D. A. Allwood et al., *Science* **309**, 1688 (2005).
A. J. Annunziata et al., *2011 IEEE International Electron Devices Meeting (IEDM) Technical Digest*, Washington, DC, pp. 3–24 (2011).
S. E. Barnes and S. Maekawa, *Phys. Rev. Lett.* **95**, 107204 (2005).
L. Berger, *Phys. Rev. B* **33**, 1572 (1986).
L. Berger, *J. Appl. Phys.* **63**, 1663 (1988).
L. Berger, *Phys. Rev. B* **54**, 9353 (1996).
J. H. Franken, H. J. M. Swagten, and B. Koopmans, *Nat. Nanotech.* **7**, 499 (2012).
S. Fukami et al., in *Symposium on VLSI Technology Digest of Technical Papers*, June 16–18, 2009, Honolulu, HI, pp. 230–231 (2009).
M. C. Gaidis et al., High density planar magnetic domain wall memory apparatus, U.S. Patent No. 8,023,305 (2011).
M. Hayashi, L. Thomas, R. Moriya, C. Rettner, and S. S. P. Parkin, *Science* **320**, 209 (2008).
M. Hayashi, L. Thomas, C. Rettner, R. Moriya, X. Jiang, and S. S. P. Parkin, *Phys. Rev. Lett.* **97**, 207205 (2006).
M. Hayashi, L. Thomas, C. Rettner, R. Moriya, Y. B. Bazaliy, and S. S. P. Parkin, *Phys. Rev. Lett.* **98**, 037204 (2007).
A. Hubert and R. Schäfer, *Magnetic Domains: The Analysis of Magnetic Microstructures*, Springer, Berlin, Germany (1998).

S. Ikeda et al., *Nat. Mater.* **9**, 721 (2010).

M. T. Johnson, P. J. H. Bloemen, F. J. A. den Broeder, and J. J.de Vries, *Rep. Prog. Phys.* **59**, 1409–1458 (1996).

J. Kim et al., *Nat. Mater.* **12**, 240 (2013).

Y. Kim et al., *IEEE Trans. Elect. Dev.* **59**, 35 (2012).

S. Aritome et al., *IEEE Trans. Elect. Dev.* **60**, 1327 (2013).

J. U. Knickerbocker et al., *IBM J. Res. Dev.* **49**, 553 (2006).

T. Koyama et al., *Nat. Mater.* **10**, 194 (2011a).

T. Koyama et al., *Appl. Phys. Lett.* **98**, 192509 (2011b).

Z. Li, S. Zhang, *Phys. Rev. Lett.* **92**, 207203 (2004).

A. P. Malozemoff and J. C. Slonczewski, *Magnetic Domain Walls in Bubble Material*, Academic Press, New York (1979).

T. R. McGuire and R. I. Potter, *IEEE Trans. Magn.* **11**, 1018 (1975).

I. M. Miron et al., *Nat. Mater.* **10**, 419 (2011).

S. Parkin, M. Hayashi, L. Thomas, X. Jiang, R. Moriya, and W. Gallagher, Emerging spintronic memories, in *Handbook of Spin Transport and Magnetism* (eds.) E. Y. Tsymbal and I. Zutic, CRC Press, Boca Raton, FL (2011).

S. S. P. Parkin, Shiftable magnetic shift register and method of using the same, U.S. Patent No. 6,834,005 (2004).

S. S. P. Parkin, M. Hayashi, and L. Thomas, *Science* **320**, 190 (2008).

K.-S. Ryu, L. Thomas, S.-H. Yang, and S. S. P. Parkin, *Appl. Phys. Exp.* **5**, 093006 (2012).

J. Slonczewski, *J. Magn. Magn. Mater.* **159**, L1 (1996).

G. Tatara and H. Kohno, *Phys. Rev. Lett.* **92**, 086601 (2004).

A. Thiaville and Y. Nakatani, in *Spin Dynamics in Confined Magnetic Structures III*, A. Thiaville and B. Hillebrand (eds.), Springer, Berlin, Germany, pp. 161–205 (2006).

A. Thiaville, Y. Nakatani, J. Miltat, and Y. Suzuki, *Europhys. Lett.* **69**, 990 (2005).

L. Thomas, R. Moriya, C. Rettner, and S. S. P. Parkin, *Science* **330**, 18 (2010).

L. Thomas and S. S. P. Parkin, in *Handbook of Magnetism and Advanced Magnetic Materials*, (eds.) H. Kronmüller and S. S. P. Parkin, John Wiley & Sons Ltd., Chichester, U.K., Vol. 2, pp. 942–982 (2007).

L. Thomas et al., *IEEE Int. Electron Devices Meeting (IEDM)*, 2011, p. 24.22.21.

E. Y. Tsymbal, *Handbook of Spin Transport and Magnetism,* Tsymbal and Zutic (eds.), CRC Press, Boca Raton, FL (2011).

J. Vogel et al., *Phys. Rev. Lett.* **108**, 247202 (2012).

D. C. Worledge et al., *Appl. Phys. Lett.* **98**, 022501 (2011).

S. Zhang and Z. Li, *Phys. Rev. Lett.* **93**, 127204 (2004).

Part V

Phase-Change Memory

Jin He, Yujun Wei, and Mansun Chan

CONTENTS

10.1 INTRODUCTION

Nonvolatile memory (NVM) plays an important role in our daily life as data storage for mobile devices, memory cards, and solid-state disks in this information era. Flash memory has been the dominating technology for the NVM markets for many years due to its high density and low fabrication cost [1–3]. A Flash memory cell is

259

made up of a transistor with a floating gate. Writing to the cell is achieved by changing the number of electrons trapped in the floating gate by Fowler–Nordheim (FN) tunneling or hot electron injection [4]. Due to the trapped electrons, the threshold voltage of the device is changed. The reading of the cell is performed by sensing the current through the memory cell under a fixed gate voltage. However, Flash memory is difficult to scale, especially beyond the sub-50 nm regime. Reduction of the device size significantly reduces the number of electrons that can be stored in the floating gate, leading to a reliability problem [5]. The small size also results in serious interference between cells because of the parasitic capacitance [6]. Much research in the NVM area is now focused on finding new types of NVM to replace Flash memory as the next-generation mainstream NVM.

Phase-change memory (PCM) may be the most promising competitor for the next-generation NVM. Compared with Flash or other existing technologies, PCM is more advantageous in reliability, high performance, and compatibility with complementary metal–oxide semiconductor (CMOS) technology [7,8]. The mechanism of the PCM is based on the thermal-driven crystal–amorphous reversible phase transition in PC materials (in particular, chalcogenides), which have different resistivity in crystal and amorphous states [9]. Figure 10.1 shows the general structure of a PCM cell that can switch between two different states: the set state, which exhibits low resistance, and the reset state, which exhibits high resistance. The switching between these states can be achieved by applying different programming pulses as shown in Figure 10.2 [10]. In the set operation, a current pulse with long duration and low amplitude is applied to keep the PC material in between the glass transition point and the melting point for crystallization. All the PC materials will be annealed to the crystalline phase, which has low resistivity. In the reset operation, a short-duration but high-amplitude current pulse is used to heat the cell to a temperature beyond the melting point followed by a cool-down process in a very short time to create a

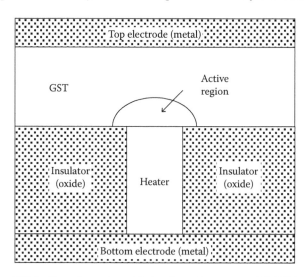

FIGURE 10.1 PCM cell configuration with the phase change material, GST, placed between the top electrode and the heater.

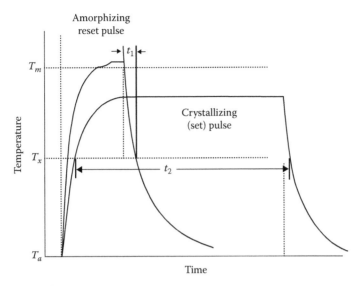

FIGURE 10.2 Different temperature responses of the PCM cell under different programming pulses.

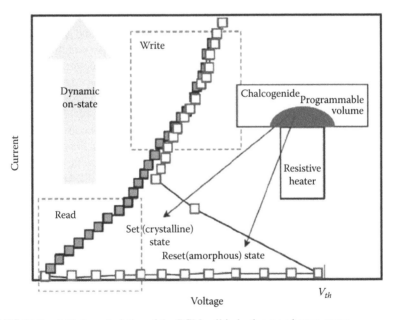

FIGURE 10.3 *I–V* characteristics of the PCM cell in both set and reset states.

portion of the amorphous phase inside the PC material, which has resistivity several orders higher than that of the crystalline phase, above the bottom electrode contact (BEC). Figure 10.3 shows the *I–V* characteristics of PCM in both states [11].

Researches and applications of PCM are in the epoch before booming. It has been reported that a PCM cell with a μ-trench structure using $Ge_2Sb_2Te_5$ material [12] can achieve low program voltage and high program velocity (erasing time is 100 ns, writing time is tens of nanoseconds). Its data retention time can achieve 10 years at 85°C with 10^8 programming cycles. In addition, only a few more masks are needed on top of a standard CMOS process for PCM integration [24], making it cost-effective for manufacturing. By using diodes as driving devices, a high-density array can be achieved to improve the yield. It is predicted that PCM can be batch-produced under a 25 nm node, while traditional NOR and NAND Flash will face significant challenges in the 32 nm node.

Although PCM has many attractive features, its adoption also has many challenges to overcome, especially in respect to the scaling limit and cell array reliability. One of the most important issues is how to reduce the programming currents, which is limited by scaling of the memory driver. In fact, compared with NAND Flash, the area of PCM cell must be reduced to $4F^2$ to keep its competitive advantages in density. In the 25 nm node, the driving diode can supply less than 150 µA driven current [13]. Current industrial test results from the 90 nm node show that it requires a programming current of several milliamperes for proper function. Therefore, estimating PCM scaling behavior and designing low-power PCM cells are some of the important issues for farther research [14–21]. On the other hand, to ensure that PCM technology meets the batch production requirement, problems of unit and array reliability must be addressed [22], especially the cross talk induced by thermal disturbance between adjacent units, unit and array data retention, and programming–read cycles. To explain the experimental data of PCM samples, designing new cell structure [23], optimizing structure, and predicting scaling effects, a physics-based model and numerical simulation tools are required.

Numerical simulations are often employed to predict device performance and optimize device structure [25]. The simulation should be able to predict the performance of a PCM cell for different materials and structures, including I–V characteristics, distribution of electric field, temperature profile, phase distribution, and resistance distribution. As for the reliability of a PCM cell, the impact of material fatigue driven by electric–thermal–stress coupling is involved to predict the lifetime and data. For a complete array, cross talk is the most important issue for reliability, which is necessary to evaluate the impact of space between the cells. Numerical simulation is established based on solving a coupling system, including an electrothermal module and PC dynamics. The numerical simulation needs the partial differential equation (PDE) solving process, which requires a lot of computational resource for predicting the device performance for the goal of structure optimization. However, numerical simulation is not suitable for array simulation including peripheral circuits. Therefore, it is necessary to develop a simulation program with integrated circuit emphasis (SPICE) model for array-type simulations with peripheral circuits. In the following sections, the numerical model, the SPICE model, and the reliability model are discussed based on research work from our group.

10.2 NUMERICAL MODEL

Numerical simulation of a PCM cell can be divided into two parts: (1) ovonic threshold switching (OTS), which is a process in which resistance sharply decreases when the electric field of amorphous materials reaches a critical value; and (2) ovonic memory switching (OMS), which is a thermal-driven PC process.

10.2.1 RESISTANCE-BASED NUMERICAL MODEL

Currently, some companies and research institutes are concentrated in developing a practical thermal-induced phase transition model. Energy Conversion Devices Inc is the first group to develop a numerical simulator for PCM. The simulator includes three parts—phase change model, electric model, and thermal model: (1) the phase change model is described by the Johnson–Mehl–Avrami–Kolmogorov (JMAK) equation and Arrhenius's law, where the former is used for modeling the set process whereas the latter is used for the reset process; (2) the electric model can be obtained by Laplace's equation; and (3) the thermal part is modeled by the heat conduction equation. To solve the aforementioned coupled system, the finite difference method (FDM) is utilized. The equation system is as follows:

$$\nabla \cdot (\sigma \nabla V) = 0 \tag{10.1}$$

$$\frac{\rho c \partial T}{\partial t} = Q + \nabla \cdot (K \nabla T) \tag{10.2}$$

where Equation 10.1 is the Laplace equation and Equation 10.2 is the heat conduction equation; it's worth noting that the OTS effect is ignored here, while the amorphous and crystal germanium–antimony–tellurium (GST) I–V characteristics are approximately linear in the set state, so the Poisson equation is replaced by the Laplace Equation 10.1. The JMAK equation is used here for the PC model:

$$V_c = 1 - \exp\left(-\left(\frac{t}{t_0}\right)^p \exp\left(\frac{-E_a}{k_B T}\right)\right) \tag{10.3}$$

where
  V_c is a fraction of the crystal volume
  E_a is active energy
  p is the nucleation rate

After solving the thermoelectric coupled equation, temperature distribution is put into the JMAK equation, and a fraction of the crystal volume is obtained; then an equivalent resistance is obtained by a fraction of the crystal volume based on the percolation theory.

Figures 10.4 through 10.6 show programming current, fraction of crystal volume, and algorithm scheme obtained by the three respective models. Through the algorithm, the electrical field, temperature, and phase distribution are obtained, respectively. According to the phase distribution, the corresponding resistance is calculated. Except for using the JMAK equation to model the crystallization process, an erasable DVD's crystallization process is simulated using the nucleation/growth (N/G) theory and a good result is achieved. A classic N/G crystallization theory is proposed by Turnbull and Fisher based on the Becker–Doring theorem [25], and summarized by Kelton [26].

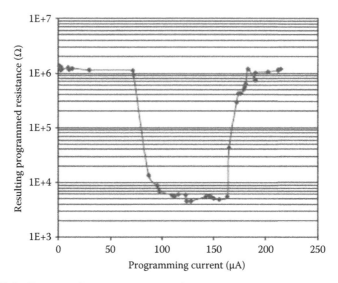

FIGURE 10.4 Programming current versus resistance.

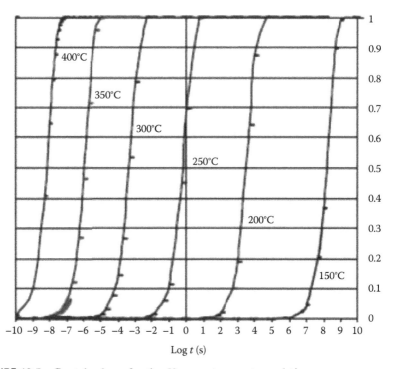

FIGURE 10.5 Crystal volume fraction V_c versus temperature relation.

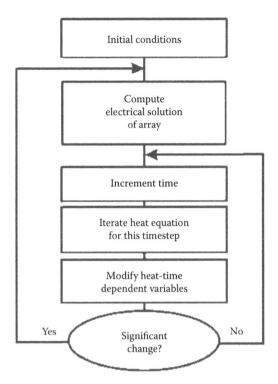

FIGURE 10.6 Algorithm scheme of the electrothermal simulation.

Stable crystal rate I^{ss} is ensured by Equation 10.4, which represents the nucleus number of unit time and volume. Nucleus size is a critical size n obtained by Equation 10.5 [27–28]:

$$I^{ss} = \frac{24Dn^{2/3}N_A}{Vol_{mol} \cdot \lambda^2} \left(\frac{|\Delta g|}{6\pi k_B Tn} \right)^{0.5} \exp\left(\frac{-\Delta G_n}{k_B T} \right) \tag{10.4}$$

$$n = \frac{32}{3} \frac{Vol_{mon}^2 \sigma^3}{\Delta g^3} \tag{10.5}$$

$$\Delta G_n = 4\pi r^2 \sigma - n\Delta g \tag{10.6}$$

$$\Delta g = Vol_{mon} \cdot \Delta H_f \frac{T_m - T}{T_m} \cdot \frac{7T}{T_m + 6T} \tag{10.7}$$

$$D = \frac{K_B T}{3\pi \lambda v} \tag{10.8}$$

The model is proposed on the assumption that the crystal nucleus is a sphere, where Δg is bulk-free energy difference of two phase molecules, σ is surface energy,

r is bulk radius, and n is number of molecules contained in crystal beam. Assuming that the crystal beam is a sphere, the total surface energy is $4\pi r^2\sigma$ and the total free energy is $n\Delta g$. So the total energy of the new phase is

$$\Delta G = 4\pi r^2\sigma - n\Delta g \tag{10.9}$$

Δg can be ensured by the empirical formula proposed by Singh and Holz:

$$\frac{\Delta g}{Vol_{mon}} = \Delta H_f \frac{T_m - T}{T_m}\left[\frac{7T}{T_m + 6T}\right] \tag{10.10}$$

where
T_m is the GST melt temperature
ΔH_f is the melting enthalpy

The smaller the bulk energy is, the more stable is the substance; therefore, the nucleation critical radius is obtained by making Equation 10.6 take on the maximum value:

$$\frac{d\Delta G}{dr} = 8\pi r\sigma - \frac{4\pi r^2\Delta g}{Vol_{mon}} = 0$$

The critical radius is obtained:

$$r_c = \frac{2\sigma Vol_{mon}}{\Delta g}$$

Using the aforementioned formula, ΔG is obtained as

$$\Delta G = \frac{16\pi}{3}\frac{Vol_{mon}{}^2\sigma^3}{\Delta g^2}$$

The expression of the crystal number of unit time and volume can be obtained by using ΔG in Equation 10.4, which is the expression of the nucleation rate. Figure 10.7 is the curve of the nucleation rate as a function of temperature from 300 to 850 K, and from the calculated result, we can see that the nucleation rate achieves a maximum when the temperature is 710 K.

Nuclei begin to grow after they are generated, as the temperature rises. Growth is a reversible process: some nuclei keep growing after they reach their critical volume if some molecules enter the nuclei; some nuclei disappear if some molecules escape from the nuclei. Thereby, the average growth rate is shown in Equation 10.11.

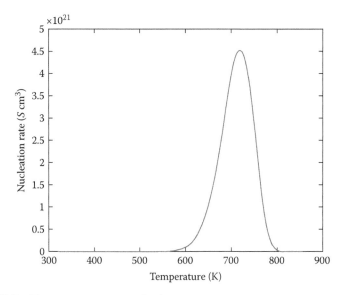

FIGURE 10.7 Temperature versus nucleation rate.

It is seen from Equation 10.11 that the growth rate is not only a function of the temperature, but also a function of the bundle radius r:

$$V_g = \frac{dr}{dt} = \frac{16D}{\lambda^2}\left(\frac{3Vol_{mon}}{4\pi}\right)^{1/3} \sin h\left[\frac{1}{2k_BT}\left(\Delta g - \frac{2\sigma}{r}Vol_{mon}\right)\right] \qquad (10.11)$$

Figure 10.8 is a curve that shows growth rate changes with temperature when the bundle radius is 1 nm. From Figure 10.8, we can observe that after the crystallization

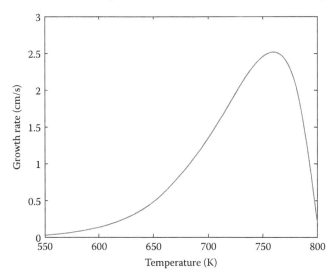

FIGURE 10.8 Growth rate versus temperature at $r = 1$ nm.

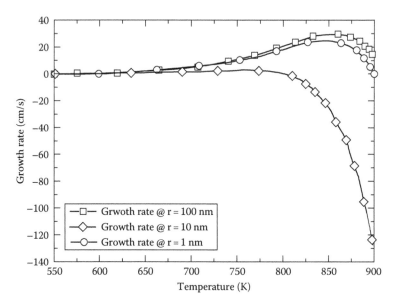

FIGURE 10.9 Growth rate versus temperature with different sizes.

rate achieves its maximum, the growth rate attains its maximum at about $T = 750$ K as the temperature increases. Figure 10.9 shows curves that demonstrate growth rate under different bundle radii. We can thus conclude that growth rate increases as bundle radius grows. At the same time, when $r = 1$ nm and the temperature exceeds 820 K, the growth rate becomes negative, which means that the nuclei begin to melt. According to the results obtained, it is important to choose an appropriate programming pulse to achieve appropriate temperature so that the PCM cell obtains a high programming speed.

Samsung's team [29] as well as the IMEC group [30] obtained PCM cell temperature distribution, phase distribution, and reset/set current curve as shown in Figures 10.10 through 10.12 by combining the N/G theory and the electrothermal model.

10.2.2 NUMERICAL MODEL BASED ON DRIFT-DIFFUSION THEORY

10.2.2.1 OTS Simulation

In the numerical model, the OMS effect has been modeled by the JMAK equation and the N/G theory coupled with the electrothermal model, but the OTS effect has been ignored. The OTS effect is performed by the filament effect based on statistics. These researches hold that current passing through amorphous GST material produces Joule heat that makes temperature rise; so very few nuclei appear in the amorphous material, and these nuclei cause a hot filament effect [31,32] and then the OTS effect occurs. But currently, the mechanism of OTS is an open issue. Alder [33] considered that OTS can be explained by the generation–recombination mechanism, ignoring the thermal effect. Pirovano's group [34] explained and verified Alder's view. This model is based on the impact ionization mechanism

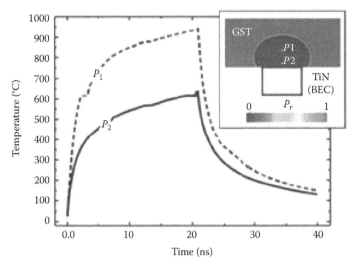

FIGURE 10.10   Temperature distribution.

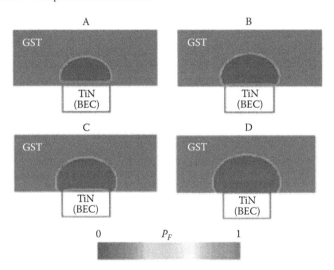

FIGURE 10.11   Phase distribution.

and the Shockley-Read-Hall (SRH) recombination mechanism. For amorphous GST, Poisson's equation and the drift-diffusion equation are

$$\nabla^2 \Psi = \frac{-q}{\varepsilon}\left(n - p + N_d - N_a\right) \tag{10.12}$$

$$-q\frac{\partial n}{\partial t} + \nabla \cdot \vec{J}_n = q\left(R_n - G_n\right)$$

$$q\frac{\partial p}{\partial t} + \nabla \cdot \vec{J}_p = -q\left(R_p - G_p\right) \tag{10.13a}$$

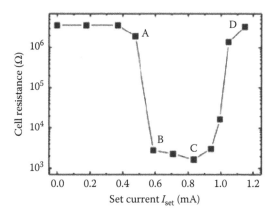

FIGURE 10.12 Set current.

$$c\frac{\partial T}{\partial t} = \nabla(k \cdot \nabla T) + Q \tag{10.13b}$$

$$Q = (\vec{J_n} + \vec{J_p}) \cdot \vec{E} \tag{10.13c}$$

where J_n and J_p are the current density formed by an electron and hole, respectively, and can be expressed by the following equations:

$$\vec{J_n} = -q\mu_n n \nabla \phi_n$$
$$\vec{J_p} = -q\mu_p p \nabla \phi_p \tag{10.14}$$

In Equation 10.14, ϕ_n and ϕ_p represent the electron and hole quasi-Fermi potential, respectively:

$$\phi_n = \frac{E_i}{q} + \frac{kT}{q}\ln\frac{n}{n_i}$$
$$\phi_p = \frac{E_i}{q} - \frac{kT}{q}\ln\frac{p}{n_i} \tag{10.15}$$

In Equations 10.12 and 10.13, G_n, R_n and G_p, R_p represent the electron and hole generation rate and recombination rate, respectively.

The amorphous GST subthreshold characteristics can be explained with the generation–recombination mechanism in a semiconductor: the semiconductor carrier generation uses the impact ionization model, and the impact ionization carrier generation rate is proportional with the electric field strength; carrier recombination uses the indirect composite model (RSH model), according to the energy band structure of A-GST. There are p-type and n-type traps that are formed by the defects in the band gap, and these traps form the RSH recombination centers of the RSH recombination mechanism. When the bias voltage is relatively low, the impact ionization generates a small number of carriers, all of which can be recombined, resulting in a balance between generation and recombination; meanwhile, as a localized state exists, the carrier mobility rate is relatively low, and thus current passing through A-GST is relatively low, although A-GST is in a high impedance state. When the voltage increases, carriers generated by impact ionization begin to increase gradually, and the current begins to increase as well. When the number of carriers generated begins to exceed the density of recombination centers, there are not enough recombination centers to keep the balance between generation and recombination. The generation rate has to reduce, that is, the voltage must reduce till it reaches the threshold voltage in the process of phase change of A-GST. The carrier generation rate and recombination rate can be determined by the following formula:

$$G_n = \alpha_n \frac{|\vec{J}_n|}{q}$$

$$G_p = \alpha_p \frac{|\vec{J}_p|}{q}$$

(10.16)

$$R_n = \frac{np - n_i^2}{\tau_{1p0}(n + n') + \tau_{1n0}(p + p')}$$

$$R_n = \frac{np - n_i^2}{\tau_{2p0}(n + n') + \tau_{2n0}(p + p')}$$

(10.17)

where α_n and α_p are electron and hole ionization coefficient, respectively, associated with the electric field strength, which can be determined by

$$\alpha_n = A_n \exp\left[-\left(\frac{B_n}{E}\right)^{\beta_n}\right]$$

$$\alpha_p = A_p \exp\left[-\left(\frac{B_p}{E}\right)^{\beta_p}\right]$$

(10.18)

where
  E is the electric field strength
  $A_n, A_p, B_n, B_p, \beta_n, \beta_p$ are constants associated with the material

For the following device structure (Figure 10.13 from the Master's thesis of Kailiang Lu of HKUST [35]), we obtain a V–I curve as shown in Figure 10.14.

The material parameters are taken from Pirovano et al. [34], without consideration of the electrothermal effect. But from the conclusion of these authors [34] we can see that the temperature has little effect on the subthreshold characteristics; thus, the threshold switching in the subthreshold region can be seen as a product in the dynamic balance of the competing generation and recombination carriers in a semiconductor,

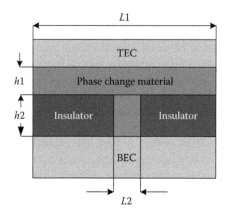

FIGURE 10.13 Device structure diagram.

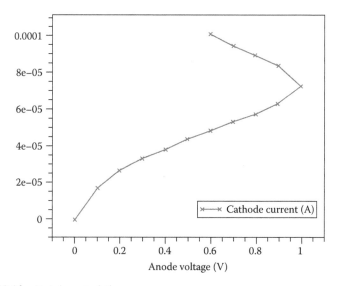

FIGURE 10.14 *V–I* characteristics curve.

which can replace the filament model [31,32], and avoid statistical analysis in modeling the A-GST subthreshold region. This model is more intuitive and can explain the threshold switching based more on physical characteristics. On the basis of the threshold model, a research group in Milan University of Milano [36] obtained a numerical simulator containing the OMS and OTS effects by self-consistently solving the drift-diffusion equation, the heat transfer equation, and the N/G theory.

10.2.2.2 OMS Simulation

Synopsys Inc. is currently working with Ovonyx Company to launch an TCAD tool [37,38] for PCM cell simulation. In Sentaurus Version A-2008.09, a resolution model of phase space mode was supplied. This model allows phase change simulation in more than two phases.

Suppose that there are N phases in a material; its probabilities s_1, s_2......s_N of becoming one certain phase satisfies the following equation:

$$\sum_i s_i = 1 \qquad (10.19)$$

For any two phases i and j, it allows any number of phase changes (in the use of capture rate and emission rate; i.e., the reaction rate between the various phases), and the dynamic equation is

$$\dot{s}_i = \sum_{j \neq i} \sum_{t \in T_{ij}} c_{ij} s_j - e_{ij} s_i \qquad (10.20)$$

where T_{ij} represents the phase change between phase i and phase j. In this phase change system, c and e represent capture rate and emission rate, respectively, which are a pair in the reciprocal process. Following the order $c = c_{ij} = e_{ji}$, $e = e_{ij} = c_{ji}$, the dynamic equation (10.20) can be written as

$$\dot{s} = Ts \qquad (10.21)$$

where
T is the total phase change matrix
s is the phase probability of each phase

According to the mechanism of PCM, one can consider that there are three phases: crystalline phase, amorphous phase, and liquid phase. The probability of their formation is, respectively, s_c and, s_a, s_m; thus

$$s_a = c_{ac} s_c - e_{ca} s_a + c_{am} s_m - e_{ma} s_a \qquad (10.22)$$

$$s_c = c_{ca} s_a - e_{ac} s_c + c_{mc} s_c - e_{cm} s_m \qquad (10.23)$$

$$s_m = 1 - s_a - s_c \qquad (10.24)$$

where the process of amorphous to crystalline phase change can be modeled by the N/G theory, the process of change from liquid to amorphous phase can be modeled by a linear relationship associated with a temperature change rate due to its fast response, and the liquid to crystalline phase change can be modeled by the Arrhenian law [39].

To model the thermal electric coupling effect between crystalline GST and amorphous GST, a heat-transfer model is used. Carrier distribution, and temperature distribution of crystalline GST and amorphous GST are obtained by self-consistently solving the Poisson equation, the continuity equation, and the heat conduction equation. Substituting the temperature distribution into the analytical phase change model, we get phase distribution and the equivalent resistance. As energy band structures of crystalline GST and amorphous GST are different, we establish an apparent band-edge shift (ABES) energy band model aiming at the crystalline and amorphous phase change to determine the carrier concentration.

According to Pirovano et al. [34], GST can be regarded as a p-type semiconductor, and the hole concentration can be described by the mobility of the valence-band edge:

$$p = N_v F_{1/2}\left(E_v - E_{F_p} - \Lambda\right) \tag{10.25}$$

$$\Lambda = \begin{cases} \lambda_0 \dfrac{T_g - T}{T_g - T_0}(1 - s_c), T \le T_g \\ 0, T > T_g \end{cases} \tag{10.26}$$

where
λ_0 is the band gap difference between amorphous and crystalline GST
T_g is the glass transition temperature (about 800 K)

Because hole is the majority carrier in GST, we can only consider the hole current [36], and the hole carrier mobility can be described by a linear function:

$$\mu_p^*(T) = s_c\mu_p^c(T) + \mu_p^a(T)(1 - s_c) \tag{10.27}$$

Using the algorithm in Equation 10.27, the phase distribution dependence on temperature and the transient response of phase distribution in reset, set, and read processes can be obtained.

Figures 10.15 through 10.17 show the steady-state and transient distribution of phase and temperature, respectively, when a reset pulse is applied. From Figures 10.15 and 10.16, we can see that the temperature reaches a maximum value of 1100 K, which is above the melting temperature, at the interface between heater and GST. In the melting region, the amorphous volume fraction reaches 0.98. Figure 10.17 shows the dynamic phase distribution when a 3 mA current pulse has a duration of 20 ns. From Figure 10.17, it is seen that the GST liquid phase fraction starts to increase at 2 ns, which means the internal part starts to melt and the volume fraction of the crystalline phase starts to decline. When the pulse ends and the falling time is 1 fs to reach the quenching rate of amorphization, the liquid fraction starts to decline, the amorphous phase fraction starts to increase, and the reset process is completed.

Note that because the model assumes that the initial GST is crystalline, even the reset process is completed. The volume fraction of the amorphous phase approaches 1 in the active region but far less than 1 (approaches 0.3 as showed in Figure 10.17) for the whole GST as a result of the fact that the active zone is concentrated in the interface of heater and GST, where the temperature reaches its maximum.

When the reset process is completed, we use the results (the phase distribution at the end of the reset process) as the input to start the modeling process; a pulse with an

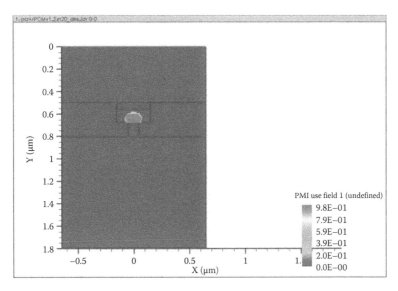

FIGURE 10.15 Amorphous phase distribution obtained at the end of a reset pulse.

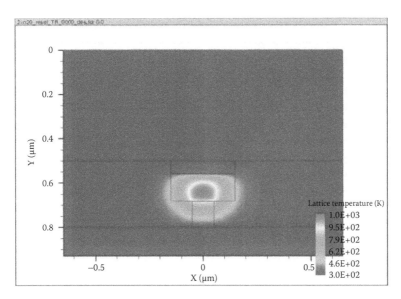

FIGURE 10.16 Temperature distribution obtained at the end of a reset pulse.

amplitude of 1.5 mA and a width of 1 μs is loaded on the top electrode. Figure 10.18 shows the phase distribution at the end of the pulse. As shown in Figure 10.19, the rates of the amorphous phase and crystalline phase decrease and increase, respectively, over time. When the pulse ends, the volume fraction of the crystalline phase is slightly smaller than 1 at about 0.925 due to residual amorphous GST in the reset process. To transform the GST completely into crystalline GST, a programming

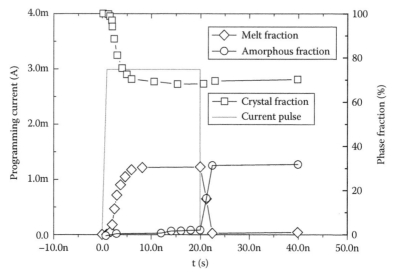

FIGURE 10.17 Dynamic distribution of the internal phase of GST obtained by applying a reset pulse.

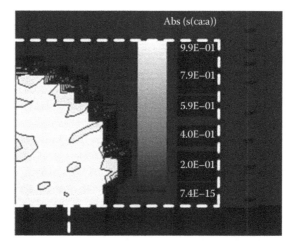

FIGURE 10.18 Amorphous phase distribution at the end of a set pulse.

pulse with a larger amplitude or width can be applied to increase the temperature of the active region to crystallize the residual amorphous GST completely. However, even without complete crystallization of the active region, the conductivity distribution also fully meets the needs of OMS. Figure 10.20 shows the resistance distribution, and 1.5 mA programming current can make a difference of up to two orders of magnitude between set and reset conductivity.

After the programming pulse is removed, a read pulse with an amplitude of 0.1 mA and a width of 1 μs is loaded. Figure 10.21 shows the temperature distribution

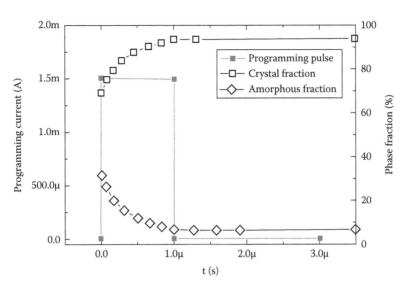

FIGURE 10.19 Dynamic phase distribution obtained by applying a set pulse.

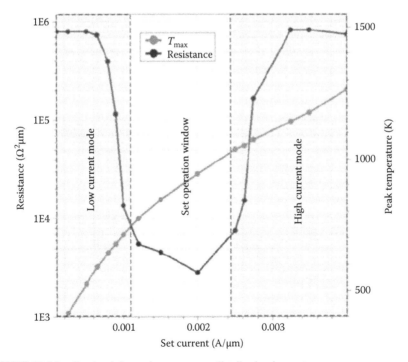

FIGURE 10.20 Conductivity and temperature distribution in a set process.

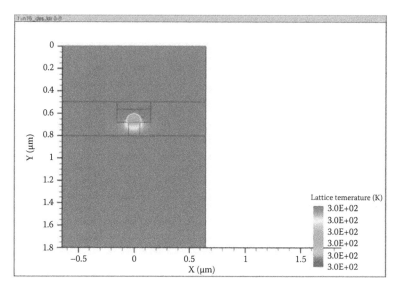

FIGURE 10.21   Temperature distribution in a read process.

obtained by applying a read pulse, and we can see that the read pulse is not enough to increase the temperature of the PCM cell. As a result, crystallization and error induced by the read pulse are avoided. Figure 10.22 shows the *I–V* feature curves in set, reset, and fully crystalline states. Under the same current, the voltage of the set state is larger than that of the fully crystalline state, indicating that the residual amorphous GST of the set state leads to an increase of resistance. The voltage of the

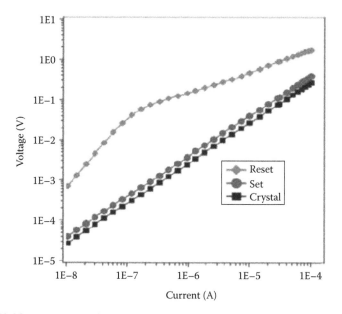

FIGURE 10.22   *I–V* curves of reset, set, and crystal processes.

TABLE 10.1
Comparison of the Three Models

Models	Electrical and Thermal Model	Crystallization Model	Phase Change Model
Level 1 (Ovonyx)	Laplace equation and heat conduction equation	JMAK equation	Percolation theory
Level 2 (Samsung)	Laplace equation and heat conduction equation	N/G theory	Percolation theory
Level 3 (Synopsys)	Drift-diffusion equation and heat conduction equation	N/G theory	Analytical phase-space model

reset state is two orders of magnitude larger than that of the set state, indicating that after the reset and set processes, OMS occurs and the difference in resistance meets the storage needs.

10.2.2.3 Summary

Based on the discussion in the previous sections, the modeling methods utilized in the three core models are summarized in Table 10.1.

The three models, from the electrical and thermal modeling, can be divided into two categories: an electrical and thermal coupling model based on electrical conductivity and a coupling model based on the drift-diffusion theory. The former considers the relationship between V and I as approximately linear. As a result, the Laplace equation rather than the Poisson equation is utilized. While the electrical and thermal coupling model based on the drift-diffusion theory is in strict accordance with the semiconductor theory, and the self-consistent solution to the Poisson equation, the current continuity equation and the heat conduction equation are obtained, where the latter is more accurate than the former in terms of electrical and thermal model. The crystallization model includes the JMAK equation and the N/G theory. However, the model does not well reflect the effect of OTS, and is entirely a thermally induced crystallization model. OMS is controversial in terms of numerical simulation of the current PCM cell, and is one main contribution of our research work.

10.3 PCM CELL COMPACT/SPICE MODEL

The establishment of an accurate and effective cell model is key for PCM to achieve the circuit-level design simulation. In recent years, despite many PCM research directions, the SPICE models proposed are mostly based on behavior characteristics and only a few models are recognized by the industry. Therefore, the research on models, especially the analytical model based on physical significance, will be the focus of future research.

A practical device model should include both the reading and writing features when working and the storage features when out of work. The reading and writing features include the respective I–V characteristics of crystalline and amorphous phases and the phase transition properties, and should cover electrical, thermal, and crystallization kinetics and other factors. According to the study of mechanism of the device, the models can be divided into analytical models and behavioral models.

According to the measured device characteristics curve, behavioral models obtain their results from the perspective of behaviors, and only a few parameters are related to the device physics, while most depend on the results of a particular process, so they do not have good adaptability. However, behavioral models can be quickly established, and it is possible to create a high-quality model for industrial applications by adding some physical significance to the following research. At present, there is a lot of research on semibehavioral models with a certain physical meaning, such as the Italy STMicroelectronics group [40], University of Exeter, UK [41], Japan OSAKA University group [42], Taiwan Ilan University group [43], etc. We should also recognize that behavioral models must be built on an experimental basis, and cannot physically reflect the relationship between the material properties, device structures, and electrical characteristics, thus limiting their use.

The analytical models explain the physical behavior of the device, that is, its operating characteristics including parameters and expression with a real physical meaning. The core of these models does not depend on process variation, and consistent models with good adaptability can be quickly obtained according to the parameters extracted from the relevant process. However, due to the amorphous semiconductor involved in PCM, no accurate interpretation of its physical properties has been given, and only a few reports are available about analytical models. Intel and STMicroelectronics research groups alone have a number of useful studies in this respect [44]. To establish analytical models, studies on the electrical and thermal properties of crystalline and amorphous GST and on the characteristics of the phase transition are necessary.

We can study the characteristics of analytical models from the aspects shown in Figure 10.22.

10.3.1 IMPEDANCE MODEL

The impedance model is mainly used to study the DC and AC *IV* characteristics of a PCM cell. The DC characteristics are the basic characteristics of PCM devices. Since Ovshinsky proposed the concept of PCM and gave the basic DC *IV* characteristics in 1968 [3], study on the DC model of PCM has been in progress. The most basic DC feature is the *IV* characteristics, and the experimental results are shown in Figure 10.23.

In the crystalline set state, the resistance is small and the current increases linearly with voltage. But in the amorphous reset state, the initial resistance is large. In the set state, with lower resistance, current increases linearly with voltage. However, in the reset state, when initial resistance is higher, current shows little change with varying voltages until the voltage rises to the threshold voltage V_{th}, when the device will undergo threshold conversion. At the same time, the current increases to the holding current I_h, and the voltage drops quickly to the keeping voltage V_h to achieve stability, and this is the so-called OTS effect. Thereafter, the voltage of the material continues to increase, with increasing current, till it reaches the molten state. Two conditions of the *IV* curve can overlap in the molten state. As considered in the programming process, controlling the programming pulse amplitude and pulse

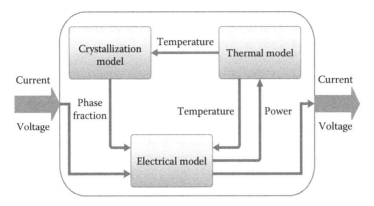

FIGURE 10.23 Research directions of PCM analytical models.

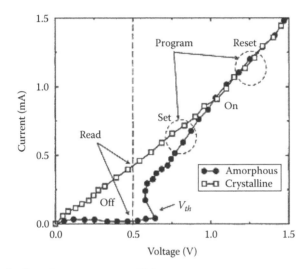

FIGURE 10.24 Experiment to measure *IV* data.

width can achieve the switching between two states. Because of the clarity of the DC characteristic of the crystal semiconductor, we lay emphasis on the switching voltage, holding current I_h, and holding voltage V_h of the amorphous state (Figure 10.24).

As for the OTS effect, there are some forming mechanisms that need explanation, including the current filament theory [46], collision ionization theory [35], carrier generation [47], recombination theory, and deep tunneling in carrier theory [45]. But none can interpret the OTS effect perfectly, as most models still remain at the numerical simulation level. Only the tunneling mechanism can offer an incomplete analytical model. In this theory, with the combination of a high electric field, the Poole–Frenkle effect [48], and the carrier hopping theory [49], the threshold *I–V* characteristics are achieved:

$$I_{tot} = I_f - I_r = 2qAN_T \frac{\Delta z}{\tau_0} e^{-\frac{E_C - E_F}{kT}} \sin h\left(\frac{q\Delta z}{kT} \frac{V}{u_a}\right) \tag{10.28}$$

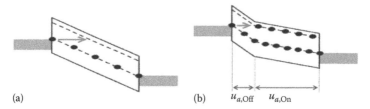

FIGURE 10.25 Potential profile and energy distribution in low-current state (a) and high-current (b).

When the OTS effect occurs, as showed in Figure 10.25, carriers at deep levels tunnel to the shallow energy level, forming a high carrier concentration conductivity, which is called the on area. But in the off area, the current is sustained by both tunneling current and initial hopping current:

$$I = 2qAN_T \frac{\Delta z}{\tau_0} \exp\left(-\frac{E_C - E_F}{kT}\right) \sin h\left(\frac{qF}{kT} \frac{\Delta z}{2}\right) \left[1 + \frac{G_{TUN}\tau_n}{N_T} \exp\left(\frac{E_{T2} - E_F}{kT}\right)\right] \quad (10.29)$$

Based on this conduction theory, the data of device size and parameter, and amorphous *I–V* testing from Shanghai microsystem research institution, we have completed the reset state DC modeling work. As showed in Figure 10.26, the simulation results fit the measured data well.

Although DC characteristics of the mechanism are not clear enough to meet the production needs of the industry, the main research has set up a file in the unit model's specific electrical properties, but not at the complex base of the physical mechanism, which is too much of a struggle. However, we still have much work to do to continue our study of the physical mechanism, and we will need DC-related parameters of the actual measurement results as well as the variation trend with the physical properties to have a reliable forecast, especially in aspects of AC reading and writing characteristics [49].

AC characteristics of the research include the delay time of the threshold switching *td*, the switching time *ts*, and the recovery time after switching *tr*, as well as the noise characteristics interference to the read operation. All these have a very big effect on the devices' work speed.

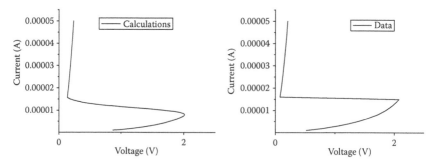

FIGURE 10.26 Simulation data and measured data.

As early as Ovshinsky put forward his finding that threshold switching in PCM devices undergoes a certain delay time, Adler also described the threshold switching delay phenomenon in his research [33]. The experimental results show that the delay time *td* is of inverse proportion function when the outer voltage ΔV is higher than the threshold voltage V_{th}. And *td* disperses greatly when the outer voltage is removed from the threshold voltage. As shown in Figure 10.20, although the outer voltage is lower than V_{th}, it can also trigger the switching.

Lacaita's group further discussed the physically based delay time in the International Electron Devices Meeting (IEDM) recently [43]. They proposed to explain the V_{th} delay phenomenon by the noise statistics theory, and values far higher than V_{th} were explained by load RC delay, thus finally unifying the fittings. As in Figure 10.27, switching time τ_s and recovery time τ_r still need further investigation, and we can only provide some experimental parameters obtained from the test (Figure 10.28).

Research of transient parameters is helpful to identify devices' work speed. We can see that the order of magnitude of the delay time is small, but as the speed

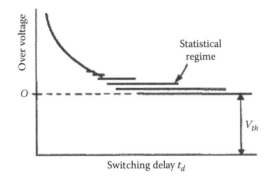

FIGURE 10.27 Relations between delay time and voltage.

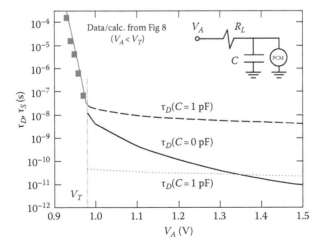

FIGURE 10.28 OTS delay time and switching time.

requirements of the device improve, its effects will have more and more influence on the whole programming time. Moreover, the switching time and keeping time, which affect the relationship between the work and the device's error correction ability, also need much attention in our research.

10.3.2 TEMPERATURE DISTRIBUTION MODEL AND HEAT GENERATION TRANSMISSION MODEL

This part is related to the thermal model and focuses on studying devices with the electric heat coupling effect and the conduction of temperature distribution.

The calculation of temperature is always done using the joule theory and the heat transfer equation, through a numerical analysis for the temperature distribution of the device. But, in fact, because the coupling of the electric heat is too complex to operate, as especially after involving phase change it brings new crystallization and uncertainty factors, there is never an ideal solution. We will divide the thermal model into a temperature distribution model and a heat producing transmission model to make the research more effective. The heat producing transmission model is just a method to obtain the temperature distribution in the analytical model, which is the ultimate goal.

There is generally no dispute in heat generation. As long as we get the current and response state of the resistance value, the value of heat can be generated by the Joule theory. The problem lies in the quantity of heat conduction, and studies show that there is thermal boundary resistance (TBR) between the GST materials and the heater and oxide [50], which will affect the conduction of the heat and result in a change of temperature distribution, as shown in Figures 10.29 and 10.30.

Obviously, considering the TBR, the temperature distribution changes significantly, and this will have a significant impact on the crystal situation. At the same time, it will also change the current programming thermal efficiency. At present,

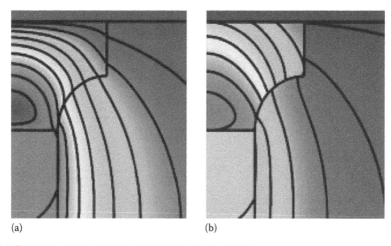

(a)   (b)

FIGURE 10.29   (a) No TBR included (b) including TBR.

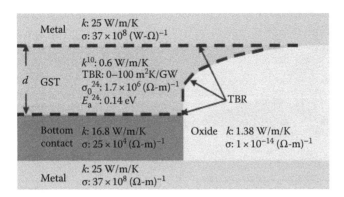

FIGURE 10.30   TBR between different boundaries.

it is known that TBR can decrease the reset current [51]. However, as shown in Figure 10.28, the value of the TBR ranges between 1 and 100 $m^2 \cdot K/GW$. The method to get the exact TBR values and analysis is to conduct further research into the impact factors.

The research methods are mostly based on numerical analysis, but the numerical model is still needed in the actual application to achieve higher efficiency, which is also one of the ultimate goals. IBM put forward an analytical numerical model aimed at the reset process [52], where the model chooses a planar structure and makes a semicircle approximation of the phase change area to obtain a three-dimensional temperature distribution, as showed in Figure 10.31. It also analyzes the dynamic resistance and reset by the influence of the temperature of current. However, this model does not include the influence of TBR, so the temperature distribution is not reliable enough. A team from the University of Illinois put forward a TBR containing the effect of temperature distribution based on the theory of the nanowire

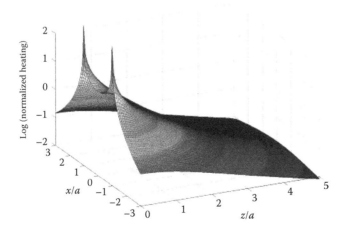

FIGURE 10.31   Temperature distribution of planar structure.

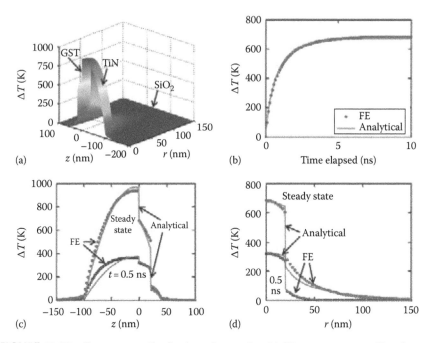

FIGURE 10.32 Temperature distribution of nanowire. (a) 3D temperature profile of nanowire structure PCM cell, (b) transient temperature profile at hot point in GST, (c) Axial-direction temperature profile along the symmetry axis at $t=0.5$ ns, and (d) radial-direction temperature profile at the GST-heater interface.

analytical model [53]. As shown in Figure 10.32, they also analyzed the effect of relative current and resistance.

Overall, there is still plenty of work to be done on the temperature distribution in an analytical model, and the current goal is to combine the generation of heat transmission to get a general structure of the temperature analytical model of the PCM unit taking the TBR effect into consideration.

10.3.3 CRYSTALLIZATION MODEL

The crystallization model is included in the heat distribution model on the basis of the analysis of the phase transition process of the PCM device, with the aim of getting results of the OMS effect.

As to the crystallization process, we follow the N/G theory and the JMAK theory. The N/G theory is the crystal growth theory, where according to the corresponding temperature, one must calculate every change in speed of crystallization and nucleation, in order to get a more precise phase change distribution. This kind of calculation process is, however, relatively complex, and most of these methods are used for numerical analysis. The JMAK equation uses temperature, time, and heat to activate energy to calculate the proportion of material crystallization, in order to obtain phase change statistical distribution, which is based on the composition of amorphous state and amorphous calculated resistance value. The advantage of this method is that it can be calculated easily from the overall consideration of getting the device's general crystallization fraction. The principle of the numerical simulation has been already stated in the mentioned session, so it is not discussed here.

Most compact models utilize the JMAK theory [54,55]; only the OSAKA university found a compact model based on the crystal growth theory [42]. In the current models, a common device is recognized as a whole. We process the components by the temperature unity, but this method ignores the internal phase change of the distribution. Our work focuses on the analytical model of temperature to get the distribution of the analytical results by phase change. The influence of resistivity and heat resistance conditions will be a feedback to the impedance model and the heat transfer model.

This work is related to the physical level of a steady-state solution for crystallization while in the practical application of the device, the OMS effect mainly concerns the reset current value, the programming process time, and the final resistance. The reset current Ireset in the PCM cell is the largest in the current working devices, and this value directly affects the device's power consumption; lowering the Ireset is also a requirement for the development of PCM. But reducing the Ireset in general will reduce the heat efficiency and extend the programming time. The programming time of the device will influence the write speed, which is critical in high-speed applications for the near future. At the other end of the programming, the current passes from a high value down to 0 in the process of crystallization, causing heavy crystallization of GST materials. If the quenching time tq is too long, and the finally formed amorphous resistance is insufficient, the value of Rreset/set decreases, which may cause programming failure and also affect the development of multiple values stored for the future.

10.3.4 MEMORY MODEL

The storage model mainly concerns the device's data-keeping stability and the problems about the reliability of the device's life. Due to the instability of the amorphous structure, its resistance can change. The results obtained from a burn-in test at high temperature are showed in Figure 10.31 [56]. One can observe that the reset resistance increased slightly in the initial period, and then gradually declined until failure. Current research thinks that the change in amorphous resistance is caused by structure relaxation (SR) and crystal growth (N/G), which work together [50,51,66,67].

SR is the concentration of the trap phenomenon that tends to reduce in amorphous GST materials because of the requirements of free energy reduction. Because the amorphous current conduction mainly relies on the trap carriers in the jumping conduction, reducing the concentration of the trap can cause the concentration of the charged carriers to reduce, thus increasing the average distance between traps. As a result, the current declines. The relationship between the current and voltage changes from the Poole distribution to the PF distribution. [52]. According to Figure 10.33, the equation of the resistance is

$$R = R_0 \left(\frac{t}{t_0 t_0} \right)^\nu \tag{10.30}$$

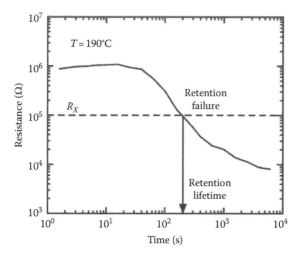

FIGURE 10.33 Data-keeping curve.

The crystal growth (N/G) theory has been discussed in the related part of thermology. During the analysis of data-keeping issues, we mainly focused on the data failure mechanism under different growth rates Vg. As shown in Figure 10.34, the crystallization is dominated at low vg. Large nuclei are formed in the amorphous GST state. The current transmission is carried out abiding the percolation theory; at medium vg, the crystallization and growth occur simultaneously; when vg becomes higher, growth plays a major part, and the device resistance is equivalent to connecting the crystalline and amorphous states in series as shown in Figure 10.35.

SR and crystal growth are a pair of opposite physical mechanisms: the former increases resistance while the latter decreases resistance. Generally speaking, SR has lower activation energy and occurs more easily at room temperature, while crystal growth requires higher activation energy and lasts for a long time. Therefore, the

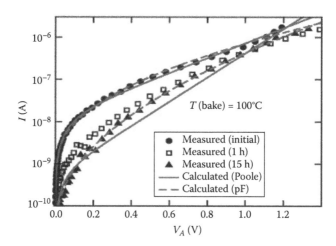

FIGURE 10.34 SR effect of current.

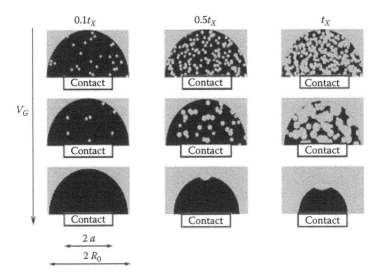

FIGURE 10.35 Crystal profile with different growth rate.

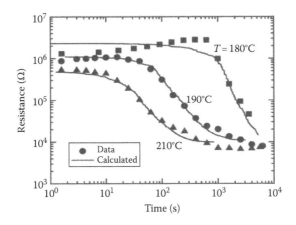

FIGURE 10.36 Data retention curves at growth rates.

device's initial resistance is increased with the effects of SR, while in the last stage of the process, resistance degradation becomes slower. However, when the temperature is high, crystal growth will have a stronger offset on SR, as shown in Figure 10.34; the SR phenomenon at high temperature is not obvious (as shown in Figure 10.36).

Temperature change in the device under nonphase conditions is mainly caused by the read current; the greater the current, the higher is the temperature. According to the data-maintaining theory, we need to optimize read current data; while maintaining data stability, we should be aware that drift does not occur, and at the same time ensure an adequate working life. Therefore, our model should be based on the corresponding theory of storage to meet the prediction of the device's life, and provide methods to optimize its lifetime. As the data retention and reliability issues are closely related, more in-depth theoretical analysis will be discussed in the reliability model section.

The models discussed previously are not entirely independent of each other, because each model parameter change will affect the results of the other models. Our work aims to first establish an analytical model, and then unify its parameters to get a self-consistent compact model.

10.3.5 MODEL DESCRIPTION AND RESULTS

The material of the PCM cell is chalcogenide Ge2Sb2Te5 with two different reversible states. One is the set state exhibiting low resistance, while the other is the reset state exhibiting high resistance. Different programming pulses are able to induce the transformation between the two states by Joule heating. A high and short current pulse forms a reset state while a low and long current pulse forms a set state. Hence, the electrical model is divided into two parts: set and reset, which are connected to each other in series.

The set state is mainly made up of polycrystalline GST material, whose electrical characteristics can be calculated as for a conventional crystalline semiconductor. The structure of crystalline Ge2Sb2Te5 is proven to be face-centered cubic (FCC) structure, in which 20% of Ge-Sb atomic sites are vacancies, forming accepter-like traps. So the material is seen as the p-type semiconductor and the current in the electrical field is dominated by hole transport. The current under a voltage V_c is

$$I_c = \frac{S}{L}\sigma_c V = \frac{S}{L}qN_v\mu_h V_c \tag{10.31}$$

where
  N_v is the hole state concentration relative to the vacancies
  μ_h is the mobility of holes
  S and L are cross section area and length, respectively

The PCM cell in the reset state is mainly made up of amorphous GST material with a high density of defects, for which the Fermi level EF is pinned in the midgap. Considering the high density of traps and few free carriers, the hopping theory is introduced to explain the transport of carriers in the amorphous state. Ielmini's work has proved [3] that carriers hopping forward and backward under a bias voltage leads to a hyperbolic sine I–V relationship:

$$I_a = 2qSP_t\frac{R}{t_0}\exp\left(-\frac{\Delta E_F}{kT}\right)\sin h\left(\frac{qV_a}{kT}\frac{R}{2L}\right) \tag{10.32}$$

where
  P_t is the total concentration of traps
  R is the average distance between traps
  t_0 is the free time of located carriers
  ΔE_F is the height of the barrier

The I–V relationship exhibits linearly under low voltage and exponentially under higher voltage in the subthreshold region. When the electric field reaches the critical field Ft, threshold switching happens and resistance decreases to a small value.

10.3.5.1 Temperature Model

The temperature model is used to detect the peak temperature of a PCM device. The temperature is calculated by the margin of generation and dissipation powers [4]. The generation power is the sum of the set part and reset part powers, with an effective coefficient β, given as

$$W_g = \beta(I_a V_a + I_c V_c) \tag{10.33}$$

The dissipation power is calculated by the thermal conduction equation, given as

$$W_d = \frac{\partial Q}{\partial T} = \frac{k\Delta T}{R_t} S_t \tag{10.34}$$

where
   k is the thermal conductivity
   R_t is the average distance from the heating point to room temperature environment, which is linear to the temperature
   S_t is the thermal cross section relative to R_t

The temperature increment ΔT from the environment is

$$\Delta T = \int_{t0}^{t1} \frac{W_g - W_d}{C V_0} dt = \frac{W_g R_t}{\kappa S_t}\left[1 - \exp\left(-\frac{\kappa S_t}{R_t V_0 C} t\right)\right] \tag{10.35}$$

where
   C is the heat capacity of the GST material
   V_0 is the volume of the effective heating region

10.3.5.2 Crystallization Model

The crystallization model decides the fraction of crystalline and amorphous region, and feeds it back into the electrical model. In this model, the JMA equation is used to calculate the fraction of crystalline composition in the active region by temperature and time [5]. The crystalline fraction is given as

$$C_{rate} = 1 - \exp\left[-\frac{t}{t_0}\exp\left(-\frac{E_a}{kT}\right)\right] \tag{10.36}$$

where
   E_a is the activation energy
   t_0 is the time constant

The crystallization process takes place only when the temperature is between glass point T_g and molten point T_m. The material state remains unchanged when the

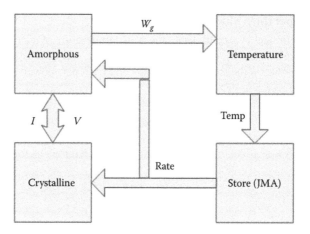

FIGURE 10.37 Complete model of the PCM cell.

temperature is lower than T_g, while it changes into molten state when the temperature is higher than T_m. As crystallization occurs during the process when the temperature decreases from higher than T_m to lower than T_g, rapid cooling is needed in reset programming in order to form the amorphous state.

10.3.5.3 Model Integration

The complete model is shown in Figure 10.37, combined with each of the modules discussed previously and implemented in Verilog-A code. The circuit-level simulation is done by Hspice, and the results are given in the following section.

10.3.5.4 Results

The DC analysis gives out the I–V characteristics of this model shown in Figure 10.38. On the conditions of a different crystalline fraction, an ~0–50 µA current is applied. We can see the different threshold voltages in each condition.

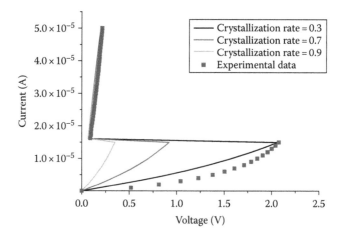

FIGURE 10.38 I–V characteristics of the model with different crystalline fractions.

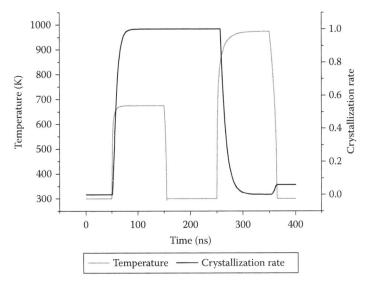

FIGURE 10.39 Temperature and crystallization rate in the set and reset programming processes.

In order to emulate the transient programming operation, the set and reset pulses are 1.25 and 2.25 mA, respectively, with the same width 100 ns, and are applied on the model in succession. The initial state is set be amorphous. From Figure 10.39, we can see the temperature and crystallization rate change as time goes by.

During the set and reset programming processes, the temperature and crystalline fraction is shown in Figures 10.39 and 10.40, respectively. In the set process, we can see that the temperature is maintained between T_g and T_m, and a higher set current

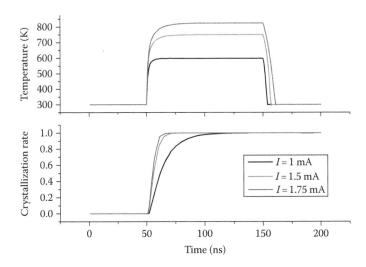

FIGURE 10.40 Temperature and crystallization rate under different set currents.

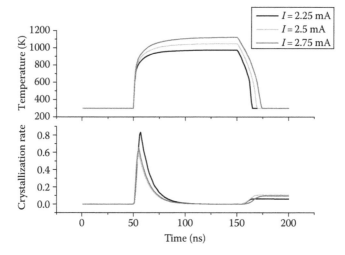

FIGURE 10.41   Temperature and crystallization rate under different reset currents.

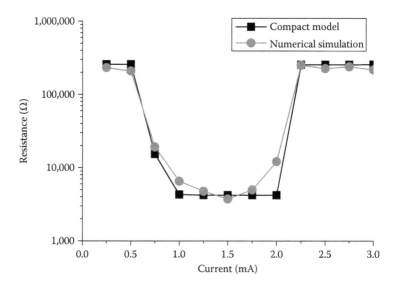

FIGURE 10.42   Resistance–current curve under different programming currents.

leads to faster crystallization. In the reset process, when the temperature exceeds T_m, the crystalline fraction drops to 0, but recrystallization occurs as the dropping temperature crosses the range between T_m and T_g. So a little crystalline material still exists in the cell, but does not affect the resistance much (Figure 10.41).

The R–I curve is shown in Figure 10.42. Because of partial crystallization, the resistance decreases gradually as the set current increases. The simulation result has already been calibrated by experimental data and numerical simulations.

10.4 RELIABILITY MODEL

Theoretically, PCM has a unique advantage over the traditional Flash in speed, power, read and write cycles (endurance), and data retention (retention) because of its compatibility with the standard CMOS process and it could be proportionally scaled down, thus making it an expected replacement for Flash memory in the discrete and embedded NVM market and could gradually enter the mainstream.

Nevertheless, in the actual research process, PCM reliability is indeed a challenging problem, as it is not yet a practical physics-based reliable model that can ensure data retention, read and write reliability, and life prediction, so a strict, precise (mostly based on physical) reliable model is of great significance.

In general, PCM reliability mainly involves three aspects: (1) data retention; (2) read and write reliability (reading and programming disturb); and (3) read and write cycles (endurance).

10.4.1 DATA RETENTION MODEL

The University of Milan and STMicroelectronics have done a lot of meaningful work on the reliability of the PCM [36]. In the field of data retention, they used high temperature to accelerate failure mechanisms, and combining the Arrhenius distribution, they predicted a 10-year life of the working temperature, with favorable results. However, for large multimegabit memory cell arrays, it is worth studying the statistical distribution of retention [57,58]; we can also use the statistical distribution to study the data-holding time.

According to the PCM storage mechanism, we could use different conductivity of phase change materials in the crystalline phase and amorphous phase to obtain data retention. Take, for example, the GST material. When c-GST's Gibbs free energy is minimum, its structure is the most stable, and according to the principle of minimum free energy, the structure of a-GST is in its metastable state, so when the PCM cell is in the reset state, the a-GST will always have a tendency of converting to c-GST. Therefore, PCM data retention is actually about the amorphous state phase-change material's converting problem under the influence of accumulation time and temperature. This research can be divided into two areas: first, the crystallization of the a-GST region, that is, at a specific temperature condition, there will be the creation and growth of nuclei with the time accumulated. So PCM data retention could be studied by the law of percolation theory and statistics [59,60]. The second method could be by studying the SR effect and its impact on data retention [50,51]. According to the principle of minimum free energy, the atomic arrangement in amorphous semiconductors has gradually fallen into order with the cumulative effect of time, so that the elevated levels of atoms arranged in amorphous semiconductors result in a lower internal density of trap states. According to the Poole–Franke theory, the decrease in trap density could result in decreased conductivity, and this has also become an influential factor in PCM data retention. The impact of crystallization and growth on data retention can be discussed under the frame of N/G theory and percolation theory. Ugo Russo [61] has studied the data retention problems caused by crystallization/growth and proposed a physics-based data retention model. Crystallization is a process of nucleation and growth of the nuclei. Nucleation and growth is a function of time and temperature. At a specific temperature, the crystalline phase and amorphous phase change in proportion over time, resulting in decreased conductivity; when conductivity drops to a certain point, this will lead to 1 bit data failure. The data retention problem is essentially about the dynamic calculation of the proportion of the crystalline phase and amorphous phase.

In short, the source of the problem in PCM data retention lies in its metastable amorphous properties. Two related continuous changes of mechanism are the crystallization/growth and the SR. The SR is the atomic-level reconstruction resulting in short-term order, but long-term disorder. Therefore, the SR effect is reversible atomic level transition resulting in a transition from short-term disorder to short-term order, and the crystalization is lattice restructure from short-term disorder to long-term disorder. Reliability issues caused by crystal growth/SR effect are essentially related to the band structure and the material itself; the greater the band gap, the greater is the energy required to stimulate the local atomic reorganization, which means crystallization becomes more difficult, indicating that the phase change material is more reliable, but this also leads to the corresponding increase in the programmed current. By changing the stoichiometry of GST, and appropriately increasing the high activation energy (band gap can also be considered noncrystalline) components, we can extend the data-maintaining life. So we need to compromise between reliability and power cost. The development of new materials as well as the modification of existing materials have been reported at present [19,63].

10.4.2 READ AND WRITE RELIABILITY MODEL

Read reliability is the relationship between the threshold voltage and the read pulse. We need to prevent the misuse of the phase change unit by the read signal such as noise-induced threshold conversion. Write reliability is mainly about the programming efficiency affected by delay time and phase transition time. For the reset process, the phase change unit, when coupled with a pulse, goes through three periods: the delay time (τ_D), the conversion time (τ_S), and the recovery time (τ_R), as shown in Figure 10.43.

τ_D is closely related to reliability; if the reading pulse width is smaller than τ_D, we could avoid the misuse caused by noise, as shown in Figure 10.44 [56]. The relation between τ_D and the voltage applied has been obtained by using statistical laws and experiment results. The greater the voltage, the smaller is τ_D. At the same time, τ_D is also connected with the circuit parasitic parameters; in nonideal circumstances, the bit line will have parasitic capacitance, and τ_D will increase with parasitic capacitance. As shown in Figure 10.45, however, there is no effect of τ_S on parasitic capacitance. τ_R is the recovery time, which is the required time to reach a steady-state equilibrium in the phase-change material at the end of the pulse. During this time, parasitic capacitance will discharge and read is not allowed. Before the balance is reached, a small proportion of the amorphous phase still exists, resulting in a drop of threshold voltage and decrease in electrical resistivity, as shown in Figure 10.46. If a reading operation is loaded at this moment, the probability of the misoperation of

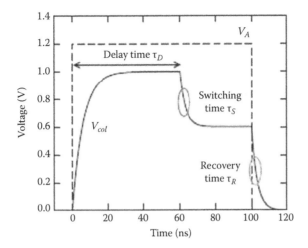

FIGURE 10.43 Write pulse and phase-change time.

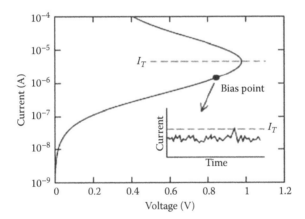

FIGURE 10.44 Impact of noise on the read signal.

reading is increased. When the bias voltage exceeds the threshold voltage, the PCM unit of the bias voltage is suddenly decreased, and the parasitic resistance discharges to the memory cell, and then enters the recovery process. Such a repeat leads to the oscillation phenomenon of conversion to restore balance, as shown in Figure 10.47.

τ_S and τ_D also determine the PCM cell's programming speed, and therefore the study is not only related to reliability but also plays an important role in programming speed optimization in PCM cells. However, the mechanism is still not clear to establish a model of τ_D, so further research is needed in this area. Reifenberg et al. [50] describe the reasons for τ_R and its impact on the reliability of reading, but there is no further description of the mechanism. The formation of τ_R is related to the dynamic balance between the crystalline phase and the amorphous phase. Therefore, phase space model as well as the N/G model theory and Arrhenius's law can be used to build the model and we could base our discussion on the crystalline/amorphous phase/liquid phase equilibrium in phase space.

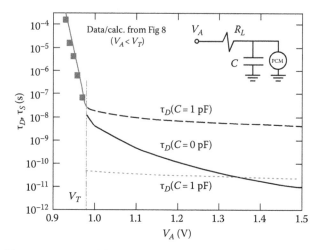

FIGURE 10.45 τ_D versus parasitic capacitance.

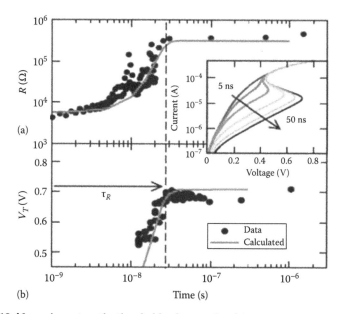

FIGURE 10.46 τ_R impact on the threshold voltage and resistance.

The study on read and write reliability was based on PCM cell transient simulation. It is completely possible to integrate read and write reliability models into numerical simulation models and achieve an analysis of read and write reliability from the structure to the circuit in the PCM cell.

10.4.3 READ AND WRITE CYCLES MODEL

Currently, the forecast for read and write cycles is based on experimental data. The PCM cell is put through repeated read and write operations, and when the failure

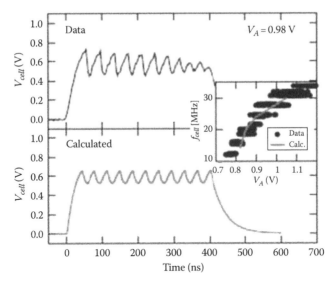

FIGURE 10.47   SR oscillation.

rate reaches a certain value, the number of read and write operations is considered as the number of read and write cycles. However, this method is based on a large number of experiments, which are not only greatly discrepant, but also very time-consuming. More importantly, there is no effective bridge connecting read-write cycles and device materials and structures. Therefore, there is a need to establish a model of read and write cycles.

The read and write cycles are mainly associated with material fatigue and related processes and structures. In the set and reset processes of PCM, the structure undergoes high-temperature annealing and quenching, which entails a slow rise in temperature followed by a sharp temperature decline. After these severe temperature changes, thermal stress will be produced within the material. There is as yet no model simulating the thermal stress–induced fatigue effects on the read and write cycles. With regard to the electric–thermal stress coupled simulation, Gille and coworkers [68] carried out an electric–thermal stress coupled simulation on the horizontal structure of PCM. The simulation results show the existence of >5% volume change during phase transitions, which will introduce 1.5 Gpa stress in theory and will continue to increase over time. The trend of volume change is shown in Figure 10.48, and the change of stress with time is shown in Figure 10.49. There is a large stress and high frequency load for the cell to bear during the PCM life cycle. The fatigue issue is an important factor in read–write cycles, so the establishment of a thermal stress failure model is the next major task.

While considering the device structure and process, thermal stress also plays an important role in the hot electrode heater and the contact quality of the phase-change GST material, as well as the GST interdiffusion with neighboring materials to avoid instability—in other words, the PCM circuit failure (open-mode failure) and PCM device breakdown or short circuit fault (short-mode failure). In addition, read and write reliability is mainly related to the thermal cross talk during the writing process

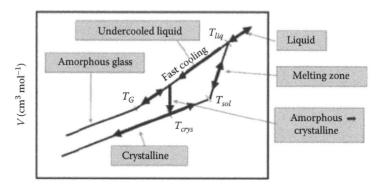

FIGURE 10.48   Volume change process.

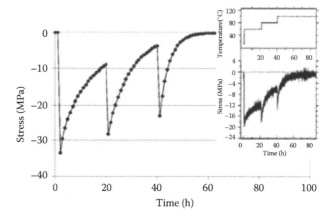

FIGURE 10.49   Changes of stress with time.

(cross talk) and the phase transition caused by local excessive heat in the reading stage. Thermal cross talk is mainly affected by technology and feature size. There are reports stating that thermal cross talk is negligible under the 65 nm node [64], but thermal cross talk could become a prominent issue above the 32 nm node [65].

10.5   CONCLUSION

Though PCM has a bright future as the candidate for next-generation NVM, it faces challenges ranging from an understanding of its physics to engineering application. From the perspective of device physics, the mechanism of current transport is not clearly known, and it relies on an empirical model. Moreover, research has been done on individual aspects only. From the point of view of engineering application, the instability of the process makes the properties of the materials unstable; as a result, the devices and circuits become very unstable and reliability becomes an important issue. All these aspects need further research, and the development of a model and simulation will decide whether PCM can be marketized and successfully replace Flash, thus becoming the new criterion in NVM.

As for simulation, there have been three different methods whose differences mainly lie between an electrothermal coupling model and the kinetics of a phase transition model. The electrothermal coupling model relies mainly on the models based on conductivity and drift-diffusion. The kinetics of phase transition model is based on the JMAK theory, the N/G theory, and the analytical phase transition model. Different electrothermal models coupled with different phase transition models form three different simulation methods. The three methods focus mainly on the process of OMS, but simulations for OTS and threshold voltage still remain to be made.

The process of OMS can be comprehended as a process of heat-induced phase transition, and the process of OTS can only be attributed to an electrical process. Amorphous GST contains many traps that are charged or neutral, formed by lone pairs; at low bias, the carriers are recombined by traps. As the bias increases, the carrier concentration increases gradually, and when the bias reaches the critical electric field, it begins the process of impact ionization. The carrier concentration is exponentially dependent on the bias, and the traps are filled, resulting in a sharp increase in current. At the same time, the recombination rate cannot balance generation any more, and the only way to reach a new steady-state condition is by decreasing the voltage drop across the device, causing a snap-back effect. The numerical models discussed focus mainly on OMS and don't apply to OTS, so we need to set up a complete numeric simulator that contains the effects of OTS.

With the aim of solving some defects in the numerical simulation, we introduce the concept of classification simulation. Classification simulation in microelectronic devices is a methodology that is mainly useful for different choices of equations in different areas with different points of focus that have not received enough attention. For example, it contains only the hole continuity equation or electrical continuity equation, in which we divide the threshold into two levels: a subthreshold area and an overthreshold area. In the threshold area, we have a self-consistent solving method for the Poisson equation and a current continuity equation, so we can import the indirect composite theory and collision ionization theory to simulate the OTS effect. Then we can obtain a full bias range of *I*–*V* characteristics.

There are already some SPICE models for the macro model behavior level. These models can produce the phase-change behavior characteristic curve, but cannot reflect key size or give material parameters dependent on the electrical characteristics of the device. The model is commonly based on the characteristics of a measured curve on the general device, from the point of view of producing relevant results of behavior. The device physics are related to only a few parameters, as it depends more on the results of the specific process, which does not have very good adaptability. Therefore, it is necessary to set up a model with strong adaptability based on the physical reaction to the device key size and material parameters.

The aim of our research is to develop a perfect SPICE model, which includes working and nonworking hours of storage in the read and write characteristics. The read and write characteristics include the amorphous state and the amorphous state under their current voltage characteristics, as well as switching between the two phases. We need to cover electrical, thermal, and crystallization kinetics and also factors that exist in SPICE models and the impedance model. On the basis of all this work, we need to further develop the thermometer heat transfer model, the crystal model, and the storage model, and study their mutual influence and effect to obtain a combined and efficient compact model. At present, we have a conductivity model of amorphous GST based on the Poole–Franke theory, and an OTS effect of the analytical model based on FN tunneling and the use wear theory. We hope to obtain an OMS model in the future. The OMS model poses difficulty in obtaining an analytical solution for temperature distribution. At present, the solutions for the model consider phase change as isothermal, with the aim of getting isothermal temperature according to energy conservation laws. But, in fact, there was a big temperature gradient in the internal phase-change area, so the assumption of an isothermal body could introduce many errors. Most of the time, OMS is a process for thermally induced phase switching, in which temperature plays an important role. Therefore, the modeling of the temperature distribution in a PCM cell will be a significant step for the whole model, and this will be the main content of our research. For the foundation of the whole model, only after establishing the temperature model, the impedance model, and the phase change dynamics model, can we use Verilog-A to realize the whole model.

REFERENCES

1. J.-H. Lee, Y.M. Kim, S.-H. Bae, and K.-R. Han, Cell devices for high density flash memory, *IEEE, ICSICT*, 2008, Beijing, China, pp. 819–822.
2. S.-M. Jung, J. Jang et al., Three dimensionally stacked NAND flash memory technology using stacking single crystal Si layers on ILD and TANOS structure for beyond 30 nm node, *IEEE, IEDM*, 2006, San Francisco, CA, pp. 1–4.
3. R. J. McPartland and R. Singh, 1.25 Volt, low cost, embedded flash memory for low density applications, *IEEE, VLSI Circuits*, 2000, pp. 158–161.
4. P. Pavan, R. Bez, P. Olivo, and E. Zanoni, Flash memory cells-an overview, *IEEE, Proceedings of the IEEE*, 85(8), 1248–1271.
5. P. Cappelletti, Flash memory reliability, *Microelectronics Reliability*, 1998, 38(2), 185–188.
6. J.-D. Lee, S.-H. Hur, and J.-D. Choi, Effects of floating-gate interference on NAND flash memory cell operation, Electron Device Letters, *IEEE*, 23(5), 264–266.
7. S. Lai, Current status of the phase change memory and its future, *2003 IEEE International Electron Devices Meeting, Technical Digest*, Washington, DC, 2003, pp. 255–258.
8. G. Atwood and R. Bez, Current status of chalcogenide phase change memory, *Device Research Conference Digest, 2005. DRC '05. 63rd*, Santa Barbara, CA, 2005, pp. 29–33.
9. S. R. Ovshinsky, Reversible electrical switching phenomena in disordered structures, *Physical Review Letters*, 1968, 21, 1450.
10. Ovynx Company, www.ovonyx.com/technology/technology.com, 2001.
11. Y.-T. Kim, K.-H. Lee et al., Study on cell characteristics of PRAM using the phase-change simulation company, *SISPAD*, Athens, Greece, September 2001, pp. 211–214.
12. F. Pellizzer, A. Benvenuti et al., A 90 nm phase change memory technology for stand-alone non-volatile memory applications, *VLSI Technology, 2006. Digest of Technical Papers. 2006 Symposium on*, 2006, pp. 122–123.
13. A. L. Lacaita, D. Ielmini et al., Status and challenges of phase change memory modeling, *Solid-State Electronics*, 2008, 52, 1443–1451.
14. Schrott, A., H.-L. Lung; Happ, T.; Chung Lam, Phase-Change Memory Development Status, *VLSI Technology, Systems and Applications, 2007. VLSI-TSA 2007. International Symposium on*, pp. 1–2.

15. S. J. Ahn, Y. J. Song et al., Highly manufacturable high density phase change memory of 64 Mb and beyond, *IEEE International Electron Devices Meeting 2004, Technical Digest*, San Francisco, CA, 2004, pp. 907–910.
16. S. L. Cho, J. H. Yi et al., Highly scalable on-axis confined cell structure for high density PRAM beyond 256 Mb, *2005 Symposium on VLSI Technology, Digest of Technical Papers*, 2005, pp. 96–97.
17. S. J. Ahn, Y. N. Hwang et al., Highly reliable 50 nm contact cell technology for 256 Mb PRAM, *2005 Symposium on VLSI Technology, Digest of Technical Papers*, 2005, pp. 98–99.
18. Y. N. Hwang, S. H. Lee et al., Writing current reduction for high-density phase-change RAM, *2003 IEEE International Electron Devices Meeting, Technical Digest*, Washington, DC, 2003, pp. 893–896.
19. H. Horii, J. H. Yi et al., A novel cell technology using N-doped GeSbTe films for phase change RAM, *VLSI Technology, 2003. Digest of Technical Papers. 2003 Symposium on*, 2003, pp. 177–178.
20. N. Matsuzaki, K. Kurotsuchi et al., Oxygen-doped gesbte phase-change memory cells featuring 1.5 V/100-/spl mu/A standard 0.13/spl mu/m CMOS operations, *IEEE International Electron Devices Meeting, 2005. IEDM Technical Digest*, Washington, DC, 2005, pp. 738–741.
21. Y. C. Chen, C. T. Rettner et al., Ultra-thin phase-change bridge memory device using GeSb, *International Electron Devices Meeting, 2006. IEDM'06*, San Francisco, CA, 2006, pp. 1–4.
22. Y. J. Song, K. C. Ryoo et al., Highly reliable 256 Mb PRAM with advanced ring contact technology and novel encapsulating technology, *VLSI Technology, 2006. Digest of Technical Papers 2006 Symposium on*, 2006, pp. 118–119.
23. N. Takaura, M. Terao et al., A GeSbTe phase-change memory cell featuring a tungsten heater electrode for low-power, highly stable, and short-read-cycle operations, *IEEE International Electron Devices Meeting, 2003. IEDM'03 Technical Digest*, Washington, DC, 2003, pp. 37.2.1–37.2.4.
24. Happ, T. D.; Breitwisch, M.; Schrott, A.; Philipp, J. B.; Lee, M. H.; Cheek, R.; Nirschl, T.; Lamorey, M.; Ho, C. H.; Chen, S.-H.; Chen, C. F.; Joseph, E.; Zaidi, S.; Burr, G. W.; Yee, B.; Chen, Y. C.; Raoux, S.; Lung, H. L.; Bergmann, R.; Lam, C., Novel One-Mask Self-Heating Pillar Phase Change Memory, *VLSI Technology, 2006. Digest of Technical Papers. 2006 Symposium on*, pp. 120–121.
25. D. Turnbull and J. C. Fisher, Rate of nucleation in condensed systems, *Journal of Chemical Physics*, 1949, 17, 71–73.
26. K. F. Kelton, Crystal nucleation in liquids and glasses, *Solid State Physics-Advances in Research and Applications*, 1991, 45, 75–177.
27. H. B. Singh and A. Holz, Stability limit of supercooled liquids, *Solid State Communications*, 1983, 45, 985–988.
28. S. Senkader and C. D. Wright, Models for phase-change of Ge2Sb2Te5 in optical and electrical memory devices, *Journal of Applied Physics*, 2004, 95, 504–511.
29. D.-H. Kim, F. Merget et al., Three-dimensional simulation model of switching dynamics in phase change random access memory cells, *Journal of Applied Physics*, 2007, 101, 064512.
30. T. Gille, K. De Meyer et al., Amorphous-crystalline phase transitions in chalcogenide materials for memory applications, *Phase Transitions*, 2008, 81, 773–790.
31. C. Popescu, The effect of local non-uniformities on thermal switching and high field behaviour of structures with chalcogenide glasses, *Solid-State Electronics*, 1975, 18, 671–681.
32. A. E. Owen, J. M. Robertson et al., The threshold characteristics of chalcogenide-glass memory switches, *Journal of Non-Crystalline Solids*, 1979, 32, 29–52.

33. D. Adler, H. K. Henisch et al., The mechanism of threshold switching in amorphous alloys, *Reviews of Modern Physics*, 1978, 50, 209.

34. A. Pirovano, A. L. Lacaita et al., Electronic switching in phase-change memories, *IEEE Transactions on Electron Devices*, 2004, 51, 452–459.

35. K. Lu, Phase-change memory study, in The Department of Computer and Electronic Engineering. Thesis of Master of Philosophy: HKUST, 2007.

36. A. L. Lacaita, A. Radaelli et al., Electrothermal and phase-change dynamics in chalcogenide-based memories, *IEEE International Electron Devices Meeting 2004, Technical Digest,* San Francisco, CA, 2004, pp. 911–914.

37. http://synopsys.mediaroom.com/index.php?s=43&item=610.

38. B. Schmithusen, P. Tikhomirov et al., Phase-change memory simulations using an analytical phase space model, *SISPAD: 2008 International Conference on Simulation of Semiconductor Processes and Devices*, Hakone, Japan, 2008, pp. 57–60.

39. Synopsys TCAD Seminar 2008, 2008, p. p33.

40. P. Fantini, A. Benvenuti et al., A compact model for phase change memories, *2006 International Conference on Simulation of Semiconductor Processes and Devices*, Monterey, CA, 2006, pp. 162–165.

41. R. A. Cobley and C. D. Wright, Parameterized SPICE model for a phase-change RAM device, *IEEE Transactions on Electron Devices*, 2006, 53, 112–118.

42. K. Sonoda, A. Sakai et al., A compact model of phase-change memory based on rate equations of crystallization and amorphization, *IEEE Transactions on Electron Devices*, 2008, 55, 1672–1681.

43. Y. B. Liao, Y. K. Chen et al., An analytical compact PCM model accounting for partial crystallization, *EDSSC: 2007 IEEE International Conference on Electron Devices and Solid-State Circuits, Vols 1 and 2, Proceedings*, Tainan, Taiwan, 2007, pp. 625–628.

44. D. Ielmini and Y. G. Zhang, Analytical model for subthreshold conduction and threshold switching in chalcogenide-based memory devices, *Journal of Applied Physics*, 2007, 102, 054517.

45. G. C. Wicker, A comprehensive model of submicron chalcogenide switching devices. PhD thesis, Wayne State University, Detroit, MI, 1996.

46. A. Redaelli, A. Pirovano et al., Electronic switching effect and phase-change transition in chalcogenide materials, *IEEE Electron Device Letters*, 2004, 25, 684–686.

47. A. K. Jonscher and C. K. Loh, Poole-Frenkel conduction in high alternating electric fields, *Journal of Physics C: Solid State Physics*, 1971, 4, 1341–1347.

48. A. K. Jonscher, Energy losses in hopping conduction at high electric fields, *Journal of Physics C: Solid State Physics*, 1971, 4, 1331–1340.

49. I. Karpov, M. Mitra et al., Temporal changes of parameters in phase change memory, *2008 International Symposium on VLSI Technology, Systems and Applications (VLSI-TSA), Proceedings of Technical Program*, 2008, pp. 140–141.

50. J. Reifenberg, E. Pop et al., Multiphysics modeling and impact of thermal boundary resistance in phase change memory devices, *2006 Proceedings 10th Intersociety Conference on Thermal and Thermomechanical Phenomena in Electronics Systems*, 2006, 1 and 2, pp. 106–113.

51. J. P. Reifenberg, D. L. Kencke et al., The impact of thermal boundary resistance in phase-change memory devices, *IEEE Electron Device Letters*, 2008, 29, 1112–1114.

52. B. Rajendran, J. Karidis et al., Analytical model for RESET operation of phase change memory, *IEEE International Electron Devices Meeting 2008, Technical Digest,* San Francisco, CA, 2008, pp. 551–554.

53. I. R. Chen and E. Pop, Compact thermal model for vertical nanowire phase-change memory cells, *IEEE Transactions on Electron Devices*, 2009, 56, 1523–1528.

54. X. Q. Wei, L. P. Shi et al., Universal HSPICE model for chalcogenide based phase change memory elements, *2004 Non-Volatile Memory Technology Symposium, Proceedings*, San Diego, CA, 2004, pp. 88–91.

55. Y. B. Liao, J. T. Lin et al., Temperature-based phase change memory model for pulsing scheme assessment, *2008 IEEE International Conference on Integrated Circuit Design and Technology, Proceedings*, Grenoble, France, 2008, pp. 199–202.

56. S. Lavizzari, D. Ielmini et al., Transient effects of delay, switching and recovery in phase change memory (PCM) devices, in *IEEE International Electron Devices Meeting 2008, Technical Digest*, San Francisco, CA, 2008, 215–218.

57. B. Gleixner, A. Pirovano et al., Data retention characterization of phase-change memory arrays, *2007 IEEE International Reliability Physics Symposium Proceedings—45th Annual*, Phoenix, AZ, 2007, pp. 542–546.

58. A. Chimenton, C. Zambelli et al., Set of electrical characteristic parameters suitable for reliability analysis of multimegabit phase change memory arrays, *2008 Joint Non-Volatile Semiconductor Memory Workshop and International Conference on Memory Technology and Design, Proceedings*, Monterey, CA, 2008, pp. 49–51.

59. U. Russo, D. Ielmini et al., Intrinsic data retention in nanoscaled phase-change memories—Part I: Monte Carlo model for crystallization and percolation, *IEEE Transactions on Electron Devices*, 2006, 53, 3032–3039.

60. A. Redaelli, D. Ielmini et al., Intrinsic data retention in nanoscaled phase-change memories—Part II: Statistical analysis and prediction of failure time, *IEEE Transactions on Electron Devices*, 2006, 53, 3040–3046.

61. U. Russo, D. Ielmini et al., Analytical modeling of chalcogenide crystallization for PCM data-retention extrapolation, *IEEE Transactions on Electron Devices*, 2007, 54, 2769–2777.

62. Y. C. Chen, C. T. Chen et al., 180 nm Sn-doped Ge2Sb2Te5 chalcogenide phase-change memory device for low power, high speed embedded memory for SoC applications, *Proceedings of the IEEE 2003 Custom Integrated Circuits Conference*, San Jose, CA, 2003, pp. 395–398.

63. S. Lavizzari, D. Ielmini et al., Reliability impact of chalcogenide-structure relaxation in phase-change memory (PCM) cells—Part II: Physics-based modeling, *IEEE Transactions on Electron Devices*, 2009, 56, 1078–1085.

64. A. Pirovano, A. L. Lacaita et al., Scaling analysis of phase-change memory technology, *2003 IEEE International Electron Devices Meeting, Technical Digest*, Washington, DC, 2003, pp. 699–702.

65. W. Feng, Nano-size structure for reading reliability in phase change memory, in *Materials Science and Engineering*, Thesis of Master of Science: UCLA, 2007.

66. D. Ielmini, D. Sharma et al., Reliability impact of chalcogenide-structure relaxation in phase-change memory (PCM) cells—Part I: Experimental study, *IEEE Transactions on Electron Devices*, 2009, 56, 1070–1077.

67. D. Ielmini, S. Lavizzari et al., Physical interpretation, modeling and impact on phase change memory (PCM) reliability of resistance drift due to chalcogenide structural relaxation, *2007 IEEE International Electron Devices Meeting*, Washington, DC, 2007, vols. 1 and 2, pp. 939–942.

68. T. Gille, J. Lisoni et al., Modeling of the mechanical behavior during programming of a non-volatile phase-change memory cell using a coupled electrical-thermal-mechanical finite-element simulator, *EUROSIME 2007: Thermal, Mechanical and Multi-Physics Simulation and Experiments in Micro-Electronics and Micro-Systems, Proceedings*, London, 2007, pp. 283–288.

11 Phase-Change Memory Devices and Electrothermal Modeling

Helena Silva, Azer Faraclas, and Ali Gokirmak

CONTENTS

11.1 CURRENT STATUS OF DATA STORAGE

Demand for data storage is increasing along with computational power, accessibility to high-speed Internet, availability of high-resolution video and mobile devices, and high volume of data generated in 3-D medical imaging. Currently, the main data storage media are the following:

- Magnetic tapes: highest volume and lowest cost, no random access, typically used for backup
- Magnetic hard drives: high density, high volume (~2 TB), low cost, common main storage for PCs

- Optical storage: medium density, low cost, 50 Gb chalcogenide-based rewriteable disks available
- Flash memory[1]: charge-based storage, very high density (~128 Gb/chip), moderate cost, highest speed, no moving parts, lightweight, typically used for mobile applications

Innovations in fabrication processes and devices are continuing to fuel all competing technologies. Increased storage capability with reduced costs, significantly higher-speed random access, and lightweight have pushed flash memory into competition with hard drives for notebook computers and high-performance systems and has been one of the enabling technologies for lightweight low-power tablet PCs. Advances in lithography, materials, and device engineering have been enabling the scaling of flash memory. Intel/Micron has announced products using high-density NAND flash using 20 nm node technology,[2] and Hynix has announced 15 nm prototypes.[3] NAND flash memory, using a single-transistor cross bar array of gate-stack lines (word) and active region lines (bit) in ~15 nm length scale, is one of the smallest devices (of any sort) that can be made using today's technology. Scaling of charge-based storage to these dimensions had been deemed questionable in past decades due to reliability concerns, and this had sparked investigations into alternative technologies. Currently, any competing solid-state memory technology has to either outperform flash memory in its own memory segment—which is difficult in terms of density unless multi-bit per cell operation is achieved—or offer higher performance.

Among the current alternative technologies, phase-change memory[4-7] (PCM, or PRAM) and other resistive RAM (RRAM) technologies (such as electrolyte memories and metal-oxide memories[8,9]) are seen as the most promising candidates for future storage-class memory (nonvolatile, high density, low cost, high speed).[10,11] RRAM approaches, relying on reversible resistance changes in a material, have been demonstrated to operate at significantly higher speeds and can potentially be scaled to <10 nm. There is therefore strong interest in exploring RRAM concepts to replace DRAM (the main computer memory), which typically uses a one-transistor–one-capacitor cell (1T–1C) to store one bit of information and requires periodic refreshing of the cells. Successful implementation of RRAM as main computer memory would significantly reduce the energy requirements for computation but requires extremely high endurance (>10^{16} cycles) and high speed (<10 ns).[10]

As device-level research progresses toward future resistive memories, significant research at the architecture level is also taking place to take advantage of these new high-density, two-terminal memory devices.[12,13] It is believed that RRAM technologies may replace NOR flash memory (for significantly higher speed) as well as DRAM (for significantly lower power). There is still no potential candidate to replace SRAM, which currently delivers sub-ns programming times, and it is rather challenging to achieve densities higher than NAND flash (1T cell). Nevertheless, NOR flash and DRAM are very large memory segments, and improved technologies that can replace these would have significant impacts.

PCM is currently the most mature of the RRAM technologies.[6,14] Samsung has announced 8 GB PCM chips on 20 nm technology node[15] and introduced PCM to the mobile phone market. Micron has started volume production of 1 GB PCM modules using 45 nm technology node (July 2012)[16] as high-performance nonvolatile memory for tablet PCs and smart phones. PCM benefits from single-bit alterability (no need to erase or initialize before write) resulting in higher-speed operation and no need for a buffer. These advantages enable faster and more energy-efficient architectures that can handle power interruptions with minimal loss of data. Hence, PCM has entered the market as a potential replacement for NOR flash as of 2012. The requirement of relatively large current levels, resulting in constraints on the access devices and higher-than-desired energy consumption, as well as endurance and reliability concerns, limit PCM density (module capacities) and the possibility of DRAM-like usage at this point.

Scaling down to ~10 nm regime significantly reduces the current requirements but may worsen reliability and endurance. Rigorous device models are needed for significant improvements in device performance, energy consumption, and reliability and endurance. Modeling of PCM devices at these scales is rather challenging, while being very interesting, since some of the electrothermal processes at these extreme length scales—which give rise to extreme thermal gradients (>1 K/nm) and heating rates (>100 K/ns)—are not well understood yet. PCM may turn out to be one of the first mainstream technologies that truly utilize physical phenomena that become significant at nanometer scale and is also a unique system for such fundamental studies.

11.2 PHASE-CHANGE MEMORY

The operation of PCM is based on the reversible phase transition of a material (typically a chalcogenide) between the resistive amorphous and conductive crystalline states through self-heating. Depending on the chalcogenide material used and the cell geometry, a resistance contrast of a few orders of magnitude can be realized between the different states[5,6]; such contrast is the key for reliable data storage. $Ge_2Sb_2Te_5$ (GST) is the most studied PCM material due to its stability and high resistivity contrast between the crystalline (c-) and amorphous (a-) phases.[7] In terms of device structure, mushroom cells (and their variations) are the most common PCM cell geometries due to high packing density and relatively easy process integration (though recent array demonstrations use the more confined pore[6] or dash-type cells[17–19]). Each individual PCM cell in the memory array is addressed through an access device, usually a field effect transistor (FET), bipolar junction transistor, or a diode, depending on the density and power requirements.[20] The *reset* current requirements determine the dimensions of the access device, as well as the power consumption, with the access device size typically determining the array density.

The mushroom device consists of a narrow bottom contact—*the heater*, a thin film of phase-change material, and a shared planar top contact (Figure 11.1). A nanoscale semispherical volume over the bottom electrode (*mushroom*) is amorphized (*reset*) by rapid cooling after melting (melt-quenched) with a large-amplitude short-duration electrical pulse ($T_{melt} \sim 873$ K for GST). The rapid cooling prevents rearrangement of the atoms in a crystalline structure. Crystallization (*set*) is achieved by heating the amorphized mushroom above the crystallization temperature ($T_{cryst} \sim 425–625$ K for GST) for a sufficient period of time (~crystallization time) using a longer-duration, lower-amplitude pulse; alternatively, a melting pulse with a longer fall time that allows for slower cooling can be used for crystallization. Since the resistivity of amorphous GST

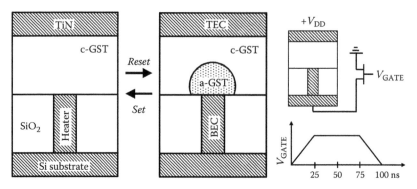

FIGURE 11.1 PCM mushroom cell schematic. A small volume of GST above the heater is switched between crystalline (*set*) and amorphous (*reset*). The top electrode contact (TEC) and bottom electrode contact (BEC) are both TiN in these simulations. Temperature boundary conditions of 300 K are applied on the top of the TEC and bottom of the Si substrate, and electrical boundary conditions are applied at the top of TEC and the bottom of BEC. All other boundaries are thermally and electrically floating. The access transistor is connected through the BEC as shown (for the positive polarity). V_{DD} is held at 1.75 V while the gate is pulsed with a peak voltage of 1.75 V. Typical resistance ratios for PCM cells are on the order of 10^3 with switching times of 50/120 ns for the *reset* and *set* operations, respectively.

(a-GST) is very large, self-heating for *set* requires sufficiently high-amplitude pulses that initiate electrical breakdown (threshold switching of the amorphous material into a conductive state[21,22]). The mushroom size (hence energy required for the switching) is determined by the bottom electrode (heater) area, which must be covered to achieve *reset*, and by the electrical pulse conditions. Various sub-lithographic techniques for the definition of the heater have been proposed, including side-wall spacer approaches,[6] use of carbon nanotubes as heaters,[14] and formation of a nanoscale conductive filament in an oxide layer atop the heater via electrical breakdown (rupture oxide).[23-26]

The phase-change materials used in PCM cells dynamically transition between the amorphous, nanocrystalline (nc-) (nanoscale crystalline grains embedded in an amorphous matrix), and polycrystalline phases (adjacent crystalline grains). The crystallinity of the material is determined by the nucleation and growth dynamics, which depend on how long the material experiences any given temperature as well as on element size, interfaces, mechanical stress, and possibly electric field conditions. The temperature-dependent material parameter functions of the phase-change materials strongly depend on crystallinity, the amorphous phase being the most sensitive to temperature. The deposition process conditions determine the initial phase of the materials. The subsequent *set/reset* operations (or any initializing electrical or thermal stresses applied) determine the phase of the active region of the device and its surroundings, which experience most heating and largest current densities. Hence, the pulse conditions to achieve a successful *reset* operation are determined in part by the cell history, with the most recent *set* pulse being the most dominant contributor. A successful *set* operation requires dielectric breakdown of the amorphous region (plug) that isolates the bottom contact. The thickness of the amorphous region is determined by the last *reset*. Hence, the *set* voltage requirements of a reset cell also depend on the last *reset*. Stable *set/reset* device operation has been demonstrated up to ~10^{10} cycles,[27] and early products (1 GB modules) guarantee ~10^6 cycles.[28] The *set* pulse duration and/or shape can be adjusted to *set* the cell to different resistivity levels, enabling controlled multilevel operation (multi-bit per cell for high-density storage).[29]

Complex dynamics in PCM devices also arise from the fact that the *set* and *reset* pulses can be applied with sub-ns rise/fall times, a time scale comparable to the thermal diffusion time scales at these dimensions (~ns). Thermal carrier generation leads to an exponential reduction in resistivity as a function of temperature for the amorphous and crystalline (*fcc*) phases. This strong negative temperature coefficient of resistivity (TCR) leads to thermal runaway and filament formation in the current paths carrying the maximum current. This effect is exacerbated for faster rise times, since conductive filaments can form prior to significant lateral heat diffusion during the transient period, and determine the final amorphization profile.

11.3   ELECTROTHERMAL MODELS FOR PCM DEVICES

Here, we describe our current electrothermal models for PCM devices and present example simulations using these models that illustrate the critical dependences on varying material parameters such as electrical resistivity and the role of thermoelectric effects in the heating distribution within these nanoscale structures.

A mushroom cell using GST as the phase-change material, TiN as the top and bottom electrode contacts (TEC and BEC), and SiO_2 as the insulating material is simulated using 2-D rotational symmetry using a commercial finite-element tool (COMSOL MultiPhysics). A SPICE model of an access transistor in series with the PCM element is integrated with the finite element models in the simulations. The simulated geometry and bias conditions are shown in Figure 11.1.

The electrothermal transport within the PCM element is modeled by solving the current continuity (1) and heat transfer (2) equations self-consistently including thermoelectric effects (coupling between charge and heat flow).[30] In homogeneous non-isothermal materials, diffusion current is due to the temperature gradient and can be expressed as $J_{\mathrm{diff}} = -\sigma\, S\nabla T$, where σ is the electrical conductivity and S is the Seebeck coefficient or thermoelectric power[31,32]:

$$\nabla \cdot J = -\underbrace{\nabla \cdot \sigma(T)\nabla V}_{\text{Drift current}} - \underbrace{\nabla \cdot \sigma(T)S\nabla T}_{\substack{\text{Thermoelectric} \\ \text{diffusion current}}} = 0 \tag{11.1}$$

$$\underbrace{dC_P(T)dT/dt}_{\text{Heating}} - \underbrace{\nabla(\kappa(T)\nabla T)}_{\text{Heat diffusion}} = J \cdot E - J \cdot \nabla(ST) = \underbrace{J^2/\sigma(T)}_{\text{Joule heating}} - \underbrace{J \cdot T\nabla S}_{\text{Thomson heat}} \tag{11.2}$$

where
  J is the electrical current density
  V is the electric potential
  T is the temperature
  κ is the thermal conductivity
  d is the mass density
  C_P is the heat capacity

In open-circuit conditions, a temperature gradient results in an open-circuit voltage due to drift-diffusion balance (Seebeck effect). In the presence of an electric current, thermoelectric heat takes two forms: *Thomson heat*, which results from a Seebeck coefficient gradient within a current carrying homogenous material (which arises from a temperature gradient) and leads to a redistribution of heat; and *Peltier heat*, which results from a difference in Peltier coefficient ($\Pi = ST$) at a junction of two different materials and results in net heating or cooling at this junction. The direction of current determines whether the result is local heating or cooling.

Although the thermoelectric terms in (1) and (2) include all thermoelectric effects (Seebeck, Peltier, and Thomson effects) depending on particular conditions,[31] the finite element tool used cannot compute the derivatives across the different material domains, and an additional surface heat source is added at the GST/TiN interfaces (both top and bottom contacts) to account for Peltier heat at this junction[24]:

$$q_{Peltier} = -J\Delta\Pi = -J\Delta(S \cdot T) = -JS_{GST}T_{GST} + JS_{TiN}T_{TiN} \qquad (11.3)$$

(S_{GST}, T_{GST}) and (S_{TiN}, T_{TiN}) are probed at either side of a virtual layer that is added at this interface (Figure 11.2) to account for thermal boundary resistance (TBR, discussed later). These are average values over the whole interface (one temperature and one Seebeck coefficient value for either side of the modeled interface). This additional Peltier heat source is added within the GST, 0.5 nm away from a virtual 1 nm TBR layer, as Peltier heat is expected to be absorbed or released mostly in the GST where generation and recombination occur within a few carrier mean free paths of the interface. Peltier heat is included only at the GST/TiN interfaces as it is not expected to be significant anywhere else. In these simulations, J is defined along the cylindrical axes; thus in the positive polarity, J flows in the $-z$ direction from TEC to BEC and is considered negative.

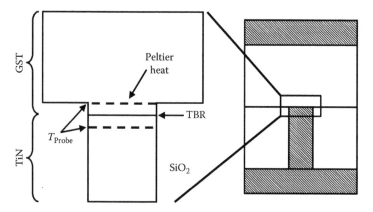

FIGURE 11.2 Peltier heat is applied 0.5 nm above the boundary between GST and TiN. This requires temperature and Seebeck coefficient values from both sides of the interface, which are probed at 0.5 nm from either side (dashed lines).

11.3.1 MATERIAL PARAMETERS

GST undergoes a first phase transition, from amorphous to crystalline-*fcc* at ~425 K, and a second phase transition, from crystalline-*fcc* to crystalline-*hcp* at ~625 K. Likely due to multiphase films (amorphous/*fcc*/*hcp*, with varying granularity) and varying deposition conditions, there is a wide range of reported electrical resistivity ($\rho = 1/\sigma$) for GST.[33–36] Variations in this and other material parameters greatly impact the amorphization and crystallization dynamics of the cells. Most importantly in PCM devices, energy dissipated in the active region is directly dependent on its electrical resistivity, and the way this heat is distributed in the active region depends on the thermal conductivity. Thermoelectric properties, TBRs, and material density all also impact device operation and performance.

11.3.1.1 Electrical Conductivity

The temperature-dependent electrical resistivity functions used in the simulations are shown in Figures 11.3 and 11.4a and are based on our measurements of nanoscale GST structures up to ~620 K and in liquid state.[37] These resistivity data were obtained through slow measurements (variable temperature chuck, heating rate ~1 K/min), which allow the GST to change its phase during the heating process since the crystallization times are much faster than these time scales. During *set* and *reset* pulses in PCM devices, however, where melting can be reached in nanoseconds, GST is not expected to undergo intermediate phase transitions before it reaches

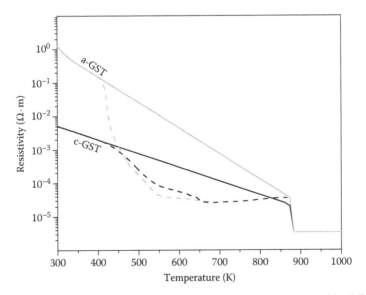

FIGURE 11.3 Resistivity as a function of temperature for a pulse-amorphized GST line structure, $L \times W \times t = 460$ nm $\times$ 160 nm $\times$ 50 nm, using a 500 ns, 2.2 V pulse at room temperature (gray dashed line), and an as-fabricated *fcc* GST line structure, $L \times W \times t = 460$ nm $\times$ 255 nm $\times$ 50 nm (black dashed line). (Heating rate of 1 K/min. From Cil, K. et al., *IEEE Trans. Electron Devices*, 60, 433, 2013.) Assumed metastable resistivity functions for each phase are shown with solid lines.

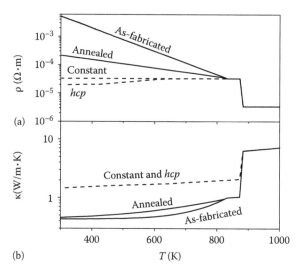

FIGURE 11.4 (a) Metastable electrical resistivity for *as-fabricated fcc* and *annealed fcc* models (solid line). Measured resistivity for *hcp* and a hypothetical constant resistivity function (dashed lines). *as-fabricated fcc* curve is measured from a GST line structure, $L \times W \times t = 460$ nm $\times$ 255 nm $\times$ 50 nm (same as in Figure 11.3). *Annealed fcc* curve is the average of the measurements of 33 different devices after being held at 450 K for ~1 h with resistance measured during cooling. All curves are linearly transitioned to the liquid value between 873 and 883 K. (b) Thermal conductivity functions for each electrical resistivity model (constructed functions for the *as-fabricated* and *annealed fcc* models, and an experimental thermal conductivity curve from slow measurements (From Lyeo, H. K. et al., *Appl. Phys. Lett.*, 89, 151904, 2006.) extrapolated to high temperature for the *constant* and *hcp* models).

the molten state. To model these metastable crystalline and amorphous phases, the exponential behavior for the *fcc* and amorphous GST resistivity at low temperature is extrapolated to the melting temperature (883 K) (Figure 11.3). A measured value of 3.4 $\mu\Omega \cdot$ m was used for the liquid phase.[28] All material parameters are linearly transitioned from their solid-state values to the liquid values in a 10 K temperature range (873–883 K) to model the solid–liquid transition (in addition to the latent heat of fusion). All the example simulations presented are for *reset* pulses (switching from crystalline to amorphous), so only the metastable crystalline parameters are used.

Various $\rho(T)$ functions for crystalline GST were implemented in the model (Figure 11.4a) for the *reset* simulations to demonstrate the effect of electrical resistivity of GST on the cell performance. Two *fcc* functions (*as-fabricated fcc* and *annealed fcc*) extrapolated from different measurements are compared with measured *hcp* data and a hypothetical *constant* resistivity function representing a mixed *fcc/hcp* phase. A constant resistivity could be achieved in mixed phase *fcc* (which has a strong negative TCR) and *hcp* (with a weak positive TCR) in the appropriate proportions by adjusting the anneal temperature.

Our measurements of GST structures with TiN contacts show an ohmic behavior, suggesting no significant contact resistance[37]; hence, we do not include any electrical boundary resistance between GST and TiN in the model.

11.3.1.2 Thermal Conductivity

The thermal conductivity of *fcc* GST was estimated based on the phonon (κ_{ph}) and the electronic (κ_{el}) contributions such that $\kappa_{Total} = \kappa_{ph} + \kappa_{el}$. Phonon conduction is assumed to dominate at low temperatures and decay as the temperature increases (dropping to negligible values upon melting). The number of broken bonds increases exponentially with temperature. Likewise, the electrical conductivity and the electronic thermal conductivity are assumed to increase with temperature mostly due to the generation of free charge carriers with the breaking of each bond. Heat transfer due to convection in liquid state is neglected. κ_{el} is obtained from Wiedemann–Franz (W–F) law:

$$\kappa_{el}(T) = L \cdot \sigma(T) \cdot T \tag{11.4}$$

where L is the Lorenz number (2.44×10^{-8} W·Ω·K^{-2}). Electrical conductivity is low below the melting temperature for *fcc* GST and increases significantly at melting; hence, electronic thermal conductivity contribution is expected to be negligible in solid GST but becomes important as GST becomes metallic upon melting. Based on these assumptions, a simplified phonon contribution to thermal conductivity $\kappa_{ph}(T)$ is constructed starting at experimentally measured κ at room temperature and decaying with the number of broken bonds, $n(T)$:

$$\kappa_{ph}(T) = \kappa(300 \text{ K})\left(1 - \frac{n(T)}{n(\text{melt})}\right) \tag{11.5a}$$

$$n(T) = \frac{\sigma(T)}{\mu \cdot q} \tag{11.5b}$$

$$\mu \approx \frac{\sigma(300 \text{ K})}{n(300 \text{ K}) \cdot q} \tag{11.5c}$$

The electron mobility, μ, is assumed to be constant as a function of temperature to eliminate μ in (11.5a) and obtain a simplified function for κ_{ph} that varies from $\kappa(300 \text{ K})$ to zero in the liquid state:

$$\kappa_{ph}(T) = \kappa(300 \text{ K})\left(1 - \frac{\rho(\text{melt})}{\rho(T)}\right) \tag{11.5d}$$

For the *as-fabricated fcc* and *annealed fcc* $\rho(T)$ functions, $\rho(\text{melt})/\rho(300 \text{ K})$ is ~0.01, and the constructed function for κ_{ph} gives the approximately correct value at room temperature, $\kappa_{ph}(300 \text{ K}) \approx \kappa(300 \text{ K})$. For the *constant* and *hcp* models, however, $\rho(\text{melt})/\rho(300 \text{ K}) \sim 0.1$, and the constructed function for κ_{ph} would result in a $\kappa_{ph}(300 \text{ K}) \sim 0.9\kappa(300 \text{ K})$. Since the electrical resistivity of GST for the *constant* and *hcp* models does not change greatly with temperature before melting, an extrapolated function based on slow measurement data of thermal conductivity of crystalline *hcp* GST[35] is used instead of the modeled function. Measurements of thermal conductivity of liquid GST are not available, hence the W–F law is used for all models to obtain $\kappa_{el,liquid}$, and thermal conductivity in the liquid state is assumed to be due only to electronic conduction. The $\kappa(T)$ functions used for each model are shown in Figure 11.4b.

11.3.1.3 Thermal Boundary Resistance

The expected inefficient coupling between electron and phonons at interfaces between materials where one or the other dominate thermal conduction results in TBRs[38,39] that impede the flow of heat through PCM structures. This is beneficial in PCM operation since the heat generated through self-heating in the active region of the devices (Joule heating and thermoelectric heat) is better contained. TBR reduces the effect of the heater acting as a heat sink, sapping heat from the active region into the large bottom contact and substrate. In our model, TBR is accounted for by introducing a virtual layer with a given thermal conductivity at the interface.

At the metallic TiN and nonmetallic GST interface, thermal transport is modeled assuming weak electron–phonon coupling (negligible κ_{ph}) at low temperatures, but improved coupling with increasing temperature, due to increasing free electron concentration in the GST. A temperature-dependent thermal boundary conductivity (TBC) function for the interface between GST and TiN was generated by adding $\kappa_{el}(T)$ (using the W–F law) to the room temperature TBC of 0.05 W/m · K.[40] As TBC increases, its effect on thermal transfer across the interface is lessened. After melting, the room temperature component becomes negligible, and the 1 nm virtual layer acts as one additional nanometer of GST. (It is important to note the difference between TBR and TBC; TBR is the total thermal resistance of the interface and TBC is the thermal conductivity of a layer with a given thickness that is used to model the thermal resistance of the interface [see Figure 11.5a]).

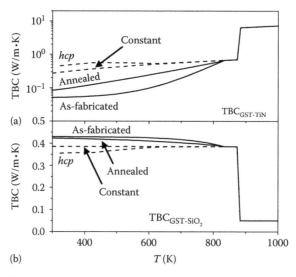

FIGURE 11.5   Thermal boundary conductivity of a 1 nm virtual layer used to model thermal boundary resistance for the GST/TiN interface (a) and GST/SiO$_2$ interface (b), for the four resistivity models.

The boundary between GST and SiO_2 is modeled in a similar manner. Heat flow in SiO_2 is due to phonon conduction, so we assume that heat transfer between the GST and SiO_2 is more efficient than that between GST and TiN. Below melting, the constructed $\kappa_{ph}(T)$ for GST is used as the TBC of a 1 nm virtual layer used to model the GST/SiO_2 interface, which acts just as an additional 1 nm layer of GST. After melting, however, the boundary layer is assigned a constant value of κ of 0.05 W/m·K, assuming that the coupling between liquid GST and SiO_2 is just as inefficient as the coupling between room temperature GST and TiN. The melting temperatures of TiN and SiO_2 are significantly above the operation temperatures of PCM devices, and a constant TBC between TiN and SiO_2 of 0.05 W/m·K is assumed.

11.3.1.4 Heat Capacity and Latent Heat of Fusion

For simplicity, the latent heat of fusion for GST ($L_f = 126 \times 10^3$ J/kg[41]) is accounted for using a 10 K wide spike in the heat capacity starting at 873 K[42,43] (rather than basing it on the energy required for the exponential increase in the number of broken bonds as a function of temperature, which peak at melting).

11.3.1.5 Seebeck Coefficient

The temperature-dependent Seebeck curves used in this model are based on averaged measurements of Seebeck coefficient of as-fabricated amorphous and annealed crystalline *fcc* GST thin films up to ~735 K (Figure 11.6).[44] The amorphous films crystallize during the measurements, at ~423 K, and were further annealed at

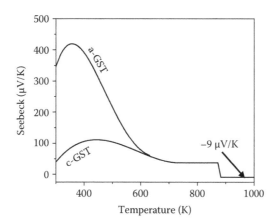

FIGURE 11.6 Seebeck coefficient for amorphous and crystalline GST as a function of temperature, obtained from averages of measurements for as-fabricated amorphous and annealed crystalline *fcc* GST thin films (100 nm) up to ~735 K.[44] The measurements were taken in 10° steps up to ~473 K and in 20° steps from 473 to 735 K. The crystalline fcc films were initially amorphous (as-fabricated) and were annealed at ~473 K for ~20–30 min. The Seebeck coefficients are assumed to remain constant (38 µV/K) in the 735–873 K range and decrease linearly upon melting (from 873 to 883 K) to match the room temperature value for TiN of −9 µV/K (modeling assumption, as GST is expected to be metallic upon melting). The Seebeck coefficient for TiN is assumed to be constant for all temperatures. (Data from Adnane, L. et al., *Bull. Am. Phys. Soc.*, 2012.)

~473 K for 20–30 min to then be measured from room temperature as the *annealed fcc* films. The Seebeck coefficients are assumed to remain constant (38 μV/K) in the 735–873 K range and decrease linearly upon melting (from 873 to 883 K) to match the TiN room temperature value of −9 μV/K, as GST is expected to be metallic upon melting. The Seebeck coefficient for TiN is assumed to be constant for all temperatures (−9 μV/K).

11.3.2 AMORPHIZATION MODEL

GST is modeled to be amorphized at any mesh point that experiences $T \geq 873$ K (onset of melting) at any time step, assuming that the entropy introduced in the material is sufficient to result in amorphization even though the material did not go through the complete solid–liquid phase transition (phase transition is modeled to be complete by $T = 883$ K after absorbing L_f). The model also assumes that cooling is fast enough to prevent crystallization (fall time of the pulse used is 25 ns). The phase of every mesh point is tracked to enable a *read/reset/read* operation sequence in a single simulation study.[45] The cells are allowed to cool down to room temperature between the *reset* and the second *read* operations. To determine the resistance state after *reset*, the room temperature value used for amorphous GST electrical resistivity in these simulations is 3.53 Ω·m. The ratio of the drain current during the first and the second *read* pulses is used to calculate the resistance contrast (ratio). The amorphized volume is visualized by mapping the conductivity in logarithmic scale (such as shown in Figure 11.10). A crystallization model for *set* operation can be implemented in a similar way, assigning the crystallized state to any mesh point that remains above crystallization temperature for a sufficient period of time (crystallization time); this can happen with slower cooling from melting, or starting from an amorphous state which requires introduction of a field-dependent electrical conductivity component to capture the threshold switching of the highly resistive amorphous material[43] (a form of electrical breakdown that enables sufficient Joule heating within practical voltages).[21,22]

11.3.3 ACCESS TRANSISTOR

The access transistor is integrated with the finite element simulations by using the basic nFET circuit model available in COMSOL.[46] The terminals of the nFET are configured as seen in Figure 11.1 for the positive polarity cases (current flowing from the top to the bottom contact). For the negative polarity cases, the PCM device is effectively "flipped" so the heater side is connected to V_{DD}, and the top contact is connected to the drain of the transistor. The transistor is turned on during *reset* by applying a 1.75 V gate voltage pulse (V_G) of 100 ns duration (which includes 25 ns rise and fall times) (Figure 11.1). The voltage applied across the access-transistor PCM-element combination (V_{DD}) is maintained at 1.75 V at all times other than during the *read* operations. A V_{DD} of 0.1 V is used to perform the *read* operations (which does not result in any significant heating of the device; $T_{max, read} \sim 313$ K for the *hcp* case where largest heating occurs). The effect of rise time, determined by the programming pulse and the cell capacitance, is not discussed here but is also a critical factor in the final temperature distributions within the PCM element.[20]

11.4 MODELING RESULTS

11.4.1 EFFECT OF ELECTRICAL RESISTIVITY

Even with the same-sized amorphous plug, the resulting resistance contrast between *set* and *reset* states (Table 11.1) should be much larger for the *hcp* model than for the *as-fabricated fcc* model just because the room temperature values for these two crystalline functions differ by 2.5 orders of magnitude. While larger resistance contrast allows for more accurate *read* operation and the potential for multiple bits storage per cell,[29] the shape of the amorphous region is also very important for device operation.

The *as-fabricated fcc* model has the largest difference between room temperature resistivity and high-temperature resistivity ($\rho(300\ K)/\rho(873\ K) = 153$). For the *annealed fcc* model, this ratio is only 6.45. When the *as-fabricated fcc* GST begins to heat, the current greatly favors the path of highest temperature, releasing more energy vertically toward the top electrode and resulting in a tall molten region, and thus a tall amorphous plug after the *reset* operation. In contrast, when *annealed fcc* GST heats, the current is able to spread out laterally, resulting in a molten region significantly larger than the heater, hence amorphization can be obtained with a lower *reset* current (Figure 11.7).

TABLE 11.1
Cell Resistances in the Crystalline (*Set*) and Amorphous (*Reset*) States, and Resistance Contrast (Ratio between the Two) for Each Resistivity Model

Resistivity Model	Set Resistance	Reset Resistance	Resistance Contrast
As-fabricated	5.07 MΩ	1.07 MΩ	212
Annealed	264 kΩ	2.30 GΩ	8.68 K
Constant	91.9 kΩ	1.91 GΩ	20.8 K
hcp	78.6 kΩ	1.95 GΩ	24.8 K

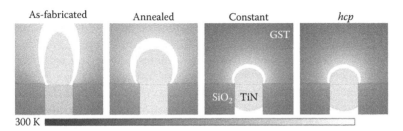

FIGURE 11.7 Peak thermal profiles (at 75 ns into the *reset* pulse) for each resistivity models. White contour lines show the 10 K transition period between solid and liquid, GST 873–883 K (within these contour lines, both GST and TiN are above 883 K). Gate and applied bias are 1.75 V, with transistor $L \times W = 22 \times 55$ nm, in the positive polarity (current flowing from the top contact to the bottom contact).

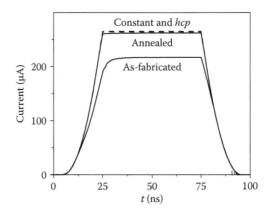

FIGURE 11.8 Current during *reset* operation for varying resistivity models. *as-fabricated* and *annealed fcc* models are shown in solid lines. *hcp* and *constant* models are shown with a dashed line. Gate and applied bias are 1.75 V, with transistor $L \times W = 22 \times 55$ nm (same for all simulations).

For the *constant* and *hcp* models, since there is no change in resistivity with temperature (or very little change, for the *hcp* model), there is no temperature-dependent preferential path for current to flow. The current spread is almost uniform from the heater–chalcogenide interface. The thermal conductivities for the *constant* and *hcp* phases are almost an order of magnitude higher than the *fcc*, allowing heat to leave the active region and flow into the surrounding GST. In addition, there is increased heat loss through the TiN heater as the higher electrical conductivity of GST leads to reduced TBR, allowing heat to escape more easily into the thermally conductive heater toward the bottom electrode and substrate. This heat loss leads to a smaller molten region in the *constant* and *hcp* cases. However, since the current spreading gives rise to a wider plug, the final resistance is approximately the same as for the *fcc* cases. Due to its higher total cell resistance, the *as-fabricated fcc* model predicts less current than any of the other models (Figure 11.8).

11.4.2 Thermoelectric Effects

In addition to Joule heating, thermoelectric heat plays an important role in the *set* and *reset* operations of PCM devices. For symmetric devices, thermoelectric heat results in asymmetric heating profiles, and for asymmetric devices (such as mushroom cells), thermoelectric heat can either assist or impede the heating of the active region of the device, depending on the voltage polarity across the cell.[47,48] By altering the temperature gradients within the device, thermoelectric heat can also affect elemental segregation that strongly impacts endurance and reliability in PCM.[49]

11.4.2.1 Thomson Heat

For a given current direction, Thomson heat leads to local heating or cooling depending on the sign of the Seebeck gradient and has a significant impact on the temperature distribution in the active region of the devices. For the positive polarity, strong Thomson heating (10^{18}–10^{19} W/m^3) is seen in the 873–883 K range where GST is

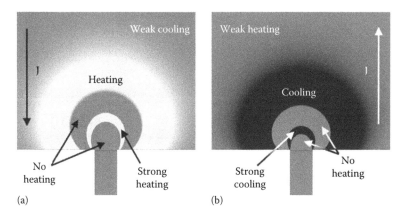

FIGURE 11.9 Thomson heat within the GST layer (volume power density) at peak heating for both (a) positive and (b) negative voltage polarities. Liquid GST regions are surrounded by strong heating or cooling for the positive or negative polarity.

modeled to transition from solid to liquid phase and the Seebeck gradient is the largest (Figures 11.6 and 11.9a). This strong Thomson heat surrounding the molten region assists its further growth. The Seebeck coefficient is modeled to be constant after melting; hence, the liquid region sees no additional thermoelectric heating from within. The same is true for GST in the temperature range of 700–873 K, where the Seebeck is assumed to be constant. From ~400 to 600 K, the Seebeck decreases with increasing temperature, leading to a second ring of significant Thomson heating. However, since the gradient is not as steep as at melting, the heating is less severe ($\sim 10^{16}$–10^{17} W/m^3). Thomson cooling takes place in the cooler outward regions of the GST layer (as in this temperature range, the Seebeck coefficient increases with increasing temperature), but since the current density, thermal gradient, and Seebeck gradient are all smaller, the cooling is small (weak cooling in Figure 11.9a). The negative polarity results in the opposite effect. As joule heating begins to create a molten GST bubble, strong Thomson cooling surrounds it, effectively impeding the growth of the molten region (Figure 11.9b). Heat is pulled away from the hot GST into the cooler regions. The result is a smaller molten region and a lower resistivity contrast for the negative polarity for the same consumed current (Figure 11.10). With positive polarity, the cell can be successfully *reset* using a smaller voltage, hence current and power, since the applied voltage in these example simulations (1.75 V for both V_{DD} and V_G) results in an *over-reset* for the positive polarity (but a "minimum" reset for the negative polarity [Figure 11.10]).

11.4.2.2 Peltier Heat

Solid GST has a much larger Seebeck coefficient than TiN. When current is flowing from the top to the bottom contact (positive polarity), $q_{Peltier} > 0$, and when current is reversed (negative polarity), $q_{Peltier} < 0$ (Figure 11.11). Like Thomson heat within the GST, Peltier heat assists in adding energy to the active region for the positive polarity, whereas in the negative polarity, it actively cools the bottom of the GST active region. When GST melts, its Seebeck coefficient is assumed to be the same

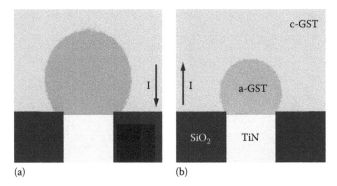

(a)   (b)

FIGURE 11.10 Conductivity profiles after a *reset* pulse (for the *annealed fcc* model) for (a) positive and (b) negative polarities. The amorphized regions are shown in dark gray. Peak currents for both polarities are 262 µA. Final resistances obtained are 2.30 GΩ for the positive polarity and 1.65 GΩ for the negative polarity.

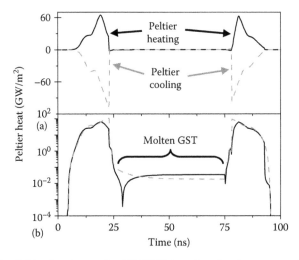

FIGURE 11.11 Peltier heating at the GST/TiN interface (surface power density) during the positive polarity pulse (solid line) and Peltier cooling during the negative polarity pulse (dashed line) in linear (a) and logarithmic (b) scales. When GST melts, the matched Seebeck coefficient (modeling assumption) and small temperature difference on either side of the GST/TiN interface result in negligible Peltier heating.

as that of TiN; further Peltier heating is only due to the small temperature difference between either side of the GST/TiN interface (Equation 11.3) and decreases by approximately an order of magnitude (Figure 11.6). In reality, a Seebeck coefficient mismatch between liquid GST and TiN at elevated temperatures is expected to lead to some Peltier heating at the liquid GST/TiN interface.

The relative power and energy contributions of Joule heating and Thomson and Peltier heating or cooling on the amorphization of the active region during a *reset* pulse are shown for both polarities in Figure 11.12 and Table 11.2. These values were calculated by integrating each heat contribution (volume power density) over

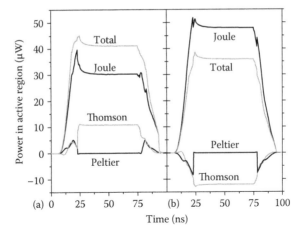

FIGURE 11.12 Joule, Peltier, Thomson, and total heat in a rectangular region tightly enclosing the active region (mushroom) during a *reset* pulse for (a) positive polarity and (b) negative polarity pulses. In the positive polarity, Peltier and Thomson heat contribute positively to the overall heating in the active region, whereas in the negative polarity, these thermoelectric heat contributions oppose the Joule heating, resulting in total power in the active region lower than the joule heating. The Joule heating is larger in the negative polarity because the current is the same and the resistance is higher (colder GST). Thomson heat increases sharply (and Peltier heat at the GST/TiN interface drops by orders of magnitude) when GST starts melting. These calculations were performed for the *annealed fcc* model, during a *reset* pulse.

TABLE 11.2

Joule, Thomson, and Peltier Heating; Total Heat Energy in a Rectangular Region Tightly Enclosing the Active Region (Mushroom); and Total Cell Energy Consumption Obtained from the Total Current and Applied Voltage

Polarity	Joule Heating (pJ)	Thomson Heating (pJ)	Peltier Heating (pJ)	Total Heat in Active Region (pJ)	Total Cell Energy (pJ)
Positive	2.25	0.66	0.07	2.98	29.42
Negative	3.30	−0.74	−0.10	2.46	29.34

Note: These calculations were performed for the *Annealed fcc* model during a *Reset* pulse.

a defined "rectangular active region" that tightly encloses the actual active region (for computational ease); these power contributions are then integrated over the *reset* pulse to calculate the total energies.

Due to significant thermoelectric transport in PCM devices that arises from the extreme thermal gradients that are established (up to ~10 K/nm), operating the cells in the positive polarity (current flowing from the mushroom to the heater) results in less energy required for operation as in this case both Thomson heat and Peltier heat help confine the heat in the active region (Figures 11.9 and 11.10). We have shown here examples of *reset* simulations, but the same polarity dependence applies to the *set* operation.

11.5 PCM DESIGN AND HIGH-TEMPERATURE NANOELECTRONICS

PCM cells are designed to confine heat to increase efficiency and reduce the thermal crosstalk between the cells, and to confine the electrical current for higher efficiency *set/reset* operations and higher resistance contrast in the *read* operation. It is desirable to have a self-limiting mechanism such as a cell reaching the resistivity level of a resistive load (reaching maximum power transfer) when the molten volume reaches the desired dimension after covering the contact interface. The shortest distance along the current path in the amorphous region determines the resistance of the *reset* state, as well as the voltage required for breakdown for the *set* operation. Hence, it is desirable to have the current confined in a quasi-1-D geometry, where both current and heat confinement can be achieved and the resistance of the *reset* state changes linearly with the amorphized volume. Line cells, recessed[50] or pore mushroom cells, and dash-type cells, like those used in recent demonstrations,[17-19] are such quasi-1-D geometries. However, line cells are also more prone to cell-to-cell variation, as the voltage required for *set* (which must result in a field above threshold switching) is a strong function of the amorphized linear distance. If the cell is over-molten during reset, *set* can become impossible in a linear structure. Mushroom geometries use a narrow point contact, and *over-reset* has only a slight impact on cell resistance and the *set* voltage requirements. Besides performance, power consumption, and density, reliability and manufacturability with minimal process variations are determining factors in device geometry choice.

The typical materials used for the electrodes have high thermal and electrical conductivities while GST has low thermal conductivity (k) in all phases, and c-GST has high electrical conductivity (σ). In conventional cells, the GST film thermally insulates the top metal electrode, aiding heat confinement. While most of the heating takes place in the GST around the heater due to current confinement, there is also some undesired cooling due to heat diffusion into this thermally conductive heater. If an interfacial layer with electrical resistivity higher than that of GST is used between the GST and the heater, substantial heating will take place right at this interface, where it is most desired, resulting in significant improvement in power efficiency and reduced peak currents. While the cell resistance in the conductive state (R_{set}) would be higher, requiring a slightly larger operation voltage, the impact on the overall cell R_{reset}/R_{set} ratio (including the access device resistance) would be relatively small since the relative increase in R_{reset} is much less significant.

Thermoelectric effects need to be considered carefully in device design. Depending on geometry, materials, and voltage polarities used for programming, these effects can be utilized for reduced power consumption, or improved performance, endurance, and reliability. The effect of Peltier heat at the critical GST/heater interface may be significant at high temperatures (in our model, we assumed the room temperature value of Seebeck coefficient of TiN for all temperatures, as well as for liquid GST) and can also be utilized for reduced power operation by choosing contact materials as to maximize this heat component (by largest Seebeck coefficient mismatch).

Conventional high-performance semiconductor devices utilize single-crystal materials, and the materials remain unchanged during normal operation conditions, maintained within a relatively small temperature window. In contrast, PCM devices utilize a phase-change material that changes dynamically in a very large temperature range, ~300–1000 K and across many phases (crystalline, amorphous, and liquid). A complete set of material parameters (electrical, thermal, mechanical) in this range and under heating rates comparable to device operation (expected to result in metastable phases as discussed earlier), as well as amorphization, nucleation, and growth models are needed; granularity in the poly/nanocrystalline phase-change material, percolation transport, and filament formation are also expected to be important and defects and local variations within the material can be a determining factor in this type of transport; changes in the material density as a function of temperature and phase induce stress at interfaces between different phases; in the liquid state, convective heat transfer may become important, and surface tension may lead to voids within the element; and electro-migration induces elemental segregation and failure of cells.[49] A complete model for PCM devices would ideally incorporate all these aspects, which is rather challenging from a computational point of view. Yet simplified models that capture the main effects observed experimentally are required for a better understanding of the PCM operation.

There are also significant gaps in knowledge related to the material parameters and electrothermal transport under extreme thermal gradients and across interfaces. When the device size is comparable to the phonon mean free path, or when the thermal gradient is significant over this distance (>1 K/nm), heat transport is expected to change from diffusive to ballistic and the Fourier heat treatment is no longer valid.[51,52] However, there is still very limited experimental data to validate proposed models. PCM devices are larger than the phonon mean free path in GST (estimated to be ~1 nm in c-GST at 300 K and to decrease with temperature[53]), but the very large thermal gradients established under typical operation, ~10–100 K/nm, are expected to result in significant ballistic thermal effects. The ballistic heat models are too complicated to be solved using finite element tools,[52] but it has been shown that the Fourier model with an effective reduced thermal conductivity gives reasonably close results.[52,54] This effective $k(T)$ can in principle be extracted using a technique similar to what we have used to extract $\rho(T)$ and *effective-$k(T)$* of silicon wires self-heated to melting (under thermal gradients ~1 K/nm) by matching simulated and experimental I–V characteristics.[55] Phase-change materials require this extraction to be done under heating rates comparable to those of device operation to capture the properties of the metastable phases.

Adoption of PCM as a main memory will require an integrated effort in design and characterization of materials, device geometries, electrical pulses and access devices, as well as fundamental studies on electrothermal transport at nanoscale. PCM operation dynamics are rather complex, and scaling will significantly benefit from physics-based models integrated with circuit simulation tools for array and module design, which can capture important transient effects. If the designs are not optimized, excessive power levels may be required to achieve acceptably low device-to-device variability, which in turn lead to reduced packing density, endurance, and reliability, and hinder multi-bit-per-cell operation.

ACKNOWLEDGMENT

The authors acknowledge the National Science Foundation for supporting part of this work through awards ECCS 0925973 and ECCS 1150960.

REFERENCES

1. J. E. Brewer and M. Gill, *Nonvolatile Memory Technologies with Emphasis on Flash*. Hoboken, NJ, John Wiley & Sons, Inc., 2008.
2. Chip Shot: Intel Micron Sample 20 nm NAND Flash, http://newsroom.intel.com/community/intel_newsroom/blog/2011/04/14/chip-shot-intel-micron-sample-20 nm-nand-flash (Accessed March 10, 2013).
3. J. Hwang, J. Seo, Y. Lee, S. Park, J. Leem, J. Kim, T. Hong, S. Jeong, K. Lee, and H. Heo, A middle-1X nm NAND flash memory cell (M1X-NAND) with highly manufacturable integration technologies, in *Electron Devices Meeting (IEDM), 2011 IEEE International*, Washington, DC, 2011, pp. 9.1. 1–9.1. 4.
4. S. Raoux, G. W. Burr, M. J. Breitwisch, C. T. Rettner, Y. C. Chen, R. M. Shelby, M. Salinga, D. Krebs, S. H. Chen, H. L. Lung, and C. H. Lam, Phase-change random access memory: A scalable technology, *IBM Journal of Research and Development*, 52, 465–480, 2008.
5. G. W. Burr, M. J. Breitwisch, M. Franceschini, D. Garetto, K. Gopalakrishnan, B. Jackson, B. Kurdi, C. Lam, L. A. Lastras, and A. Padilla, Phase change memory technology, *Journal of Vacuum Science & Technology B: Microelectronics and Nanometer Structures*, 28, 223, 2010.
6. H. Wong, S. Raoux, S. B. Kim, J. Liang, J. P. Reifenberg, B. Rajendran, M. Asheghi, and K. E. Goodson, Phase change memory, *Proceedings of IEEE*, 98, 2201–2227, 2010.
7. S. Raoux and M. Wuttig, *Phase Change Materials: Science and Applications*. US: Springer Verlag, 2008.
8. R. Waser, R. Dittmann, G. Staikov, and K. Szot, Redox-based resistive switching memories–nanoionic mechanisms, prospects, and challenges, *Advanced Materials*, 21, 2632–2663, 2009.
9. G. I. Meijer, Who wins the nonvolatile memory race? *Science*, 319, 1625, 2008.
10. International Technology Roadmap for Semiconductors (ITRS): Emerging Research Devices, 2011, http://www.itrs.net/Links/2011ITRS/2011Chapters/2011ERD.pdf (Accessed March 10, 2013).
11. G. W. Burr, B. N. Kurdi, J. C. Scott, C. H. Lam, K. Gopalakrishnan, and R. S. Shenoy, Overview of candidate device technologies for storage-class memory, *IBM Journal of Research and Development*, 52, 449–464, 2008.
12. B. C. Lee, P. Zhou, J. Yang, Y. Zhang, B. Zhao, E. Ipek, O. Mutlu, and D. Burger, Phase-change technology and the future of main memory, *IEEE Micro*, 30, 143, 2010.
13. B. C. Lee, E. Ipek, O. Mutlu, and D. Burger, Phase change memory architecture and the quest for scalability, *Communications of the ACM*, 53, 99–106, 2010.
14. F. Xiong, A. D. Liao, D. Estrada, and E. Pop, Low-power switching of phase-change materials with carbon nanotube electrodes, *Science*, 332, 568, 2011.

15. Y. Choi et al., A 20 nm 1.8 V 8 Gb RAM with 40 MB/s program bandwidth, in *2012 IEEE International Solid-State Circuits Conference*, San Francisco, CA, 2012, pp. 46–48.

16. Micron Announces Availability of Phase Change Memory for Mobile Devices: First PCM Solution in the World in Volume Production, http://investors.micron.com/releasedetail.cfm?ReleaseID=692563 (Accessed March 10, 2013).

17. D. Im, J. Lee, S. Cho, H. An, D. Kim, I. Kim, H. Park, D. Ahn, H. Horii, and S. Park, A unified 7.5 nm dash-type confined cell for high performance PRAM device, in *Electron Devices Meeting, 2008. IEDM 2008. IEEE International*, Washington, DC, 2008, pp. 1–4.

18. I. Kim, S. Cho, D. Im, E. Cho, D. Kim, G. Oh, D. Ahn, S. Park, S. Nam, and J. Moon, High performance PRAM cell scalable to sub-20 nm technology with below 4F2 cell size, extendable to DRAM applications, in *VLSI Technology (VLSIT), 2010 Symposium on*, Honolulu, HI, 2010, pp. 203–204.

19. M. Kang, T. Park, Y. Kwon, D. Ahn, Y. Kang, H. Jeong, S. Ahn, Y. Song, B. Kim, and S. Nam, PRAM cell technology and characterization in 20 nm node size, in *Electron Devices Meeting (IEDM), 2011 IEEE International*, Washington, DC, 2011, pp. 3.1. 1–3.1. 4.

20. A. Faraclas, N. Williams, F. Dirisaglik, K. Cil, A. Gokirmak, and H. Silva, Operation dynamics in phase-change memory cells and the role of access devices, in *VLSI (ISVLSI), 2012 IEEE Computer Society Annual Symposium*, Amherst, MA, 2012, pp. 78–83.

21. D. Ielmini and Y. Zhang, Analytical model for subthreshold conduction and threshold switching in chalcogenide-based memory devices, *Journal of Applied Physics*, 102, 054517, 2007.

22. D. Ielmini, Threshold switching mechanism by high-field energy gain in the hopping transport of chalcogenide glasses, *Physical Review B*, 78, 035308, 2008.

23. R. E. Scheuerlein and S. B. Herner, Non-volatile memory cell comprising a dielectric layer and a phase change material in series, US 2005/0158950, 2005.

24. B. J. Choi, S. H. Oh, S. Choi, T. Eom, Y. C. Shin, K. M. Kim, K. W. Yi, C. S. Hwang, Y. J. Kim, and H. C. Park, Switching power reduction in phase change memory cell using CVD GeSbTe and ultrathin TiO films, *Journal of the Electrochemical Society*, 156, H59, 2009.

25. B. J. Choi, S. Choi, T. Eom, S. H. Rha, K. M. Kim, and C. S. Hwang, Phase change memory cell using GeSbTe and softly broken-down TiO films for multilevel operation, *Applied Physics Letters*, 97, 132107, 2010.

26. Q. Hubert, C. Jahan, A. Toffoli, L. Perniola, V. Sousa, A. Persico, J. Nodin, H. Grampeix, F. Aussenac, and B. de Salvo, Reset current reduction in phase-change memory cell using a thin interfacial oxide layer, in *Solid-State Device Research Conference (ESSDERC), 2011 Proceedings of the European*, Helsinki, Finland, 2011, pp. 95–98.

27. A. Lacaita, Phase change memories: State-of-the-art, challenges and perspectives, *Solid-State Electronics*, 50, 24–31, 2006.

28. Micron Products: Phase Change Memory, http://www.micron.com/products/phase-change-memory (Accessed March 10, 2013).

29. T. Nirschl, J. B. Phipp, T. D. Happ, G. W. Burr, B. Rajendran, M. H. Lee, A. Schrott, M. Yang, M. Breitwisch, and C. F. e. a. Chen, Write strategies for 2 and 4-bit multilevel phase-change memory, in *Electron Devices Meeting, 2007. IEDM 2007. IEEE International*, Washington, DC, 2008, pp. 461–464.

30. D. K. C. MacDonald, *Thermoelectricity: An Introduction to the Principles*. Mineola, NY, Dover Publications, Inc., 2006.

31. L. D. Landau, L. P. Pitaevskii, and E. M. Lifshitz, Thermoelectric phenomena, in *Electrodynamics of Continuous Media*, 2nd ed. Anonymous, Burlington, MA, Butterworth-Heinemann, 1984, p. 97.

32. G. K. Wachutka, Rigorous thermodynamic treatment of heat generation and conduction in semiconductor device modeling, *IEEE Transactions on Computer-Aided Design of Integrated Circuits and Systems*, 9, 1141–1149, 1990.

33. R. Endo, S. Maeda, Y. Jinnai, R. Lan, M. Kuwahara, Y. Kobayashi, and M. Susa, Electric resistivity measurements of Sb2Te3 and Ge2Sb2Te5 melts using four-terminal method, *Japanese Journal of Applied Physics*, 49, 5802, 2010.

34. R. Fallica, J. L. Battaglia, S. Cocco, C. Monguzzi, A. Teren, C. Wiemer, E. Varesi, R. Cecchini, A. Gotti, and M. Fanciulli, Thermal and electrical characterization of materials for phase-change memory cells, *Journal of Chemical & Engineering Data*, 54, 1698–1701, 2009.

35. H. K. Lyeo, D. G. Cahill, B. S. Lee, J. R. Abelson, M. H. Kwon, K. B. Kim, S. G. Bishop, and B. Cheong, Thermal conductivity of phase-change material GeSbTe, *Applied Physics Letters*, 89, 151904, 2006.

36. D. Kim, F. Merget, M. Forst, and H. Kurz, Three-dimensional simulation model of switching dynamics in phase change random access memory cells, *Journal of Applied Physics*, 101, 064512–064512-12, 2007.

37. K. Cil, F. Dirisaglik, L. Adnane, M. Wennberg, A. King, A. Faraclas, M. B. Akbulut, Y. Zhu, C. Lam, and A. Gokirmak, Electrical resistivity of liquid based on thin-film and nanoscale device measurements, *IEEE Transactions on Electron Devices*, 60, 433–437, 2013.

38. J. Reifenberg, E. Pop, A. Gibby, S. Wong, and K. Goodson, Multiphysics modeling and impact of thermal boundary resistance in phase change memory devices, in *Thermal and Thermomechanical Phenomena in Electronics Systems, 2006. ITHERM'06. The Tenth Intersociety Conference on*, San Diego, CA, 2006, pp. 106–113.

39. J. P. Reifenberg, K. W. Chang, M. A. Panzer, S. Kim, J. A. Rowlette, M. Asheghi, H. S. P. Wong, and K. E. Goodson, Thermal boundary resistance measurements for phase-change memory devices, *Electron Device Letters, IEEE*, 31, 56–58, 2010.

40. D. Kencke, I. Karpov, B. Johnson, S. J. Lee, D. Kau, S. Hudgens, J. Reifenberg, S. Savransky, J. Zhang, and M. Giles, The role of interfaces in damascene phase-change memory, in *Electron Devices Meeting, 2007. IEDM 2007. IEEE International*, Washington, DC, 2007, pp. 323–326.

41. S. W. Lee, H. D. Cho, G. Panin, and T. Won Kang, Vertical ZnO nanorod/Si contact light-emitting diode, *Applied Physics Letters*, 98, 093110–093110-3, 2011.

42. D. Groulx and W. Ogoh, Solid-liquid phase change simulation applied to a cylindrical latent heat energy storage system, in *Proceedings of the 6th Annual COMSOL Conference*, Boston, MA, 2009.

43. A. Faraclas, N. Williams, A. Gokirmak, and H. Silva, Modeling of set and reset operations of phase-change memory cells, *Electron Device Letters, IEEE*, 32, 1737–1739, 2011.

44. L. Adnane, F. Dirisaglik, M. Akbulut, Y. Zhu, C. Lam, A. Gokirmak, and H. Silva, High temperature seebeck coefficient and electrical resistivity of GeSbTe thin films, *Bulletin of the American Physical Society*, 2012.

45. N. Kan'an, A. Faraclas, N. Williams, H. Silva, and A. Gokirmak, Computational analysis of rupture oxide phase change memory cells, *IEEE Transactions on Electron Devices*, 60, 1649–1655, 2013.

46. COMSOL-Multiphysics Modeling, http://www.comsol.com/ (Accessed March 10, 2013).

47. D. T. Castro, L. Goux, G. A. M. Hurkx, K. Attenborough, R. Delhougne, J. Lisoni, F. J. Jedema, M. A. A. 't Zandt, R. A. M. Wolters, D. J. Gravesteijn, M. Verheijen, M. Kaiser, and R. G. R. Weemaes, Evidence of the thermo-electric Thomson effect and influence on the program conditions and cell optimization in phase-change memory cells, in *IEEE International Electron Devices Meeting*, Washington, DC, 2007, pp. 315–318.

48. J. Oosthoek, K. Attenborough, G. Hurkx, F. Jedema, D. Gravesteijn, and B. Kooi, Evolution of cell resistance, threshold voltage and crystallization temperature during cycling of line-cell phase-change random access memory, *Journal of Applied Physics*, 110, 024505, 2011.

49. A. Padilla, G. W. Burr, K. Virwani, A. Debunne, C. T. Rettner, T. Topuria, P. M. Rice, B. Jackson, D. Dupouy, A. J. Kellock, R. M. Shelby, K. Gopalakrishnan, R. S. Shenoy, and B. N. Kurdi, Voltage polarity effects in GST-based phase change memory: Physical origins and implications, in *Electron Devices Meeting (IEDM), 2010 IEEE International*, San Francisco, CA, 2010, pp. 29.4.1–29.4.4.

50. A. Cywar, J. Li, C. Lam, and H. Silva, The impact of heater-recess and load matching in phase change memory mushroom cells, *Nanotechnology*, 23, 225201, 2012.

51. G. D. Mahan and F. Claro, Nonlocal theory of thermal conductivity, *Physical Review B*, 38, 1963, 1988.

52. G. Chen, Ballistic-diffusive heat-conduction equations, *Physical Review Letters*, 86, 2297–2300, 2001.

53. E. K. Kim, S. I. Kwun, S. M. Lee, H. Seo, and J. G. Yoon, Thermal boundary resistance at GeSbTe/ZnS: SiO interface, *Applied Physics Letters*, 76, 3864, 2000.

54. B. C. Larson, J. Z. Tischler, and D. M. Mills, Nanosecond resolution time-resolved x-ray study of silicon during pulsed-laser irradiation, *Journal of Materials Research*, 1, 1986.

55. G. Bakan, L. Adnane, A. Gokirmak, and H. Silva, Extraction of temperature dependent electrical resistivity and thermal conductivity from silicon microwires self-heated to melting temperature, *Journal of Applied Physics*, 112, 063527–063527-9, 2012.

Part VI

Resistive Random Access Memory

12 Nonvolatile Memory Device

Resistive Random Access Memory

Peng Zhou, Lin Chen, Hangbing Lv, Haijun Wan, and Qingqing Sun

CONTENTS

12.1 INTRODUCTION

12.1.1 RESISTIVE RANDOM ACCESS MEMORY: HISTORY AND EMERGING TECHNOLOGY

In general, nonvolatile memory (NVM) can be divided into two major parts. Most NVM devices in mobile and embedded applications today are based on charge storage, which can also be called as capacitive memories like flash, and the other NVM devices are based on various kinds of resistance switching mechanism of inorganic, organic, and molecular materials, which can also be called as resistive memories. Those capacitive memory devices have some general shortcomings, like slow programming, limited endurance, and the need for high voltages during programming and erase. Resistive switching random access memory (RRAM) devices are considered as the most attractive candidates for next-generation NVM applications, owing to their excellent merits including very low operation voltage, low power consumption, and simple device structure. The unique resistance switching behavior under the applied voltages in oxides has been found in 1960s by several groups individually [1–4]. However, with the physical limitation approaching of flash on the international technology roadmap for semiconductors, the interest on the resistance switching in oxides has been renewed in 2004 International Electron Device Meeting by Samsung [5]. Figure 12.1 shows the first complementary metal-oxide-semiconductor (CMOS) process compatible 1transistor–1resistor (1T–1R) structure based on NiO.

More candidate materials for these memories including doped perovskite $SrZrO_3$ [6], ferromagnetic materials like $(Pr,Ca)MnO_3$ [7], and transition metal oxides (BTMO) such as NiO [8], TiO_2, [9] Al_2O_3, [10], ZrO_2, [11], and CuxO [12] were proposed in past years. Compared to ternary or quaternary oxide semiconductor films such as Cr-doped $SrZrO_3$ or $(Pr,Ca) MnO_3$, binary metal oxides have the advantage of a simple fabrication process and are more compatible with CMOS process. In the last several years, binary transition metal oxide (BTMO)-based RRAM have been studied intensively by industry and universities. The number of scientific papers and contributions to conference

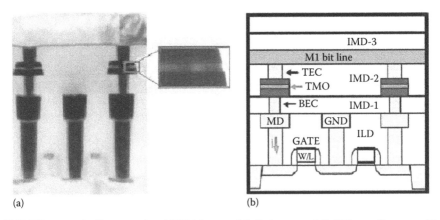

(a)   (b)

FIGURE 12.1   (a) Cross-sectional TEM image of fully integrated OxRRam cell array with magnified polycrystalline BTMO inset, and (b) corresponding schematic diagram. (From Baek, I.G. et al., *IEDM Tech. Dig.*, 587, 2004.)

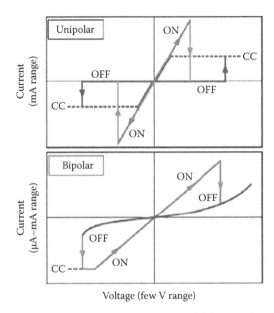

FIGURE 12.2 Unipolar and bipolar switching schemes. CC denotes the compliance current. (From waser, R. and Aono, M., *Nat. Mater.*, 6, 833, November 2007).

is continuously increasing, especially on the NiO (Samsung), TiO_2 (Seoul National University), CuOx (Spansion), WOx (Macronix), and ZrO_2 (Fudan University).

The typical memory characteristic curve and the basic definition of RRAM are shown as Figure 12.2.

When the applied voltage reached a certain value, the current of the device increased abruptly, which represented the first transition from initial state to stable low-resistance state (LRS) appears. It is well known as forming process. Sequentially, by sweeping a negative voltage, the switching from LRS to high-resistance state (HRS) occurs, exhibiting an abrupt current decrease (RESET process). Then, as the applied voltage increases from 0 V, the switching from HRS to LRS (SET process) happens. For the SET and the forming process, an appropriate current compliance should be configured for memory switching. It means that the current is limited by measurement when it reaches a preset value. In general, if the SET and RESET voltages keep the same direction like both negative voltage or both positive voltage, this resistive switching can be thought as "unipolar switching mode," and if the polarity of the SET and RESET voltages is in opposite direction like positive voltage for SET and negative voltage for RESET, this resistive switching can be thought as "bipolar switching mode."

The development of BTMO-RRAM needs reliable switching behavior and data retention property to further scale the NVM down into the sub-20 nm regime. Hence, the bottleneck challenge is the trustable electronic switching mechanism of RRAM.

12.1.2 CHALLENGE FOR RRAM ON STORAGE-CLASS MEMORY

The most possible application for RRAM on the current technology stage is the embedded application for mobile storage demand. However, if RRAM can dominate the memory market in future, it must meet the storage-class requirement. The storage-class memory (SCM) can be defined as a memory that combines the benefits of a solid-state memory, such as high performance and robustness, with the archival capabilities and low cost of conventional hard-disk magnetic storage [14].

For the SCM application, two major issues must be evaluated:

1. Performance requirement

    SCM needs highly reliable, high speed, and high density to replace the hard-disk magnetic storage and flash-based solid-state drive. The best record now is the NiO-based RRAM reported by Samsung as shown in Figure 12.3 [15]. As Figure 12.3b and c shows, a single 1.5 and 1 V pulse with a 10 ns duration has programmed the crossbar memory device successfully for SET and RESET, respectively. The write current also can be reduced to 10–30 μA with the cell size scaling to 50 × 50 nm. However, the possible switching mechanism of RRAM indicates that the reliability cannot reach the SCM demand. Actually, still not any work claims the endurance is above 10^6 times and the retention maintains 10 years under the critical condition for the same device cell. From this point of view, it is an overburden to RRAM to provide an SCM solution in the future.

2. Architecture requirement

    There are two RRAM architectures for different applications. One is 1T–1R architecture, and the other is crossbar architecture. Generally, 1T–1R is used as embedded memory in SOC or other application system; the competitors in the embedded memory application including multiple-time programmable and one-time programmable are the general floating gate–based NVM

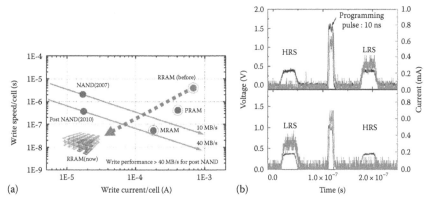

(a)   (b)

FIGURE 12.3 (a) Comparison of the cell write current vs. cell write speed for flash, PRAM, MRAM, and RRAM. Current RRAM materials using Ti-doped NiO show fast programming speeds on the order of 10 ns and low monitoring pulses with the programming pulse in between (black line), and switching from HRS to LRS induced by a single 1.5 V pulse with a 10 ns duration (red line). The cell size is 500 × 500 nm, and the LRS current of about 300 1A has been calculated by measuring the voltage across a 50 O resistor. (b) Switching from LRS to HRS driven by a single reset 1 V pulse with a 10 ns duration. (From Lee, M.J. et al., *Adv. Mater.*, 19, 3919, 2007.)

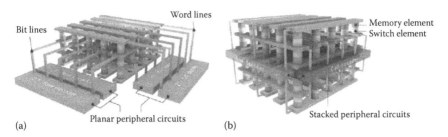

(a)   Planar peripheral circuits   (b)

FIGURE 12.4 (a) Duration of stacked memories with peripheral circuit. (b) Conceptual diagram for ideal staking structure utilizing stackable peripheral circuits. (From Lee, M.-J. et al., *Adv. Funct. Mater.,* 19, 1587, 2009.)

devices, such as Sidence's 1T-Fuse technology, Kilopass's XPM in 40 nm line, and Fujitsu's e-fuse type in 65 nm line. There are no advantages of 1T–1R RRAM for the single memory cell size, power consumption, reliability, and logic process compatibility comparing with the competitors given earlier. 1T–1R RRAM is much good at in some special applications like transparent memory or flexible memory [16]. However, the future of RRAM is dependent on the SCM development. The SCM-RRAM must use crossbar architecture in the applications, especially on the oxide-based stackable 3-D memory cells as shown in Figure 12.4 [17].

12.2 BINARY TRANSITIONAL METAL OXIDE-BASED RRAM

12.2.1 DEVICE FABRICATION AND CURRENT–VOLTAGE CHARACTERIZATION

12.2.1.1 Device Fabrication

A RRAM memory cell has a simple MIM structure that is composed of insulating or semiconducting materials sandwiched between two metal electrodes. An experimental method of fabricating an RRAM cell comprises at least three steps: forming a bottom electrode, depositing a metal oxide layer, and forming a top electrode on the metal oxide layer. Because of the simple structure, RRAM cells are easily integrated into highly scalable crossbar arrays, where the simple design reduces the cell size to 4 F^2 per bit and allows for relatively easy alignment.

A generalized crossbar array memory structure is shown in Figure 12.5a. This crossbar structure enables the circuit to be fully tested for manufacturing defects and to be subsequently configured into a working circuit [18]. Although the crossbar structure can induce high density of the memory integration, a typical problem existed that crosstalk between neighboring devices hinders the memory cells from being randomly accessed [15]. Figure 12.5b shows a typical "crosstalk" behavior of the simplest 2 × 2 cross-point cell array without switching elements. Although we want to read the information of the cell in the HRS surrounded. With three cells in the LRS, the reading current can easily flow through the surrounding cells in the LRS and thus transmit erroneous LRS information. Consequently, for any practical high-density crossbar RRAM array, elimination of crosstalk requires a rectifying element to be included in each memory cell called 1D–1R (one diode one resistor) to prevent "sneak" currents

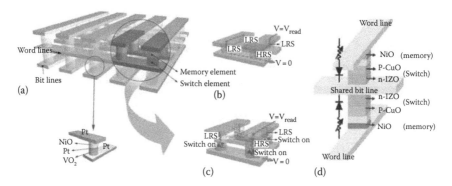

FIGURE 12.5 (a) Crossbar memory whose 1 bit cell of the array consists of a memory element and a switch element between word line and bit line. (b) Reading interference in an array without switch elements. (c) Rectified reading operation in an array with switch elements. (d) Detailed structure of a single cell consisting of a Pt/NiO/Pt memory and a Pt/VO$_2$.

from passing through nonselected cells, as shown in Figure 12.5c. The switch elements with rectifying behaviors can be fabricated by a traditional semiconductor p–n junction. However, considering the process compatibility, the oxides, p–n junctions, or rectifying elements were also fabricated, as shown in Figure 12.5d and e [19].

In order to suppress the reset current and select the operating cell in the memory arrays, integration of transistors is necessary during the fabrication to form a 1T–1R structure; Samsung has provided a new structure fabrication concept with a GaInZnO (GIZO) thin-film transistors integrated with 1D (CuO/InZnO)–1R (NiO)) structure oxide memory node element. All-oxide-based device components for high-density nonvolatile data storage with stackable structure become possible.

12.2.1.2 Current–Voltage Characterization

Based on current–voltage characteristics, the switching behaviors can be classified into two types: one is called unipolar (or symmetric) when the switching procedure does not depend on the polarity of the voltage and current signal. The other is called bipolar (or antisymmetric) when the set to an ON state occurs at one voltage polarity and the reset to the OFF state on reversed voltage polarity.

In unipolar resistive switching [20], the switching direction depends on the amplitude of the applied voltage but not on the polarity. During the I–V characterization, RRAM cell needs a so-called forming process at first. An as-prepared memory cell is in a highly resistive state and is put into an LRS by applying a high-voltage stress. After the forming process, the cell in an LRS is switched to an HRS by applying a threshold voltage, which is called "reset process." Switching from an HRS to an LRS is achieved by applying a threshold voltage that is larger than the reset voltage, which is called "set process." In the set process, the current is limited by the current compliance of the control system, or, more practically, by adding a series resistor, the current compliance can protect the device against the hard breakdown.

The bipolar switching shows directional resistance switching according to the polarity of the applied voltage. By sweeping the applied positive voltage from zero to a certain voltage with a compliance current of 10 mA, an abrupt increase in current was observed, and the LRS was reached. The device remains in the LRS after the soft breakdown of the film when it is swept back to 0 V. Subsequently, the polarity of the voltage is changed by sweeping the gate voltage from zero to negative, the resistance of the sample is abruptly increased at a certain negative voltage, and it means that the sample switches back to an HRS. Hence, reversible bipolar resistive switching was observed with an on/off resistance ratio, which provides a large enough window for readout.

For memory application, the electrical pulse characteristics are used to write/erase and read the device. Write/erase voltages should be in the range of a few hundred mV to be compatible with scaled CMOS to few V. Read voltages need to be significantly smaller than write voltages in order to prevent a change of the resistance during the read operation. The endurance and retention characteristics of the devices should be tested, as shown in Figure 12.6 [21]. A data retention time of >10 years is required for universal NVM. This retention time must be kept at thermal stress up to 85°C and small electrical stress such as a constant stream of read voltage pulses [22].

Usually, the reproducible resistive switching cycles can't be obtained through the single electrical pulse applied between the two electrodes. What's more, due to the dispersions of reset and set voltages existed during the single electrical pulse operations, there is a possibility that the device would be SET back to LRS just after reset process. Multi-pulse mode such as ramped-pulse series (RPS) is proposed to avoid the operation instability and minimize the reset voltage dispersion for the RESET operation [23]. The RPS is a kind of write–verify algorithm, which includes a series of pulses. These pulses increase from the initial (V_{start}) to the last voltage (V_{end}) with a fixed step (V_{step}). All single pulses of RPS are the same in duration but different in amplitude. The V_{end} is determined by a maximum V_{reset}. A read process is performed after each single pulse. Once the resistance reaches the reference value of HRS, the RESET process will be terminated, and the remaining pulses are canceled.

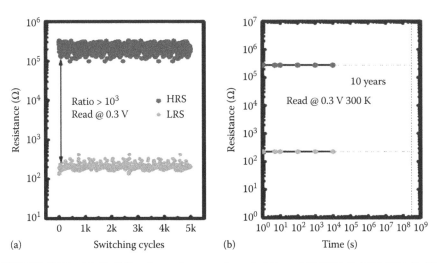

FIGURE 12.6 (a) Statistics for endurance characteristics of TaN/Al$_2$O$_3$/NbAl)/Al$_2$O$_3$/Pt RRAM cell. (b) Retention test of the device after 10^3 write/erase cycles, both.

12.2.2 BTMO-RRAM INTEGRATION FOR EMBEDDED APPLICATION ON 0.18 µm AL PROCESS AND 0.13 µm CU PROCESS

Before taking into practical application, the issue of how to integrate RRAM onto the standard process should be well solved, because this issue will not only determine the whole wafer cost, but also influence the device performance greatly. One competitive method is to integrate the RRAM structure on the backend of process line (BEOL), instead of the front end of line (FEOL), considering the low-temperature budget of RRAM fabrication and material contamination. In this chapter, the integration flow of RRAM with one transistor/one resistor (1T–1R) structure on Cu process and Al process will be discussed, with attention being focused on low cost and high reliability. CuO_x material and WO_x material will be taken as examples for switching material.

12.2.2.1 RRAM Integration on 0.18 µm Al Process

From the view of integration on Al process, WO_x is the first choice for RRAM application, because the WO_x material can be easily formed on each layer of W plug just by tuning very little process on the basis of standard production flow; thus, the research cost and time to market can be greatly reduced. In this section, the integration flow of WO_x-based RRAM will be discussed.

Figure 12.7 illustrates the fabrication process of WO_x-based RRAM with 1T–1R structure. The WO_x memory layer is formed on the contact of W plugs. With initial reference to step 1, a selective transistor and a W plug contacting the transistor are far formed after FEOL process. Thereafter, the wafer is transferred to an oxidation chamber for the growth of WO_x on W plugs, as shown in step 2. The oxidation can be performed by thermal oxidation at elevated temperatures or O-containing plasma oxidation at somewhat lower temperatures, even or by wet-chemical oxidation. By adjusting the temperature, pressure, power, and time, the quality and thickness of WO_x film can be well controlled. It should be mentioned that the W plugs contacting logic transistor are also oxidized simultaneously, which will cause underlying reliability issue on the periphery circuit, so the tungsten oxide on these parts should be cleaned completely in subsequent process. Next, with reference to step 3, a conductive layer such as TiN, Ti, Al–Cu, or W is deposited on top of the WO_x, by means of physical vapor deposition (PVD) or plasma-enhanced deposition. Using suitable photolithographic technique, this conductive layer is patterned as shown in step 4. Then, in step 5, a dry etching step is conducted by Cl-containing metal etch chemistry to remove the part of conductive layer, which is unprotected by the photo-resist (PR). Next, the PR is ashed by O_2 plasma, followed by a chemical wet clean step to remove the residual produced during etching and the WO_x layer on W plugs of logic parts. Thereafter, the RRAM device with MIM structure is accomplished, with W plug as bottom electrode, WO_x as switching layer, and patterned conductive layer as top electrode.

Next, with reference to step 6, a metal layer of Ti/TiN/Al(Cu) is provided over the resulting structure after sputtering the native oxide off the top of the electrode and is patterned as shown in step 7, using appropriate photolithographic techniques. In step 8, the common plate (M1) connecting the top electrodes of RRAM devices is formed after dry etching and PR stripping process.

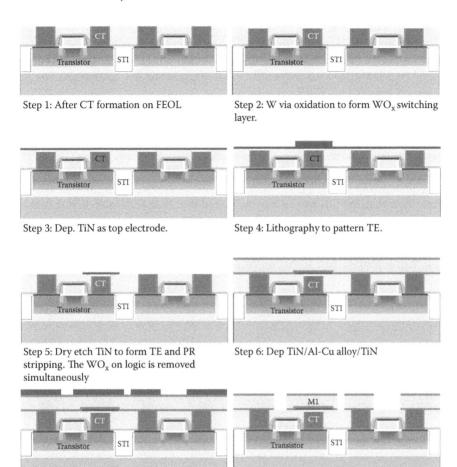

Step 1: After CT formation on FEOL

Step 2: W via oxidation to form WO$_x$ switching layer.

Step 3: Dep. TiN as top electrode.

Step 4: Lithography to pattern TE.

Step 5: Dry etch TiN to form TE and PR stripping. The WO$_x$ on logic is removed simultaneously

Step 6: Dep TiN/Al-Cu alloy/TiN

Step 7: Lithography to pattern M1.

Step 8: Dry etch to form M1 and PR stripping.

FIGURE 12.7 Integration flow of WO$_x$-based RRAM on Al process.

In this structure, each of the memory devices is in series with the selective transistor, with the gates of the transistor being the word lines and the bit lines being the common plate connecting the top electrode of RRAM. The schematic drawing of the process flow provided earlier is just a small part of the overall memory array.

12.2.2.2 RRAM Integration on 0.13 μm Cu Process

BTMOs have advantages of simple composition and good compatibility with CMOS processes. Taking CuO$_x$ for an example, the Cu material is widely used in the current art of advanced interconnect process. The fabrication of CuO$_x$ material is fully based on the apparatus of standard process, which will greatly reduce the costs for both research and production. In this section, the integration flow of CuO$_x$-based RRAM will be discussed.

Figure 12.8 shows a specified process flow for integrating CuO$_x$ RRAM onto the BEOL of Cu process with 1T–1R structure, where each of the memory devices is in series with a select transistor and the gates of the transistor are the word lines and the M2 connecting the top electrodes are the bit lines. The flow steps from substrate preparing to bit line completing. With initial reference to step 1, the substrate is formed on a semiconductor wafer after FEOL process and M1 connection. The M1 is exposed just after Cu CMP.

Next, with reference to step 2, a bilayer of thin SiN and TEOS with a thickness of 50 and 50 nm, respectively, is deposited over the M1 by plasma-enhanced deposition. By using suitable photolithographic techniques, the TEOS layer is patterned in step 3 to provide an opening area for memory cell. After lithography, the TEOS in the opening is removed by reactive ion etching with the etch stopping at the SiN layer, as shown in step 4. Next, in step 5, the PR is stripped by O_2 plasma and wet clean. After that, the SiN layer is further etched to expose Cu in step 6. It should be noted that two-step etch process is adopted to generate an opening for memory cell,

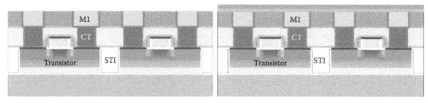

Step 1: After M1 formation on BEOL   Step 2: Dep. 50 nm SiN and 60 nm TEOS on M1

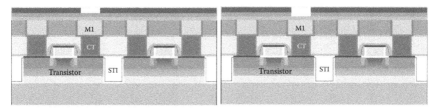

Step 3: Lithography for memory cell pattern   Step 4: Dry etch TEOS stopping at SiN layer

Step 5: PR stripping and wet clean   Step 6: Dry etch SiN and expose Cu substrate

Step 7: Cu oxidation to form CuO_x switch layer   Step 8: Dep. TaN as top electrode

FIGURE 12.8 Integration flow of CuO_x-based RRAM on Cu process.

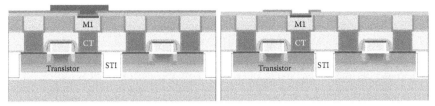

Step 9: Lithography to pattern TE

Step 10: Dry etch TaN to form TE/PR stripping

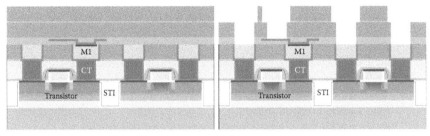

Step 11: Dep. SiN/TEOS/SiN/TEOS

Step 12: V1 and M2 opening

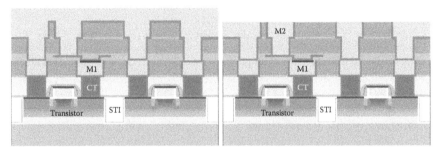

Step 13: Dep TaN/Seed Cu/ECP Cu

Step 14: CMP to form M2

FIGURE 12.8 (continued)   Integration flow of CuO_x-based RRAM on Cu process.

considering the ashing process of PR will unexpectedly oxidize the Cu substrate and have uncontrollable influence on device performance.

Next, the wafer is transferred to an oxidation chamber for the growth of CuO_x on Cu substrate, as shown in step 7. The oxidation can be accomplished by any number of means, including thermal oxidation by O_2 at elevated temperatures or reduced-pressure oxidation in an O-containing plasma at somewhat lower temperature. By adjusting the temperature, pressure, power, and time, the quality and thickness of CuO_x film can be precisely controlled.

With reference to step 8, top electrode material such as TaN, Ta, Ru, Ti, TiN, or bilayer is deposited by PVD or plasma enhanced chemical vapor deposition (PECVD) technique, with a thickness of 50 nm. Thereafter, the top electrode is patterned by lithography in step 9, followed by dry etching using typical Cl-containing metal etch chemistry in step 10. PR is stripped by a sequential O_2 plasma and organic solvent process. The RRAM device with MIM structure is thereafter formed in contact with a drain of transistor by W plug and M1. With reference to step 11, SiN capping layer, first insulating layer between metals (IMD), etching stopping layer, second IMD, and SiON antireflection layer are subsequently deposited by PECVD technique. Via 1 and trench for M2 are formed by lithography and dry etching in step 12. After that, TaN/Ta barrier layer and Cu seed layer are deposited by PVD technique after sputtering the native oxide off the tops of the electrode exposed by Via 1. Electrochemical plating (ECP) Cu is then employed to fill the Via and the trench, as shown in step 13. After a short annealing process to enlarge the crystal size of ECP Cu, a chemical–mechanical polishing step is undertaken to remove the portions overlying the IMD layer, as shown in step 14; thus, Cu plugs and M2 are formed.

This integration scheme has the following advantages:

1. High reliability.
   a. After CuO_x switching layer formation, the top electrode is deposited on it directly, avoiding unnecessary contamination on CuO_x layer.
   b. During oxidation, the logic part is protected by SiN layer, thus increasing the reliability of circuit.
2. The material for top electrode can be adjusted in a wide range.
3. Multistack structure can be realized easily.

12.2.3 DOPING EFFECT IN BTMO-RRAM

Artificially doping impurity in electron devices modifies their electronic transport and can be useful in improving their performance. The effects of impurity doping on resistive switching characteristics in binary metal oxide films have been reported in some studies [24–31].

Jung et al. [29] investigated the effects of lithium doping on bistable resistance switching in polycrystalline NiO film. They concluded that doping metallic impurity can improve the thermal stability of the off state in undoped NiO films, resulting in a much better retention property in the off state and stable on/off operation as shown in Figure 12.9. For the Li-doped device, both on and off currents were found to be stable and constant with a small value for standard deviation. However, for the undoped device, only its on current was stable.

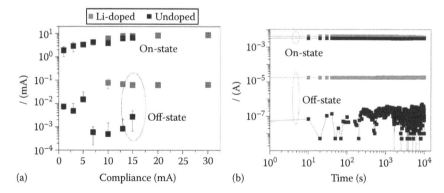

FIGURE 12.9 (a) The stability of on and off currents. (b) The measured retention properties of the on and off states. (From Jung, K. et al., *J. Appl. Phys.*, 103, 2008.)

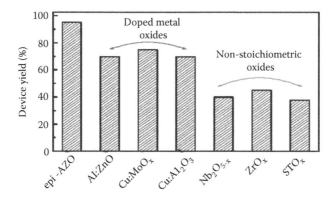

FIGURE 12.10 Comparison of several oxides with and without doping impurity on device yields.

The Dongsoo Lee group has investigated various doped metal oxides such as copper-doped molybdenum oxide, copper-doped Al_2O_3, copper-doped ZrO_2, aluminum-doped ZnO, and Cu_xO for novel resistance memory applications [31]. Compared with nondoped RRAM devices, doped metal oxides show much better device yields, as shown in Figure 12.10.

Guan et al. [28] reported a resistive switching memory device utilizing gold nanocrystals embedded in the zirconium oxide layer. They stated that the intentionally introduced golden nanocrystal, acting as the electron traps, provides an effective way to improve the device yield.

These studies suggested that this doping effect is most likely associated with the local enhancement or concentration in electric field induced by the embedded fine metallic impurity. These doped impurities or nanocrystals may provide easy path to form a fixed conducting filament in the thin films. Therefore, the fluctuation of switching parameters could be stabilized, and the devices yield can be improved through the doping method.

12.2.4 ROLE OF COMPLIANCE CURRENT

Proper compliance current is usually needed to protect the resistive switching material from permanent breakdown, for a large transient current will occur during the transition from HRS to LRS. It has been reported that I_{comp} is a key parameter that influences the resistive switching behaviors, especially for the value of the on-state resistance (R_{on}) and the reset current (I_{reset}, defined as the peak current during the reset process). The relationships of I_{reset} vs. I_{comp} (I_{reset} increases with increasing I_{comp}) and R_{on} vs. I_{comp} (R_{on} decreases when I_{comp} increases) for I_{comp} 1 mA have already been reported by several groups [9,32,33].

The RESET current was found to increase together with the SET compliance current for BTMO-RRAM. The increased SET compliance current is likely related to the density increase in conductive filament, which induces more current to generate more heat to rupture the filaments. According to the Joule heating effect–caused filament rupture model, a proper current density is needed for the RESET switching.

As the SET compliance current increases, more conductive filaments are formed and lower on-state resistance is achieved. In order to reach a certain current density to rupture these conductive filaments, larger RESET current is needed for the higher SET compliance current condition. Increasing the applied voltage can generate higher current. However, according to the Ohm's law, the lower resistance can also offer higher current in the same voltage. Considering this, the inconspicuous increase in RESET voltage can be easily understood. In "nonuniform, flawed filament" model, the rupture of filament is thought to take place only at high-resistance flaw inside the filament, because the highest temperature can be generated thereby Joule heating. As long as the critical temperature reaches at the flaw inside the filament by external current, regardless of the polarity, the RESET switching will occur.

The switching behavior of RRAM with different architecture can elucidate the compliance current effect much clearly. Figure 12.11 shows the typical bipolar resistive switching characteristics of Cu_xO-based RRAM with 1R architecture. The SET under positive voltage sweep with a compliance current, and then RESET without a compliance current. Different current compliances are used in the set processes, as shown in Figure 12.11. The resistive switching characteristics can be greatly influenced by I_{comp}, especially for the I_{reset} and R_{on}. As it can be seen from Figure 12.11, when I_{comp} is above 1 mA, I_{reset} decreases almost linearly with I_{comp}; however, when I_{comp} decreases below 1 mA from 600 to 200 μA, I_{reset} keeps stable (~1 mA). The similar phenomenon is also reported by Kinoshita et al., in which $I_{reset} \approx I_{comp}$ is observed for $I_{comp} \geq 1$ mA and I_{reset} is 2–3 mA independent of I_{comp} for $I_{comp} <$ 1 mA [33]. The relationship between R_{on} and I_{comp} can also be classified into two parts, shown in the inset of Figure 12.11: cycle endurance of 50 times is performed under DC voltage sweep mode for each I_{comp} and the 50 R_{on} values are picked up to plot the relationship between R_{on} and I_{comp}. When I_{comp} increases from 1 to 10 mA,

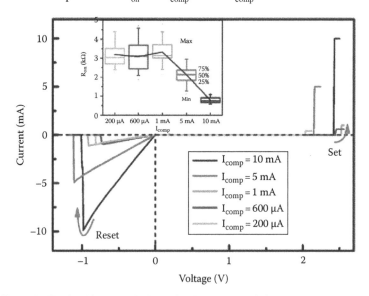

FIGURE 12.11   Typical bipolar resistive switching characteristics of RRAM memory device with 1R architecture. The inset shows the relationship between R_{on} and I_{comp}.

R_{on} decreases from ~3 to ~1 kΩ. That's to say, R_{on} has a negative relationship with I_{comp} when $I_{comp} \geq 1$ mA. However, when I_{comp} is below 1 mA, R_{on} is independent of I_{comp} and distributes around 3 kΩ.

On the other hand, in RRAM with 1T–1R architecture, the relationship of $I_{reset} \approx I_{comp}$ can be clearly observed when I_{comp} is below 1 mA, as shown in Figure 12.12. This may be caused by the excellent capability of confining I_{comp}, for the device is directly fabricated above the contact plug and connected to a transistor in 1T–1R architecture. The gate voltage (V_g) of this transistor is used to control I_{comp}. V_g is maintained at a fixed value 3.3 V (the corresponding I_{comp} is about 2 mA) during the reset process, whereas during the set process, V_g is maintained between 1.25 and 2.65 V (the corresponding I_{comp} distributes from 200 μA to 1 mA). The excellent capability of confining I_{comp} can also be seen from the almost linear relationship between R_{on} and I_{comp}, as shown in the inset of Figure 12.12. Fifty time cycles are performed under DC voltage sweep mode for each I_{comp}. R_{on} decreases from ~10 to ~3 kΩ, when I_{comp} increases from 100 to 600 μA. That's to say, the resistance value of R_{on} can be greatly improved by decreasing the value of I_{comp}, thus decreasing the value of I_{reset}. However, the I–V curve of reset process changes gradually instead of precipitating quickly when I_{reset} decreases to 200 μA or below, as shown in Figure 12.12. It is attributed that the smaller I_{reset} is, with more difficulty the conductive filament will rupture, and thus the reset speed will be negatively affected. In other words, there is a competitive relationship between power consumption and speed. High I_{reset} means not only fast speed, but also high power consumption, while low I_{reset} means low power consumption, but also slow speed. Therefore, a right balance between power consumption and speed should be struck in order to achieve an optimized resistive switching performance.

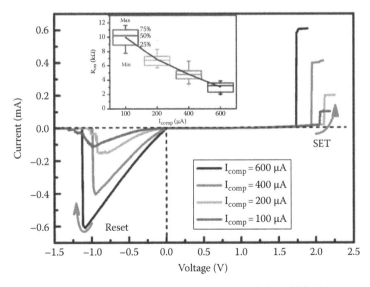

FIGURE 12.12 Typical bipolar resistive switching characteristics of RRAM memory device with 1T–1R architecture based on 0.13 μm logic CMOS technology. The inset shows between R_{on} and I_{comp}. (From Wan, J. et al., *IEEE Electron. Dev. Letts.*, 31, 246, 2010.)

To clarify the resistive switching behaviors under different compliance currents, a self-build compliance current capturing system is set up. As shown in Figure 12.13a, a Keithley 4200 SPA, a Cu_xO-based memory device with 1R architecture and a 2 kΩ sampling resistor are connected in series; an oscillograph is connected with the sampling resistor in parallel to scout the transient current flowing through

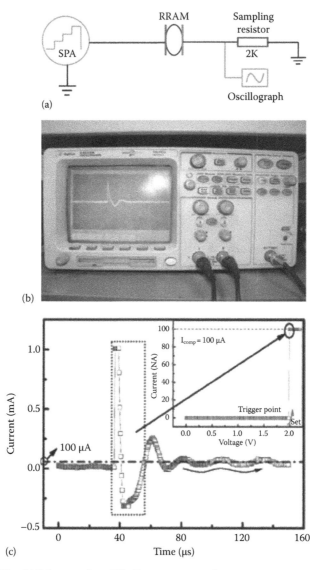

(a)

(b)

(c)

FIGURE 12.13 (a) Schemes of a self-built current capturing system set up to scout the transient current flowing through the memory device during the transition from HRS to LRS; (b) a serious compliance current overshoot phenomenon observed in memory device with 1R architecture; (c) the enlargement of the current overshoot curve in (b), the inset shows the corresponding trigger point from HRS to LRS with 100 μA compliance current.

the memory device during the transition from HRS to LRS. Surprisingly, a serious compliance current overshoot phenomenon is observed in 1R-architecture device, as shown in Figure 12.13b. This current overshoot curve is enlarged and replotted in Figure 12.13c. Although I_{comp} is set as 100 µA during the set process, a large overshoot current about 1 mA is observed at the trigger time point from HRS to LRS. This overshoot current increases quickly from ~0 to 1 mA within only 0.4 µs and then relaxes back to 100 µA in about 50 µs as shown in Figure 12.13c. The whole process happens in only about 50 µs, which is very short comparing with the DC voltage sweep speed (1 ms per step). Therefore, no compliance current overshoot is observed in the normal I–V curve during the set process shown in the inset of Figure 12.13c.

Different compliance currents, such as 200 and 600 µA, are also used in the similar capturing system; once I_{comp} is below 1 mA, the overshoot phenomenon appears and the overshoot current maintains about 1 mA. However, when I_{comp} is larger than 1 mA, the capturing current equals to I_{comp}. In other words, I_{comp} configuration is invalid when I_{comp} is below 1 mA for the existence of the compliance current overshoot phenomenon. That's why I_{reset} and R_{on} are independent of I_{comp} once I_{comp} decreases below 1 mA in 1R architecture.

Based on the earlier observations, the compliance current overshoot phenomenon with 1R architecture may be caused by the parasitic capacitance C, which exists between the external transistor in SPA and the RRAM device. At the set point, RRAM device suddenly switches from HRS to LRS; however, the parasitic capacitance C has already been charged to a certain voltage (equals to the set voltage, V_{set}) before the set transition during the DC voltage sweep process. Once the RRAM device switches from HRS to LRS, the charges stored in the parasitic capacitance C will discharge through the RRAM device and the sampling resistor, which directly induces the occurrence of the compliance current overshoot phenomenon. We can also find that the transient current fluctuates in a wave form before regressing back to 100 µA. It is attributed that, except the existence of the parasitic capacitance, the parasitic inductance L also exists between the RRAM device and SPA even though an external transistor connected between the RRAM device and the sampling resistor can control the discharging current through the sampling resistor. The stored charges in the parasitic capacitance C can still be discharged from another parasitic capacitance C_0, which exists between the RRAM device and the external transistor. Therefore, the resistive switching behaviors can still be affected by the overshoot current. Compared with 1R architecture, the memory device and the transistor are connected directly via a contact plug in 1T–1R architecture, thus the parasitic capacitance of the joint between them can be negligible. That's to say, the discharge current can be perfectly controlled by the internal transistor. Therefore, no compliance current overshoot phenomenon is observed in 1T–1R architecture, and the reduction in parasitic capacitance strongly limits the current overshoot during the set transition, thus limiting the reset current required for its subsequent dissolution. This overshoot current can remarkably affect the resistive switching characteristics in 1R-architecture RRAM, especially when I_{comp} is below 1 mA.

12.2.5   PHYSICAL MECHANISM AND ITS EVIDENCE

Various physical switching mechanisms have been proposed to clarify this important resistance change phenomenon such as (1) conductive filament formation and rupture by Joule heat–induced thermochemical reaction or charge trap/detrap process, (2) mobile anion-induced resistance change, and (3) Schottky barrier modulation by ion movement. It is noted that most models are based on the indirect I–V behavior and analytical fitting and lack of direct evidence. Here three kinds of models with three complementary views were taken to make this bottleneck problem clear.

The first view is Cheol-Seong Hwang's conductive filament model with direct evidence based on TiO_2-RRAM [35] (Figure 12.14). The second view is a theoretical approach of Jinfeng Kang's work based on ZnO-RRAM [36]. The third one is a total physical image to BTMO-RRAM based on TaN/CuxO/Cu sandwich structure [37].

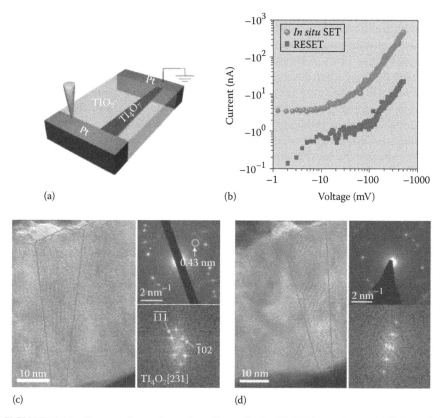

(a)   (b)

(c)   (d)

FIGURE 12.14   Structural transformation after an in situ RESET experiment. (a) Schematic to depict the experimental setup. (b) Local I–V curves in a log scale before and after RESET. The STM probe approached the top electrode, and the I–V curves represent the electrical conduction between the top and bottom electrodes. (c) High-resolution image, diffraction pattern, and fast Fourier transformed micrograph of the Magnéli structure before RESET. (d) The corresponding images after RESET. (From Kwon, D.-H. et al., *Nat. Nanotechnol.*, 5, 148, 2010; Choi, K.M. et al., *Appl. Phys. Lett.*, 91, 012907, 2007.)

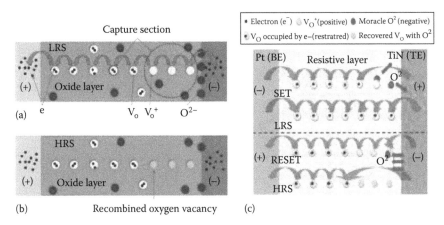

FIGURE 12.15 (a) Schematic illustration of conduction transport in LRS and the reset process of RRAM devices. (b) Schematic view for HRS. (c) Schematic views of the unified physical model for the conduction transport in and the switching processes between LRS and HRS. (From Gao, B. et al., *IEEE Electron. Dev. Letts.*, 30, 1326, 2009; Xu, N. et al., *Appl. Phys. Lett.*, 92(23), 232112, 2008; Xu, N. et al., *2008 Symposium on VLSI Technology Digest of Technical*, pp. 100–101.)

Figure 12.15 shows that oxygen vacancies rearrange to form an ordered structure and induced a stable metallic phase. After RESET, this stable Magneli phase disappeared. Although the high-resolution TEM provided by this work is convincible to TiO_2-based RRAM switching mechanism, there are still a lot of observed phenomena that cannot be understood by the same Magneli phase transformation. Kang's theoretical works also provide another view to clarify this problem.

In Kang's theory, the electron transport characteristics along the filament are calculated based on electron hopping. Current generated by hopping is calculated as $I = -e\Sigma\left[(1-f_n)W_n{}^{iC} - W_n{}^{oC}f_n\right]$, where W_n and W_o denote the electron hopping rate from electrode to oxygen vacancy Vo and from Vo to electrode, respectively. fn is the occupying probability of electron of the nth Vo along the filament. The measured temperature dependence of the reset time is (t_{reset}), where t_{reset} refers to the minimal width of pulse voltage. With increased temperature, t_{reset} is shortened due to faster transport of O^{2-}, and $log(t_{reset})$ is fitted linearly with 1/T, agreeing with the model prediction. For single-filament device, a sharp transition is observed, whereas for multiple-filament device, the transition is gradual (right column) due to different critical voltage for given filaments. Therefore, each filament is ruptured under different voltage, so a gradual transition with voltage is observed.

To give a clear physical picture of RRAM switching, a universal filament/charge trapped combined model is schematically illustrated in Figure 12.16. It is known that most of the trap centers formed by localized states and defects capable of capturing carriers distributing at the grain boundary in the oxide film. It is easy to understand in our proposed schematic model that HRS can be achieved when a portion of trap centers are empty in filaments because they capture charge carriers as shown in Figure 12.16. For the unipolar reset operation, the major contribution should be Joule heating–induced trapped charge release; in other words, the unipolar reset process is

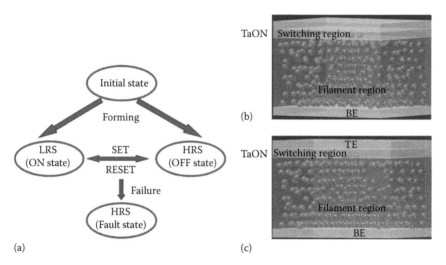

FIGURE 12.16 Physical image for filament/charge trapped combined model. (a) The four states of a normal RRAM device and relative transition process; (b) schematic diagram of the TaN/CuxO/Cu structure; it is composed of the switching region and the filament region. Here the trap centers are empty so that the system is in off state. (c) The on state of the TaN/CuxO/Cu structure. (From Zhou, P. et al., *Appl. Phys. Lett.*, 94(5), 053510, 2009; Wan, H.J. et al., *J. Vac. Sci. Technol. B*, 27, 468, 2009.)

different from the bipolar one. On the other hand, if the trapping centers are already filled with holes during the previous set pulse step or voltage sweep, the charge carriers are not influenced by these filled traps, and the LRS is obtained as shown in Figure 12.16. The conduction in filament region depends on the dynamic trap-release process of charges in neighboring trap centers; Frenkel–Poole emission and ohmic conduction are major contributions for HRS and LRS, respectively. The set and reset occurs at interface as shown in Figure 12.16b and c. It is also thought as a switching region. And reset happens when the trapped charge carriers in switching region are recombined. There are still some trapped charge carriers that cannot be released by recombination or thermal process from trap centers; it will induce failure with this kind of trap center accumulation.

12.3 MEMRISTOR

12.3.1 Leon Chua's Theory of the Fourth Fundamental Element

From the classical circuit-theoretic point of view, there are four basic circuit variable parameters, namely, the charge q, the current i, the voltage v, and the magnetic flux ϕ. Out of the six possible combinations of these four variables, five have led to well-known relationships. Among them, the physical law that relates charge and current is

$$\frac{dq}{dt} = i$$

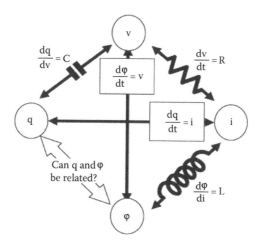

FIGURE 12.17 Possible relations among charge q, current i, voltage v, and magnetic flux φ.

Similarly, the physical law relating flux and voltage is

$$\frac{d\varphi}{dt} = \upsilon$$

These two relations are depicted in Figure 12.17. Besides, as shown in Figure 12.17, three other relationships are already given, respectively, by the axiomatic definition of the three classical fundamental circuit elements, namely, the resistor R, the capacitor C, and the inductor L.

Resistor R is defined by the relation of voltage and current:

$$\frac{d\upsilon}{di} = R$$

Capacitor C is defined by the relation of charge and voltage:

$$\frac{dq}{d\upsilon} = C$$

Inductor L is defined by the relation of magnetic flux and current:

$$\frac{d\varphi}{di} = L$$

But what about the relationship between flux φ and charge q? Can they also be related? For nearly 150 years, the known fundamental passive circuit elements were limited to the capacitor (discovered in 1745), the resistor (1827), and the inductor (1831). Then, in a brilliant but underappreciated 1971 paper "Memristor-The Missing

Circuit Element," Leon Chua, a professor of electrical engineering at the University of California, Berkeley, predicted the existence of a fourth fundamental device, which he called a memristor [38]. He proved that memristor behavior could not be duplicated by any circuit built using only the other three elements. In this paper, the relationship between flux and charge is described by a simple equation:

$$\frac{d\varphi}{dq} = M$$

where M is defined as memristance, the property of a memristor just as the resistance is the property of a resistor. With this new relationship by Chua, we will have six equations relating the four fundamental circuit parameters—R, C, L, and the newfound M.

We know that the circuit components R, C, and L are linear elements, unlike a diode or a transistor, which exhibit a nonlinear current–voltage behavior. However, Chua has proved theoretically that a memristor is a nonlinear element because its current–voltage characteristic is similar to that of a Lissajous pattern. If a signal with a certain frequency is applied to the horizontal plates of an oscilloscope and another signal with a different frequency is applied to the vertical plates, the resulting pattern we see is called the Lissajous pattern. A memristor exhibits a similar current–voltage characteristic [39]. Unfortunately, no combination of nonlinear resistors, capacitors, and inductors can reproduce this Lissajous behavior of the memristor. That is why a memristor is a fundamental element.

How do we understand the meaning of memristor? Memristor is a contraction of "memory resistor," because that is exactly its function: to remember its history. A memristor is a two-terminal device whose resistance depends on the magnitude and polarity of the voltage applied to it and the length of time that voltage has been applied. When you turn off the voltage, the memristor remembers its most recent resistance until the next time you turn it on, whether that happens a day later or a year later. In other words, a memristor is "a device that bookkeeps the charge passing its own port." This ability to remember the previous state made Chua call this new fundamental element a memristor—short form for memory and resistor.

Think of a resistor as a pipe through which water flows. The water is electric charge. The resistor's obstruction of the flow of charge is comparable to the diameter of the pipe: the narrower the pipe, the greater the resistance. For the history of circuit design, resistors have had a fixed pipe diameter. But a memristor is a pipe that changes diameter with the amount and direction of water that flows through it. If water flows through this pipe in one direction, it expands (becoming less resistive). But send the water in the opposite direction and the pipe shrinks (becoming more resistive). Further, the memristor remembers its diameter when water last went through. Turn off the flow, and the diameter of the pipe "freezes" until the water is turned back on.

Chua's memristor was a purely mathematical construct that had more than one physical realization. Conceptually, it was easy to grasp how electric charge could couple to magnetic flux, but there was no obvious physical interaction between charge and the integral over the voltage. Chua demonstrated mathematically that his hypothetical device would provide a relationship between flux and charge similar to what a nonlinear resistor provides between voltage and current. In practice, that would mean the device's resistance would vary according to the amount of charge that passed through it. And it would remember that resistance value even after the current was turned off.

After Chua theorized the memristor out of the mathematical ether, it took another 35 years for scientists to intentionally build the device at HP Labs. So let's turn to the next section.

12.3.2 HP Lab's Discovery of the Prototype-Pt/TiO$_{2-x}$/TiO$_2$/Pt Memristor

We are all familiar with the fundamental circuit elements: the resistor, the capacitor, and the inductor. However, in 1971, Leon Chua reasoned from symmetry arguments that there should be a fourth fundamental element, which he called a memristor (short for memory resistor). Although he showed that such an element has many interesting and valuable circuit properties, until now, no one has presented either a useful physical model or an example of a memristor. Here HP Labs' scientists show, using a simple analytical example, that memristance arises naturally in nanoscale systems in which solid-state electronic and ionic transport are coupled under an external bias voltage. These results serve as the foundation for understanding a wide range of hysteretic current–voltage behavior observed in many nanoscale electronic devices that involve the motion of charged atomic or molecular species, in particular certain titanium dioxide cross-point switches.

As shown in Figure 12.18, two thin layers of TiO_2 are fabricated, one is highly conducting layer with lots of oxygen vacancies $(V_O{}^+)$ and the other layer undoped, which is highly resistive [40]. Oxygen vacancies in TiO_2 are known to act as n-type dopants, transforming the insulating oxide into an electrically conductive

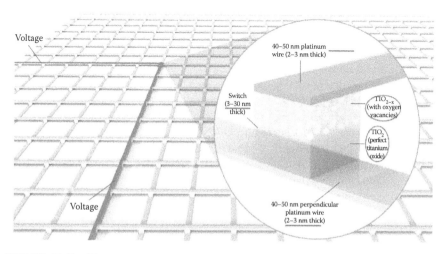

FIGURE 12.18 The crossbar architecture of Pt/TiO$_{2-x}$/TiO$_2$/Pt memristor.

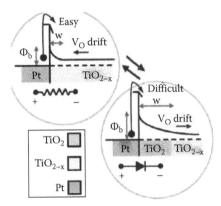

FIGURE 12.19 Metal/semiconductor contact is typically ohmic in the case of Pt/TiO$_{2-x}$ interface and rectifying (Schottky-like) in the case of Pt/TiO$_2$ interface.

doped semiconductor. Good ohmic contacts are formed using platinum (Pt) electrodes on either side of this sandwich of TiO$_2$. A switch is a 40 nm^3 titanium dioxide (TiO$_2$) in two layers: lower TiO$_2$ layer has a perfect 2:1 oxygen-to-titanium ratio, making it an insulator. By contrast, the upper TiO$_2$ layer is missing 0.5% of its oxygen (TiO$_{2-x}$), so x is about 0.05. The vacancies make the TiO$_{2-x}$ material metallic and conductive. Metal/semiconductor contacts are typically ohmic in the case of very heavy doping and rectifying (Schottky-like) in the case of low doping, as shown in Figure 12.19 [41].

The oxygen deficiencies in the TiO$_{2-x}$ manifest as "bubbles" of oxygen vacancies scattered throughout the upper layer. A positive voltage on the switch repels the (positive) oxygen deficiencies in the metallic upper TiO$_{2-x}$ layer, sending them into the insulating TiO$_2$ layer below. That causes the boundary between the two materials to move down, increasing the percentage of conducting TiO$_{2-x}$ and thus the conductivity of the entire switch. The more positive voltage is applied, the more conductive the cube becomes. When more positively charged oxygen vacancies reach the TiO$_2$/Pt interface, the potential barrier for the electrons becomes very narrow, as shown in Figure 12.20, making tunneling through the barrier a real possibility. This leads to a large current flow, making the device turn ON. When the polarity of the applied voltage is reversed, the positively charged oxygen bubbles are pulled out of the TiO$_2$. The amount of insulating resistive TiO$_2$ increases, thereby making the switch as a whole resistive. The more negative voltage is applied, the less conductive the cube becomes. This forces the device to turn OFF due to an increase in the resistance of the device and deduced possibility for carrier tunneling.

A typical memristor device structure is Si/SiO$_x$/Ti 5 nm/Pt 15 nm/TiO$_2$ 25–50 nm/Pt 30 nm, as schematically shown in the upper-left inset to Figure 12.20 [42]. All the metal layers, including Pt and Ti, were deposited via e-beam evaporation. The TiO$_2$ layers were deposited by sputtering from a polycrystalline rutile TiO$_2$ target. The Ti (1.5 nm adhesion layer) + Pt (8 nm) electrode used for the 50 nm × 50 nm nanojunctions was patterned by ultraviolet–nanoimprint lithography. The Ti (5 nm adhesion layer) + Pt (15 nm for BE and 30 nm for TE) electrodes used

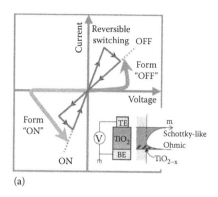
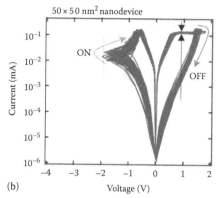

(a)   (b)

FIGURE 12.20 (a) Schematic of the forming step and subsequent bipolar reversible switching. (Inset) Polarity of switching is usually controlled by the asymmetry of the interfaces as fabricated; for all reported devices, the top interface is Schottky-like and the bottom is ohmic-like. (b) Fifty cycles of the bipolar switching for a 50×50 nm^2 nanodevice.

for the microjunctions (5×5 μm) were fabricated using a metal shadow mask. Some samples adopted a highly reduced TiO$_{2-x}$ layer.

The idealized electrical behavior of a memristive oxide switch is shown in Figure 12.20a. Repeatable ON/OFF switching follows a "bowtie"-shaped I–V curve. This repeatable switching is arrived at, however, only after an electroforming step of high positive voltage or high negative voltage changes the device from a virgin near-insulating state into an ON/OFF switching state. As shown in Figure 12.20a, opposite polarities of forming voltage and current typically produce opposite initial states of the switch. After forming, the device resistance decreases by several orders of magnitude, and the majority drop of the applied external voltage shifts from the device to the wires accordingly. After a negative voltage sweep, the device is formed in the ON state while a positive voltage sweep forms the device in the OFF state with the typical ON/OFF resistances shown in Figure 12.20b. After electroforming, the device shows repeatable nonvolatile bipolar switching up to 10^4 cycles. These devices are switched ON by a negative voltage and switched OFF by a positive voltage on the top electrodes. Polarity of switching is usually controlled by the asymmetry of the interfaces as fabricated.

What makes the memristor special is not just that it can be turned OFF and ON, but that it can actually remember the previous state that when the voltage is turned off, positive or negative, the oxygen bubbles do not migrate. They stay where they are, which means that the boundary between the two titanium dioxide layers is frozen. This is because when the applied bias is removed, the positively charged Ti ions (which is actually the oxygen-deficient sites) do not move anymore, making the boundary between the doped and undoped layers of TiO$_2$ immobile. When you next apply a bias (negative of positive) to the device, it starts from where it was left. Unlike in the case of typical semiconductors, such as silicon in which only mobile carriers move, in the case of the memristor, both the ionic and the electron movements, into the undoped TiO$_2$ and out of undoped TiO$_2$, are responsible for the hysteresis in its current–voltage characteristics.

The future application of memristors in electronics is not very clear, since we do not yet know how to design circuits using memristors along with the silicon devices, although it is not difficult to integrate memristors on a silicon chip, since we now have matured technologies to accomplish this. However, since a memristor is a two-terminal device, it is easier to address in a crossbar array. It appears, therefore, that the immediate application of memristors is in building nanoscale high-density non-volatile memristors and field-programmable gate arrays. Chua's original work also shows that using a memristor in an electronic circuit will reduce the transistor count by more than an order of magnitude. This may lead to higher component densities in a given chip area, helping us beat Moore's law.

12.4 CONCLUSION

In summary, BTMO-based RRAM has become one candidate for the next-generation NVM because of its CMOS process compatible, low program voltage and power consumption, high scaling-down ability, and low cost. There are several basic switching mechanisms related to oxygen vacancies and conductive filaments to clarify and optimize the memory device performance. The electrode interface, oxygen concentration and distribution, SET compliance current, temperature, polycrystalline grain boundary, and cell architecture dominate the memory performance jointly. For the BTMO-based RRAM development, the SCM application is the ultimate and critical direction because the embedded application has no advantages compared with the competition technologies.

REFERENCES

1. G. S. Hreynina, *Rad. Eng. Electron. Phys.* 7, 1949, 1962.
2. J. F. Gibbons and W. E. Beadle, *Sol. State Electron.* 7, 785, 1964.
3. T. W. Hickmott, *J. Appl. Phys.* 33, 2669, 1962.
4. K. L. Chopra, *J. Appl. Phys.* 36,184, 1965.
5. I. G. Baek, M. S. Lee, S. Seo, M. J. Lee, D. H. Seo, D.-S. Suh, J. C. Park et al., *IEDM Tech. Dig.* 587, 2004.
6. C. Y. Liu, P. H. Wu, A. Wang, W. Y. Jang, J. C. Young, K. Y. Chiu, and T.-Y. Tseng, *Jun. IEEE Electron. Dev. Lett.* 26, 351, 2005.
7. S. Srivastava, N. K. Pandey, P. Padhan, and R. C. Budhani, *Phys. Rev B.* 62, 13868, 2000.
8. S. Seo, M. J. Lee, D. H. Seo, E. J. Jeoung, D. Suh, Y. S. Joung, and I. K. Yoo, *Appl. Phys. Lett.* 85, 23, 2004; J.-W. Park, D.-Y. Kim, and J.-K. Lee, *J. Vac. Sci. Technol. A* 23(5), 1309, September 2005.
9. C. Rohde, B. J. Choi, D. S. Jeong, S. Choi, J.-S. Zhao, and C. S. Hwang, *Appl. Phys. Lett.* 86(26), 262907, June 2005.
10. K. M. Kim, B. J. Choi, B. W. Koo, S. Choi, D. S. Jeong, and C. S. Hwang, *Electrochem. Solid State Lett.* 9, G343, 2006.
11. D. S. Lee, H. J. Choi, H. J. Sim, D. H. Choi, H. S. Hwang, M.-J. Lee, S.-A. Seo, and I. K. Yoo, *IEEE Electron. Dev. Lett.* 26, 719, 2005.
12. T. Fang, S. Kaza, S. Haddad, A. Chen, Y. Wu, Z. Lan, S. Avanzino et al. *IEDM Tech. Dig.* 789, 2006.
13. R. Waser and M. Aono, *Nat. Mater.*, 6, 833, November 2007.

14. G. W. Burr, B. N. Kurdi, J. C. Scott, C. H. Lam, K. Gopalakrishnan, and R. S. Shenoy, *IBM J. Res. Dev.* 52(4/5), 449–464, September 2008.
15. M. J. Lee, Y. Park, D. S. Suh, E. H. Lee, S. Seo, D. C. Kim, R. Jung et al. *Adv. Mater.* 19, 3919, November 19, 2007.
16. S.-E. Ahn, B. S. Kang, K. H. Kim, M.-J. Lee, C. B. Lee, G. Stefanovich, C. J. Kim, and Y. Park, *IEEE Electron. Dev. Lett.* 30, 550–552, 2009.
17. M.-J. Lee, S. I. Kim, C. B. Lee, H. Yin, S.-E. Ahn, B. S. Kang, K. H. Kim et al., *Adv. Funct. Mater.* 19, 1587–1593, 2009.
18. A. Sawa, *Mater Today* 11, 28–36, June 2008.
19. M. J. Lee, C. B. Lee, S. Kim, H. Yin, J. Park, S. E. Ahn, B. S. Kang et al., *IEEE Int. Electron Dev. Meet. 2008, Tech. Dig.* 949, 85–88, 2008.
20. X. Wu, P. Zhou, J. Li, L. Y. Chen, H. B. Lv, Y. Y. Lin, and T. A. Tang, *Appl. Phys. Letts.* 90, 183507, April 30, 2007.
21. L. Chen, Y. Xu, Q. Q. Sun, H. Liu, J. J. Gu, S. J. Ding, and D. W. Zhang, *IEEE Electron Dev. Letts.* 31, 356–358, April 2010.
22. R. Waser, R. Dittmann, G. Staikov, and K. Szot, *Adv. Mater.* 21, 2632, July 13, 2009.
23. M. Yin, P. Zhou, H. B. Lv, J. Xu, Y. L. Song, X. F. Fu, T. A. Tang, B. A. Chen, and Y. Y. Lin, *IEEE Electron Dev. Letts.* 29, 681–683, July 2008.
24. D. Lee, D. J. Seong, I. Jo, F. Xiang, R. Dong, S. Oh, and H. Hwang, *Appl. Phys. Letts.* 90, 042107, March 19, 2007.
25. C. Schindler, S. C. P. Thermadam, R. Waser, and M. N. Kozicki, *IEEE Trans. Electron. Dev.* 54, 2762–2768, October 2007.
26. K. Tsunoda, K. Kinoshita, H. Noshiro, Y. Yarnazaki, T. Lizuka, Y. Ito, A. Takahashi et al. *IEEE Int. Electron. Dev. Meet.* 1 and 2, 767–770, 2007.
27. M. Villafuerte, S. P. Heluani, G. Juarez, G. Simonelli, G. Braunstein, and S. Duhalde, *Appl. Phys. Letts.* 90, 052105, January 29, 2007.
28. W. H. Guan, S. B. Long, Q. Liu, M. Liu, and W. Wang, *IEEE Electron. Dev. Letts.* 29, 434–437, May 2008.
29. K. Jung, J. Choi, Y. Kim, H. Im, S. Seo, R. Jung, D. Kim, J. S. Kim, B. H. Park, and J. P. Hong, *J. Appl. Phys.* 103, 034504, February 1, 2008.
30. Y. Wang, Q. Liu, S. B. Long, W. Wang, Q. Wang, M. H. Zhang, S. Zhang et al. *Nanotechnology*, 21, 29, 2010.
31. L. Dongsoo, S. Dong-Jun, C. Hye Jung, J. Inhwa, R. Dong, W. Xiang, O. Seokjoon et al. *IEEE Int. Electron. Dev. Meet.* 1–4, 2006.
32. J. W. Park, D. Y. Kim, and J. K. Lee, *J. Vac. Sci. Technol. A* 23(5), 1309, September 2005.
33. K. Kinoshita, K. Tsunoda, Y. Sato, H. Noshiro, S. Yagaki, M. Aoki, and Y. Sugiyama, *The 22nd IEEE Non-Volatile Semiconductor Memory Workshop*, pp. 66, 2007.
34. H. J. Wan, P. Zhou, L. Ye, Y. Y. Lin, T. A. Tang, H. M. Wu, and M. H. Chi, *IEEE Electron. Dev. Letts.* 31, 246–248, March 2010.
35. D.-H. Kwon, K. M. Kim, J. H. Jang, J. M. Jeon, M. H. Lee, G. H. Kim, X.-S. Li et al., *Nat. Nanotechnol.* 5, 148–153, 2010; K. M., Choi, B. J., Shin, Y. C., Choi, and C. S. Hwang, *Appl. Phys. Lett.* 91, 012907, 2007.
36. B. Gao, B. Sun, H. Zhang, L. Liu, X. Liu, R. Han, J. Kang, and B. Yu, *IEEE Electron. Dev. Letts.* 30, 1326–1328, December 2009; N. Xu, L. F. Liu, X. Sun, X. Y. Liu, D. D. Han, Y. Wang, R. Q. Han, J. F. Kang, and B. Yu, *Appl. Phys. Lett.* 92(23), 232112, June 2008; N. Xu, B. Gao, L. F. Liu, B. Sun, X. Y. Liu, R. Q. Han, J. F. Kang, and B. Yu, *2008 Symposium on VLSI Technology Digest of Technical,* pp. 100–101.
37. P. Zhou, M. Yin, H. J. Wan, H. B. Lu, T. A. Tang, and Y. Y. Lin, *Appl. Phys. Lett.* 94(5), 053510, February 2009; H. J. Wan, P. Zhou, L. Ye, Y. Y. Lin, J. G. Wu, H. Wu, and M. H. Chi, *J. Vac. Sci. Technol. B* 27, 468–2471, November/December 2009.

38. L. Chua, Memristor-The missing circuit element, *IEEE Trans. Circ. Theory* 18, 507–519, September 1971.
39. M. Jagadesh Kumar, Memristor—Why do we have to know about it? *IETE Tech. Rev.* 26(1), January–February 2009.
40. R. S. Williams, How we found the missing memristor, *IEEE Spect.* December 2008.
41. J. J. Yang, J. Borghetti, D. Murphy, D. R. Stewart, and R. S. Williams, A family of electronically reconfigurable nanodevices, *Adv. Mater.* 21, 3754–3758, 2009.
42. J. J. Yang, F. Miao, M. D. Pickett, D. A. A. Ohlberg, D. R. Stewart, C. N. Lau, and R. S. Wiliams. The mechanism of electroforming of metal oxide memristive switches, *Nanotechnology* 20, 215201, 2009.

Materials, Devices, and Circuits

Hong Yu Yu

CONTENTS

13.1 INTRODUCTION

13.1.1 LITERATURE SURVEY ON RESISTIVE SWITCHING IN SOLIDS

Resistance switching, which refers to the alternating of the conductance triggered by the external electrical field, is a very fundamental physical phenomenon of dielectrics. It is frequently observed in thin-film dielectrics, with the first experimental report from Hickmott in 1962 [1]. This observation was soon confirmed by Geppert [2]. And similar phenomena were soon witnessed on other material systems. For instance, the first observation of resistive switching in polymer was made by Mann on thin silicone polymer film [3]. Table 13.1 lists the pioneering works on resistive switching.

Resistive switching observations in a wide range of materials were reported subsequently. Majority of them were noncrystalline phase metal oxide–based systems. In 1970, one of the most important reviews on resistive switching was given by Dearnaley et al., where the concept of resistance random access memory (RRAM) was well established [12]. In 1973, a special issue of IEEE Transactions on Electron Devices (Volume 20, Issue 2) was devoted to amorphous switching devices. Despite oxide-based switching systems, excellent switching performance was also observed on other systems. For instance, Rutz et al. from IBM had reported an ultrafast (500 ps switching time) switching resistive memory with AlN active layer, and AlN-based RRAM still keeps records for its swift switching in recent publications [13,14].

However, practical memory application was not realized in the last century, particularly due to hindrance from material processing and device integration. With the advancement of CMOS technology, Beck et al. reported $SrZrO_3$:Cr as a candidate of next-generation nonvolatile memory [15]. In 2002, Zhuang et al. fabricated RRAM memory with perovskite material [16]. In 2004, NiO-based RRAM was

TABLE 13.1
Pioneering Works on Resistive Switching

Reference	Material System	Time (Submitted)
[1]	$Al/SiO_2/Au$, $Al/Al_2O_3/Au$, $Ta/Ta_2O_5/Au$, $Zr/ZrO_2/Au$, and $Ti/TiO_2/Au$	February 5, 1962
[2]	$Nb/Nb_2O_5/Hg$	October 5, 1962
[4]	$Al/Al_2O_3/Hg$	October 11, 1962
[5]	$Nb/Nb_2O_5/metal$ (metal = Ag, Al, and Au)	March 26, 1963
[6]	$Nb/Nb_2O_5/metal$ (metal = Au, Cu, and W)	September 17, 1963
[3]	Thin silicone polymer film	November 20, 1963
[7]	$Al/Al_2O_3/metal$ (metal = Au, Cu, Co, Pb, Sn, Bi, In, Al, Mg)	December 13, 1963
[8]	$Al/Al_2O_3/Se/Au$	January 23, 1964
[9]	$Ni/NiO/Ag$	March 30, 1964
[10]	$Nb/Nb_2O_5/metal$ (metal = Ag and Au), $Ti/TiO_2/metal$, $Ta/Ta_2O_5/metal$	June 10, 1964
[11]	$Nb/Nb_2O_5/Bi$	December 29, 1964

TABLE 13.2

Recent Reviews on Resistive Switching and RRAM

Reference	Material Scope	Mechanisms	Remarks
[17]	Inorganic materials	VCM and ECM	
[18]	Transition metal oxide	TCM and VCM (homogeneous)	Ti/LSMO/SRO is reported as charge injection-type RRAM also.
[19]	Oxide	VCM(homogeneous)	
[20]	Inorganic materials	TCM, VCM and ECM	
[21]	Metal oxide (binary)	TCM and VCM	
[22]	Inorganic material	TCM, VCM, ECM, phase change and spin polarization	
[23]	TiO_2, WO_3, GeSe, SiO_2, and MSQ	VCM and ECM	
[24]	TiO_2	TCM, VCM	
[25]	Inorganic materials	ECM	
[26]	Binary oxide	TCM	
[27]	Metal oxide	TCM	

demonstrated by Seo et al., which prospered subsequent studies on various transition metal oxide–based RRAM systems.

Research of RRAM has received a lot of attention and it is blooming now. The number of tracked papers in time interval from 2000 to 2011 in Compendex reaches almost 10,000. Thus, the full list of relevant publications is not feasible to be included in this chapter. Resistive switching and RRAM have recently been reviewed frequently thanks to the fast progress in technology development and underlying mechanisms investigation. Table 13.2 lists some of the recent reviews on resistive switching and RRAM.

13.1.2 RRAM Classification

A board definition refers to all types of MIM structures with electric field triggered resistance change as RRAM. Criteria to classify RRAM can be mechanisms, operation polarity, or materials.

13.1.2.1 Classification by Mechanisms

As proposed by Waser et al. and adopted by ITRS 2010, one of the ways to classify RRAMs is switching mechanisms based [20].

As shown in Figure 13.1, RRAM consists of five subcategories. They are phase-change memory (PCM), thermal chemical memory (TCM), valence change or mixed valence memory (VCM), electrochemical metallization (i.e., program metal cell or atomic switch) (ECM), and electrostatic/electronic effects memory (i.e., Ferroelectric barrier cell, Mott insulator cell, or charge trapping cell) (EEM).

In the perspective of working mechanisms, PCM primarily relies on heat-induced phase transformation. On the contrast, Electrostatic/electronic mechanism memory

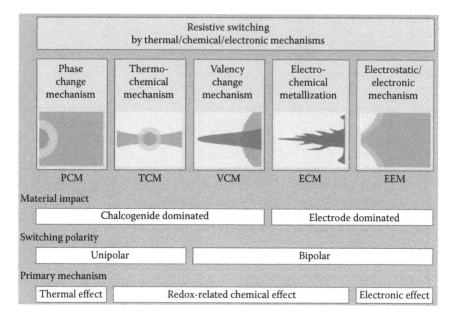

FIGURE 13.1 RRAM classification. (Reproduced from ERD/ERM Group, Assessment of the Potential and Maturity of Selected Emerging Research Memory Technologies Workshop & ERD/ERM Working Group Meeting [2010].)

is purely electronic effect. The other three are redox-related chemical effects based. And they share strong resemblance in terms of operation mechanism. Thus, TCM, VCM, and ECM are typically grouped as redox RAM, which occasionally serves as a narrow definition of RRAM. The scope of this review is limited to redox RAM.

13.1.2.2 Classification by Polarity

Two types of switching behaviors were identified depending on the operating voltage polarity, which are unipolar (or nonpolar) and bipolar operations. The rising of polarity dependence originates from the underlying mechanism as depicted by Figure 13.1.

For unipolar or nonpolar switching, SET or RESET transitions can exist with same electric field polarity (see Figure 13.2). In other words, the memory can be programmed with single polarity of voltage pulses. And typically switching can be made with either polarity of the applied field, especially in symmetrically configured material systems, which is why it is also named nonpolar operation. The SET and RESET transitions have triggered the strength of electric field.

Unipolar operations are preferred for ultimate RRAM application because of memory array integration reason, which is elaborated in the circuit section.

In bipolar resistive switching, the SET and RESET transitions desire opposite polarity of electric field as shown in Figure 13.2.

Coexistence of both unipolar operations and bipolar operations has been discovered in some material systems, since they have fulfilled the criteria for two different switching mechanisms. More detailed discussion is presented in the following material section.

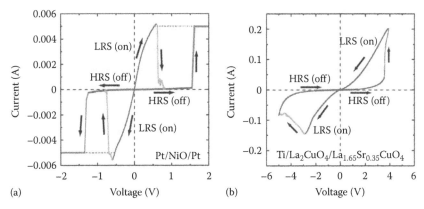

FIGURE 13.2 RRAM classification by polarity: (a) unipolar (nonpolar) RRAM and (b) bipolar RRAM. (Reproduced from Sawa, A., *Mater. Today*, 11, 28, 2008.)

13.1.2.3 Classification by Materials

A family of materials of similar chemical properties typically shows strong similarities in resistive switching if they are switchable. Therefore, it is very natural to group resistive switching systems by material families. Table 13.3 lists a classification of some frequently reported material systems.

13.1.3 GENERAL REQUIREMENTS OF RRAM

Based on IRTS 2010 update, RRAM was considered to be one of the emerging memory devices along with nanomechanical memory, polymer memory, and molecular

TABLE 13.3
RRAM Classification by Material Family

Family	TCM	VCM	ECM
Carbon and polymer	Diamond-like carbon, reduced GOa, CNTa, PVK:Au NP/PEDOT-PSSa, MSQ/Aga	GO/PrCaMnO$_3$	PI/GO-PI/PIa, GO, PEO+AgClO$_4$[a]
Chalcogenide glass			GeSe, GeTe
Metal carbide			CuC
Metal nitride	AlN, CuN		
Metal oxide	CuO, SiO$_2$, HfO$_2$, Nb$_2$O$_5$, NiO, SnO$_2$, Ta$_2$O$_5$, TiO$_2$, ZnO, ZrO$_2$	Al$_2$O$_3$, SiO$_2$, GeO$_2$, HfO$_2$, Nb$_2$O$_5$, Ta$_2$O$_5$, TiO$_2$, WO$_3$, ZnO, ZrO$_2$	Al$_2$O$_3$, CoO$_2$, CuO, SiO$_2$, Ta$_2$O$_5$, ZrO$_2$
Metal sulfide			AgS, CuS, GeS
Perovskite structure	BaTiO$_3$, PrCaMnO$_3$, SrTiO$_3$, SrZrO$_3$	LaCaMnO$_3$, PrCaMnO$_3$	

[a] GO, graphene oxide; CNT, carbon nanotube; PVK, poly(N-vinylcarbazole); NP, nanoparticles; PEDOT-PSS, poly(3, 4-ethylenedioxythiophene)-poly(styrenesulfonate); MSQ, methylsilsesquioxane; PI, polyimide; PEO, polyethylene oxide.

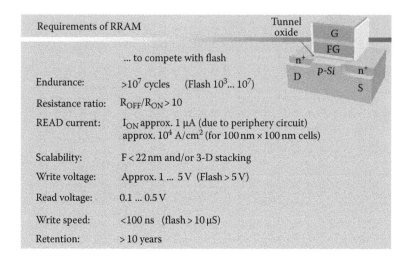

Requirements of RRAM		
	... to compete with flash	
Endurance:	>10^7 cycles (Flash 10^3... 10^7)	
Resistance ratio:	R_{OFF}/R_{ON} > 10	
READ current:	I_{ON} approx. 1 µA (due to periphery circuit) approx. 10^4 A/cm² (for 100 nm × 100 nm cells)	
Scalability:	F < 22 nm and/or 3-D stacking	
Write voltage:	Approx. 1 ... 5 V (Flash > 5 V)	
Read voltage:	0.1 ... 0.5 V	
Write speed:	<100 ns (flash > 10 µS)	
Retention:	> 10 years	

FIGURE 13.3   Requirements of RRAM. (Reproduced from ERD/ERM Group, Assessment of the Potential and Maturity of Selected Emerging Research Memory Technologies Workshop & ERD/ERM Working Group Meeting [2010].)

memory due to the long duration of technology development and polishing. Especially redox-based RRAM, which was recommended by the committee with STT-MRAM to receive additional focus in research and development to accelerate progress toward commercialization.

In order to compete with floating gate device and fulfill the circuit requirements in ultrahigh-density nonvolatile memory, performance expectations were proposed by Waser et al., which was later included in ITRS 2010 [20].

The endurance expectation of RRAM is 10^7, which is better than that of the current floating gate memory. Resistance ratio >10 is preferred to simplify and accelerate corresponding sense amplifier. Read current is preferred to be at the level of 1 µA for fast detection. And, most importantly, as a candidate of sub-22 nm node nonvolatile memory, RRAM must feature the scalability for <22 nm or feasibility of 3-D stacking (Figure 13.3).

Write voltage is preferred to be less than 5 V; this is to eliminate the V_{pp} generator circuit used for flash memory programming with 5 V USB power supply. Read voltage is around 1/10 of the write voltage, due to circuit design concern. Speed of writing is less than 100 ns, which is similar to DRAM performance. And a 10 years' lifetime is also essential.

RRAM has many inborn advantages thanks to its unique operation mechanism. For instance, filament-based RRAMs have good potential for scaling below 10 nm. And less than 10 ns switching has been demonstrated in AlN, which is at the same level of SRAM. Possible unipolar operation benefits the density of integration by exclusion of active devices (i.e., transistor) in memory matrix. But RRAM also suffers from its mechanism. One of the major tough issues comes from its poor reliability. The pros and cons of RRAM are listed in Table 13.4.

TABLE 13.4
Pros and Cons of Redox RAM

Pros	Cons
Good potential for scaling below 10 nm generation (small cell size ~ DRAM size)	Need better understanding of physical SET/RESET processes; need quantitative physical models
Fast read and write times switching <10 ns	Fractal structure; robustness of filament structure (reliability issue)
Low write current in mA range (needs further reduction for DRAM application)	Temperature dependence of read disturb (reliability issue)
Compatible with CMOS (materials, processing)	Need better endurance and retention (reliability issue)
Possible unipolar operation	Unstable unipolar operation (reliability issue)
Possible crossbar structure with potential for multiple layers	Need a reliability model
Multibit/cell storage	Currently requires an initial forming process
Reasonable endurance ~10^9 (needs improvement for SRAM application)	Per bit cost savings of 3-D stacked RRAM is questionable
Reasonable retention time @ 150°C	
Large ON/OFF ratio	

Source: Reproduced from ITRS 2010.

13.2 TYPICAL RRAM MATERIAL SYSTEMS

13.2.1 THERMAL CHEMICAL MECHANISM RRAM MATERIAL SYSTEM

13.2.1.1 General Properties

As mentioned in the previous section, one of the most important features of thermal/ chemical mechanism (or fuse/antifuse) RRAM is its unipolar or nonpolar type of operation, which distinguishes it from VCM- or ECM-type RRAM. This feature originates from the significant Joule heating effect caused by the filament in switching process. TCM RRAM also features the advantage of unipolar operation as no active elements in the memory matrix are desired.

13.2.1.1.1 Area Dependence of Resistance

In literature, the low-resistance state (LRS) resistance is typically area independent because of the filament nature of TCM cells (see Figures 13.4 and 13.5).

But for high-resistance state (HRS), both area-dependent conduction and area-independent conduction may exist (see Figures 13.4 and 13.5). This lies on the competition between carriers flowing through partially ruptured/shrunk filaments and bulk dielectric film. If ruptured/shrunk filaments dominate the current transport, the transport will lose its area dependence. In this case, HRS behaves like as a weak LRS.

13.2.1.1.2 Temperature Dependence of Resistance

In TCM RRAM, LRS is reported of very weak temperature dependence where the activation energy is typically less than 10 meV in most of the cases, as shown in Figure 13.6. This may also suggest the weak metallic nature of the filament.

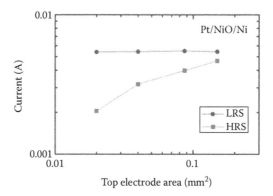

FIGURE 13.4 Area-independent LRS resistance and area-dependent HRS resistance. (Reproduced from Courtade, L. et al., Improvement of resistance switching characteristics in NiO films obtained from controlled Ni oxidation, in *Non-Volatile Memory Technology Symposium*, pp. 1–4, 2007.)

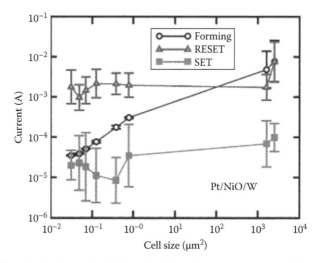

FIGURE 13.5 Area-independent LRS and HRS resistances. (Reproduced from Ielmini, D., *ECS Trans.*, 33, 323, 2010.)

Compared to LRS, HRS is typically of much larger activation energy, which is material system dependent. This may serve as evidence that the transport process is thermal facilitated (see Figure 13.6).

13.2.1.1.3 Frequency Dependence of Admittance

A very critical piece of evidence on the transport type of RRAM comes from conductance (real part of admittance) as a function of frequency. The imaginary part of admittance is correlated to the conductance by Kramer–Kronig transform.

Pristine devices typically follow the power law of polycrystalline/amorphous devices illustrated in Figure 13.7. On the contrary, LRS is of a very weak frequency dependence in low-frequency range, due to a frequency-independent

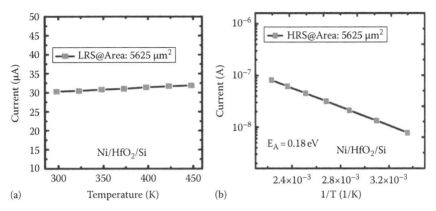

FIGURE 13.6 Temperature dependence of resistance of (a) LRS and (b) HRS of a n^+-Si/HfO$_x$/Ni RRAM cell. (Reproduced from Tran, X.A. et al., *IEEE Electron Device Lett.*, 32, 396, 2011.)

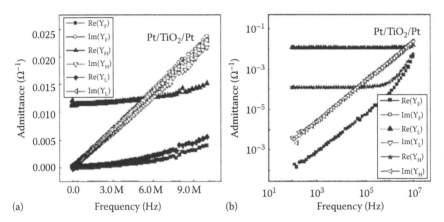

FIGURE 13.7 Frequency dependence of admittance: (a) conductance or real part of admittance and (b) imaginary part of admittance. (Reproduced from Jeong, D.S. et al., *Appl. Phys. Lett.*, 89, 082909, 2006.)

component caused by the weak metallic filament. HRS owns a curve somehow between pristine one and LRS one.

13.2.1.2 TCM Material Combinations

As mentioned in the previous section, a large variety of materials have been explored of unipolar switching. Majority of them are metal oxides. But perovskite structure materials, carbon-based materials, or polymers have also been reported with good electric performance. Table 13.5 consists of several typical unipolar operation systems reported very recently.

13.2.1.3 TCM Mechanism Study

TCM switching is filament based. With field applied, pristine device experiences dielectric breakdown due to the migration of either cations or anions. Once the

TABLE 13.5

Recent Reports on TCM-Based RRAM Systems

Type	Switching Region	Thickness (nm)	BE	TE	Device Size	Current	Voltage (V)	Speed Endurance	Retention	Remarks	Reference
Carbon and polymer	Diamond-like carbon	22	W	Pt	0.24 μm diameter	38 μA	<1.5	>10³ (SET/RESET: 50 ns/10 ns)	>10 h @ 300 C in Ar		[32]
	Reduced graphene oxide or multiwall carbon nanotube	20 or 150	ITO	Al or Ar	3 mm diameter	<1 or 4 μA	<7.5 or 15	(SET/RESET: 10 μs/10 μs)	>2 × 10³ s		[33]
	PV:Au NP/PEDOT-PSS	(120 or 360)/70	Al	Al	0.04 mm²	<10 mA	<5/<10	NS	>10⁴ s @ RT		[34]
	MSQ/Ag	150/12	Pt	Pt	0.1 × 0.1 μm² (vertical)	<0.7 mA (vertical)	<1 (vertical)	>2 × 10³ (SET/RESET: 0.1 ms/0.1 ms)	>6 × 10⁴ s @85 C		[35]
Simple metal oxide	Cu₂O	190	Pt	Cr	NS	<10 mA	<2	>10⁴	NS		[36]
	SiOx/HfSiON	2.5	Si	NiSi	0.12 μm²	<0.2 mA	<4	>50 (DC)	>10⁴ s @ 150 C	Both bipolar and unipolar	[37]
	HfO₂	30	Si	Ni	NS	<2.5 mA	<4	(SET/RESET: 500 ns/100 ns)	>10⁴ s @ 85 C		[38]
	HfO₂	5–20	Ni	TiN	0.1 × 0.1 mm²	<0.2 mA (lowest Ic)	<1.5 (lowest Ic)	>10⁴ (bipolar), 10³ (unipolar)	>20 h @ 150 C	Both bipolar and unipolar	[39]
	HfO₂	3	Si	Ni	5,625–99,225 μm²	<1 mA	<2.5	>10⁵ (SET/RESET: 50 ns/50 ns)			[30]
	HfO₂/Al₂O₃	4.2	Si	Ni	NS	<1 mA	<1.5	>10⁶ (SET/RESET: 10 ns/30 ns)	>7 × 10³ s @ 120 C		[40]

Material	Thickness	Electrode 1	Electrode 2	Area	Current	Voltage	Cycles	Retention	Notes	Ref.
InGaZnO	60	Cu	Cu	0.5 mm diameter	<1 mA	<2	>150 (DC)	NS	Flexible	[41]
NbO	40	Pt	Pt	50 × 50 μm²	<10 mA	<4	>150 (DC)	NS		[42]
NiO	35	W	Au	180 nm diameter	<4 mA	<1.5	>200 (DC)	>3 × 10⁶ s @ 125 C		[43]
NiO:W	50	Pt or TiN	Pt	40 × 40 μm²	<10 mA	<2 or <2.5	NS	NS		[44]
Al₂O₃/NiO/Al₂O₃	4/40/4	Pt	TiN	0.4 mm diameter	<10 mA	<2.5	>250 (DC)	>10⁴ s @ RT		[45]
NiO nanodot	30	Au	Au	30 nm diameter	<0.1 nA	<0.8	>200 (DC)	NS	Lateral wire structure	[46]
SiO	50	W	Cu	40 × 40 μm²	<2 mA	<6	NS	>10⁴ s @ 100 C		[47]
SiO	20	Pt	Cu	4.9 × 10⁻⁴ cm²	<1 mA	<3	NS	NS	Both bipolar and unipolar	[48]
SnO₂	50	Ti	Ag	0.635/1.778/2.413 mm diameter	<10 mA	<1.25	>30 (DC)	>10⁴ s @ RT		[49]
TaOₓ	~40	Pt	Pt	30 × 30 μm²	<0.1 A	<2.5	NS	NS	Bipolar with Ta₂O₅ layer	[50]
TiO:Gd	40	Pt	W	NS	<15 mA	<1.5	NS	NS		[51]
TiO₂	60	Pt	Pt	2 × 2 μm²	<10 mA	NS	>4 × 10³	NS		[52]
ZnO	40/90/140	Al	Al	0.2 mm diameter	<10 mA	<2/2.5/3	NS	NS		[53]
ZnO:La	200	Pt or Si	Pt	0.2 mm diameter	<20 mA	<3	>150 (DC)	>10⁶ s		[54]
ZnO, NiO/ZnO or WO₃/ZnO	50, 50/50 or 50/50	Pt/Au	Pt/Au	2 × 2–10 × 10 μm²	<10,100 or 10 mA	<2	NS	NS		[55]
Zn₀.₉₈Cu₀.₀₂O	80	ITO	Ag	1 mm diameter	<20 mA	<15	NS	>10³ s		[56]
ZrO₂/Cu/ZrO₂	20/3/20	Pt	Cu	0.1 × 0.1 mm²	<70 mA	<2.5	NS	NS	Both bipolar and unipolar	[57]

(continued)

TABLE 13.5 (continued)
Recent Reports on TCM-Based RRAM Systems

Type	Switching Region	Thickness (nm)	BE	TE	Device Size	Current	Voltage (V)	Speed Endurance	Retention	Remarks	Reference
Perovskite structure	$SrTiO_x$	60	Pt	Pt	25×25 μm²	NS	NS	NS	>60 s		[58]
	$Pr_{0.7}Ca_{0.3}MnO_3$	400	Pt	Au, Ag, Cu or Al	0.3 mm diameter	<150 mA	<7	>100 (DC)	>24 h	Both bipolar and unipolar	[59]
	$SrTiO_3$	300	Pt	Au	0.1 mm diameter	<30 mA	<1.5	NS	>14 days	Both bipolar and unipolar	[60]
	$SrTiO_3$	200	Pt	Pt	0.2 mm diameter	<50 mA	<6	NS	>10⁵ s	Both bipolar and unipolar	[61]
	$SrZrO_3$	100	Pt	Pt	0.2 mm diameter	<200 mA	<15	>600 (DC)	NS		[62]
	$BaTiO_3$:Co	400	Pt	Au	0.1 mm diameter	<30 mA	<7.5	>10⁵ (SET/RESET: 8 ns/69 ns)	>7 × 10⁴ s		[63]

filament is bridged between two electrodes, the RRAM cell is in its LRS state. However, in TCM cell, Joule heating–caused temperature gradient is a more dominant force than electric field for driving cations or anions to migrate. Thus, the filament will be ruptured by Joule heating under high current situation, and this is the origin of RESET transition. As mentioned in Table 13.2, specific TCM mechanism reviews have been given by Ielmini et al. [27,64] and Kim et al. [26].

13.2.1.3.1 Physical Characterization Study

Recently, for NiO system, Yoo et al. have reported on the direct observation of creation and rupture of a portion of a percolation network, which takes a form of grain boundary as revealed by HRTEM [65] (see Figure 13.8).

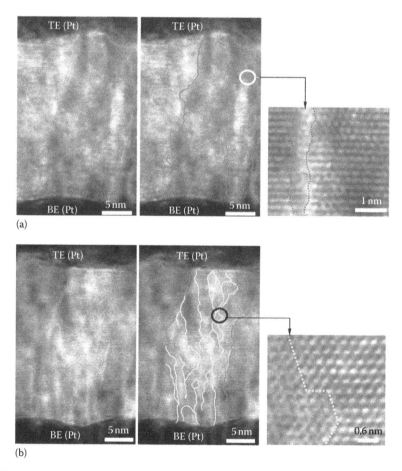

FIGURE 13.8 Observation of soft breakdown boundaries in NiO thin films during resistance switching. (a) A grain boundary is indicated by dark lines in the virgin NiO thin film. (b) New boundaries with white lines appeared after resistance switching. It is believed that conducting clusters (percolating network) are formed in both grain boundaries and breakdown boundaries during resistance switching. (Reproduced from Yoo, I.K. et al., *IEEE Trans. Nanotechnol.*, 9, 131, 2010.)

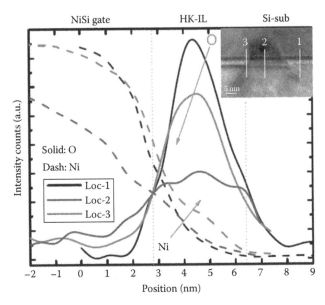

FIGURE 13.9 Ni and O EELS line profiles acquired at three locations indicated in the inserted TEM micrograph. At the filament location, O diffuses to the NiSi gate, while Ni moves into the gate dielectrics and Si substrate. The Si epitaxy region shows low Ni intensity as a result of Si protrusion from the substrate. (Reproduced from Li, X. et al., *Appl. Phys. Lett.*, 97, 202904, 2010.)

The EELS chemical component analysis of percolation path has been carried by Li et al. [66]. The weight of Ni cations and oxygen vacancies in filament region is significantly higher than that of regions nearby, which is shown in Figure 13.9.

HAXPES has been employed to identify the valence state change of Ni atoms before and after dielectric breakdown by Calka et al. [67] (see Figure 13.10). During forming, occupied bandgap states are created near NiO Fermi level (EF), which coincides with oxygen-deficient NiO bandgap structure. Besides, metallic Ni atom weight is observed to increase after forming process, but the increment is relatively small, which may be due to the reason that the filament is of small volume.

Kondo et al. have reported TEM observations of NiO breakdown also [68]. As shown in Figure 13.11, an ~400 nm diameter conductive spot was observed of significant change of surface morphology. The EDX analysis of underlying NiO reveals oxygen deficiency.

Despite Ni-based material system, another system that has been frequently reported is TiO_2. Kwon et al. have reported the observation of Magnéli phase Ti_4O_7 filament in TiO_2 unipolar RRAM by HRTEM [69] (see Figure 13.12). Ti_4O_7 is oxygen deficient compared to TiO_2 and of metallic nature above ~150 K. Both have been observed in TiO_2 RRAM. Except Ti_4O_7, another Magnéli phase Ti_5O_9 structure has also been identified in TiO_2 RRAM, as reported by Kim et al. [52] (see Figure 13.13).

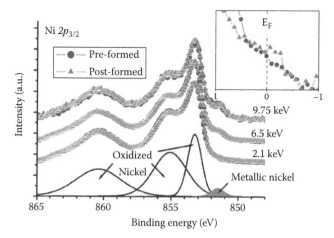

FIGURE 13.10 Ni core-level spectrum. The decomposed area for metallic peak (851 eV) increases after forming process. The inset shows the bandgap spectrum. The bandgap states are attributed to oxygen vacancies. (Reproduced from Calka, P. et al., *J. Appl. Phys.*, 109, 124507, 2011.)

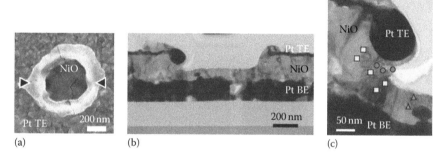

FIGURE 13.11 (a) SEM image of conductive spot, (b) and (c) corresponding TEM images. In (c), the edge of conductive spot is enlarged, where the conductive path (filament) is thought to be formed. Around this area, compositional analyses of NiO were performed by EDX. The triangles, squares, and circles correspond to positions with composition ratios of Ni:O = 1:1, 2:1, and 9:1, respectively. (Reproduced from Kondo, H. et al., *Jpn. J. Appl. Phys.*, 50, 081101, 2011.)

13.2.1.3.2 Switching and Conduction Mechanisms

Both switching mechanisms and conduction mechanisms have been proposed based on the evidence from physical characterizations.

Precise modeling of the SET/RESET process is of tremendous challenge. This lies on the atomic system of big population and severe disorder. And the exact atom configurations before and after switching are very difficult or impossible to be probed. However, for unipolar RRAM, empirical or classical treatment has been reported with success. Regarding switching mechanisms, despite the mentioned review articles, quantitative calculation model of switching properties has been reported by Russo et al. [70]. By considering temperature gradient drove ion motions and the

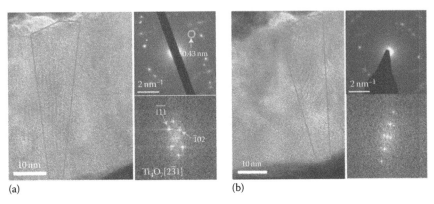

(a)   (b)

FIGURE 13.12   (a) High-resolution image, diffraction pattern, and fast Fourier transformed micrograph of the Magnéli structure before RESET. (b) The corresponding images after RESET. The diffraction spot (marked as a circle in c) from the Magnéli structure disappeared after RESET. (Reproduced from Kwon, D.-H. et al., *Nat. Nano.*, 5, 148, 2010.)

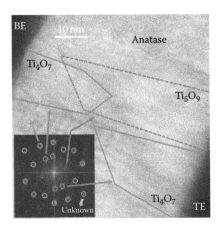

FIGURE 13.13   HRTEM image of the hourglass-shaped conductive filament. Inset shows the fast Fourier transformed diffraction patterns. (Reproduced from Kim, G.H. et al., *Appl. Phys. Lett.*, 98, 262901, 2011.)

redox reactions, Bocquet et al. have simulated the I–V and sweep rate vs. V_{SET} by solving the partial differential equation [71]. The simulated SET and RESET transitions are shown in Figure 13.14.

Similarly, with Matsui–Akaogi potential, Zhao et al. have performed the molecular dynamics simulation of Ti and O ions' distribution subjecting to external electric field [72]. The temperature profile has been calculated, and the filament melting was imitated (see Figure 13.15).

For NiO system, Chien et al. have proposed a novel model for switching interpretation based on Ni_2O_3 [73]. The proposed model is that, in HRS state, switching region is of the form NiO. NiO transforms to Ni and Ni_2O_3 if the temperature is sufficiently high by Joule heating. Metallic Ni will be responsible for the large current

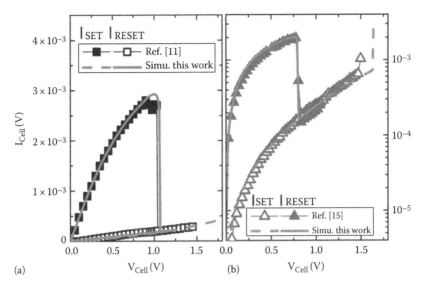

FIGURE 13.14 Experimental and simulated SET and RESET operations of (a) single Pt/Nio/Pt RRAM cell (Ref. [11]) and (b) Pt/Nio/Pt RRAM cell in a ID-IR structure (Ref. [15]). (Reproduced from Bocquet, M. et al., *Appl. Phys. Lett.*, 98, 263507, 2011.)

flow in LRS state. However, Ni_2O_3 is not thermally stable also. It may easily decompose to NiO by the release of oxygen ions at even higher temperature. The released oxygen ions oxidize metallic Ni atoms and thus filament is ruptured. Their claim is supported by photoemission spectrum shown in Figure 13.16.

In the field of carrier transport, metallic impurity band due to oxygen vacancies is proposed by several groups for interpreting the weak metallic nature of LRS. For instance, Park et al. have reported that aligned neutral oxygen vacancy chains in TiO_2 may form delocalized impurity bands due to overlapping of t_{2g} orbitals [74] (see Figures 13.16 and 13.17).

Similar calculation has also been implemented on NiO by Magyari et al. [75] (see Figure 13.18). And calculated formation energy of oxygen vacancies indicates that neutral oxygen vacancy is preferred if Fermi level of external electron reservoirs aligns at the middle of the bandgap.

For HRS state, Ielmini et al. have found the impact of oxygen vacancy/defect concentration on activation energy [64]. A gradual change can be seen in Figure 13.19. The role of the oxygen vacancies is similar to dopants. Therefore, in LRS, large local density of oxygen vacancies causes metal–insulator transition in a similar manner with the degenerately doped semiconductors.

13.2.2 Valence Chemical Mechanism RRAM Material System

13.2.2.1 General Properties

VCM switching can be further divided into filament type and homogeneous type based on experimental observations. Filament VCM memory shows very similar properties with TCM memory. In fact, as mentioned in the previous section, several

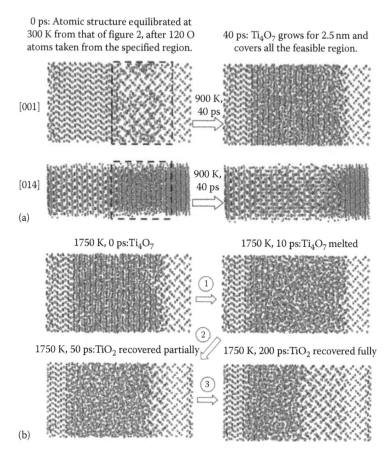

FIGURE 13.15 Atom distribution in the (a) SET simulation and (b) RESET simulation. (Reproduced from Zhao, L. et al., *IEEE Electron Device Lett.*, 32, 677, 2011.)

systems exhibit both unipolar and bipolar switching properties. This is due to the reason that Joule heating and electric field are two competing forces in RESET processes. For TCM cells, the RESET actions are mainly thermal driven and thus unipolar. While in filament VCM RRAM, the anions, which are typically oxygen ions, were drifted by electric fields and therefore show bipolar switching.

13.2.2.1.1 Area Dependence of Resistance

Most cases of VCM RRAM are filament based, showing very similar properties as TCM cases that LRS typically has area-independent resistance. But HRS may either be area independent or dependent based on whether the current contribution from ruptured/shrinked filament can dominate or not as depicted in Figures 13.20 and 13.21.

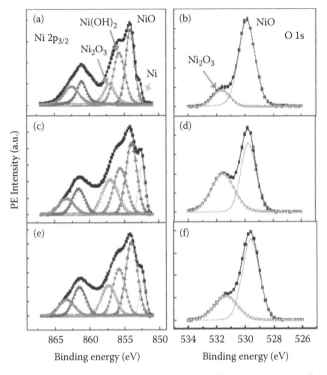

FIGURE 13.16 Photon emission spectra of Ni $2p_{3/2}$ and O 1s taken in the original NiO film (a) and (b). The ON spot (c) and (d), and the OFF spot (e) and (f). (Reproduced from Chien, F.S.-S. et al., *Appl. Phys. Lett.*, 98, 153513, 2011.)

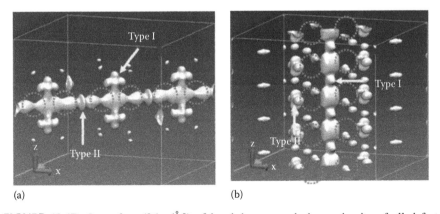

FIGURE 13.17 Isosurface (0.1 e/Å3) of band-decomposed charge density of all defect states. (a) [110] vacancy chain. (b) [001] vacancy chain. (Reproduced from Park, S.-G. et al., *IEEE Electron Device Lett.*, 32, 197, 2011.)

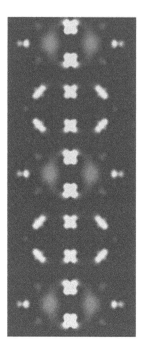

FIGURE 13.18   Partial charge density within (E_F, E_F + 0.3 eV) in the (100) plane including oxygen vacancies and a Ni metal chain. (E_F is the Fermi level) (Reproduced from Magyari-Köpe, B. et al., *Nanotechnology*, 22, 254029, June 24, 2011.)

The other type of VCM RRAM, the interface switching memory, is experiencing homogeneous change of interface properties during switching process. This makes both LRS and HRS scale with area (see Figure 13.22).

13.2.2.1.2   Temperature Dependence of Resistance

For VCM RRAM, in majority cases (either filament oriented or homogeneous interface oriented), the observed LRS resistance shows very weak temperature dependence. Along with its ohmic field dependence, LRS is attributed to weak metallic nature, which is the same as the TCM case. And similarly, HRS is temperature sensitive, suggesting a thermal-facilitated transportation type (see Figure 13.23).

However, there is also report of ultralow conductance LRS VCM RRAM in which the LRS behaves like typical HRS. And the HRS is of even poor conductance, which is close to that of pristine device (see Figure 13.24).

13.2.2.1.3   Frequency Dependence of Admittance

Similar to TCM cells, real part of admittance of pristine VCM RRAM cells is typically following the universal power law of polycrystalline/amorphous semiconductors. On the contrary, LRS is normally dominated by a strong frequency-independent component in low-frequency range, probably because of the metallic filament. And HRS is of a state in between pristine state and LRS (see Figure 13.25).

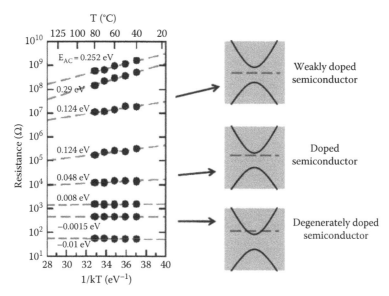

FIGURE 13.19 Arrhenius plot of resistances for different nanofilament states (left) and corresponding schematic band structures (right). LRS display a metallic conductivity behavior, where resistance increases with T, as revealed by the negative apparent activation energy. This suggests that the Fermi level is pinned in the conduction (or valence) band as for a degenerately doped semiconductor. For increasing resistance, the nanofilament behavior becomes increasingly semiconductor-like, showing an increasing activation energy. The corresponding physical picture is a semiconductor with a different Fermi level with respect to the band edge. (Reproduced from Ielmini, D. et al., *Nanotechnology*, 22, 254022, 2011.)

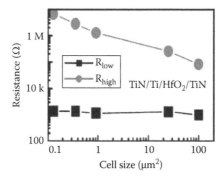

FIGURE 13.20 Area-independent LRS resistance and -dependent HRS resistance. (Reproduced from Lee, H.Y. et al., Low power and high speed bipolar switching with a thin reactive Ti buffer layer in robust HfO2-based RRAM, in *IEDM Technical Digest*, pp. 1–4, 2008.)

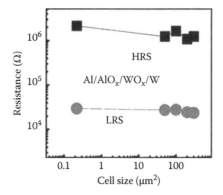

FIGURE 13.21 Area-independent LRS and HRS resistances. (Reproduced from Song, Y.L. et al., *IEEE Electron Device Lett.*, 32, 1439, 2011.)

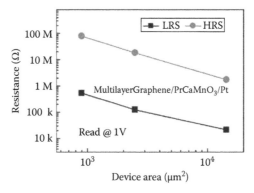

FIGURE 13.22 Device area dependence of resistance values in HRS and LRS. (Reproduced from Lee, W. et al., *Appl. Phys. Lett.*, 98, 032105, 2011.)

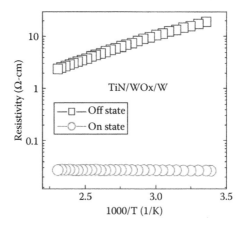

FIGURE 13.23 Temperature-independent LRS resistance and -dependent HRS resistance. (Reproduced from Ho, C. et al., A highly reliable self-aligned graded oxide WO$_x$ resistance memory: Conduction mechanisms and reliability, in *VLSI Symposium Technology Digest*, pp. 228–229, 2007.)

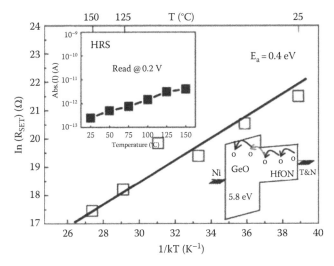

FIGURE 13.24 Temperature-dependent LRS and HRS resistances. (Reproduced from Cheng, C.H. et al., *IEEE Electron Device Lett.*, 32, 366, 2011.)

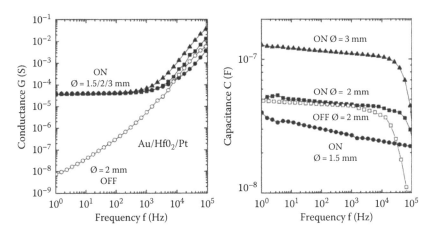

FIGURE 13.25 Frequency dependence of the C and G in the ON and OFF states, as a function of the top electrode size. (Reproduced from Gonon, P. et al., *J. Appl. Phys.*, 107, 074507, 2010.)

13.2.2.2 VCM Material Combinations

Similar to TCM, VCM switching is observed in a large variety of material families, including carbon-based material, polymer-based material, metal nitride, metal oxide, and perovskite structure material. Table 13.6 lists several typical examples of VCM RRAM from very recent publications.

13.2.2.3 VCM Mechanism Study

Similar to TCM RRAM, the forming process and SET transition in VCM RRAM are the results of migration of cations or anions in both filament oriented device

TABLE 13.6

Recent Reports on VCM-Based RRAM Systems

Type	Switching Region	Thickness (nm)	BE	TE	Device Size	Current	Voltage (V)	Speed Endurance	Retention	Remarks	Reference
Carbon and polymer	GO/PCMO	30/25	Pt	Pt	NS	<3 mA	<1	>125	>10^4 s @ 85 C		[82]
Metal nitride	AlN	150	Pt	Ti	50–200 μm diameter	<10 μA	<2	>10^8 (SET/RESET: 10/10 ns)	>10^5 s @ 85 C		[14]
	AlN	130	ITO	ITO	0.2 mm diameter	<10 mA	<3	>10^8 (SET/RESET: 10/10 ns)	>10^5 s @ 85 C		[83]
	Cu_xN	~80	Cu	Ni	0.1 mm diameter	<0.2 mA	<0.8	>50 (DC)	NS		[84]
Metal oxide	Al_2O_3/RuNCs/Al_2O_3	10/3/10	Pt	TaN	0.1 mm diameter	<100 mA	<1.5	>10^3	>10^5 s @ RT	NS	[85]
	AlO_x/WO_x	5/20	W	Al	0.22 × 0.22–300 × 300 μm^2	<30 μA	<2	>10^3 (SET?RESET: 1/1 μs)	NS		[77]
	$CoSiO_x$	30	Pt	TiN	1–64 μm^2	<10 mA	<2	>100 (DC)	NS		[86]
	GeO_x or GeO_x/$SrZrO_x$	5, 8, 12/5 or 12/8	TaN	Ni	11,304 μm^2	<1 μA (for $SrZrO_x$, GeO_x, 12/8 nm)	<1.5 (for $SrZrO_x$, GeO_x, 12/8 nm)	>10^6 (SET/RESET: 50/50 ns)	NS		[87]
	GeO_x/HfON	8.5/7.5	TaN	Ni	11,300 μm^2	<0.1 μA	<3	>10^6 (SET/RESET: 20s/20 ns)	NS		[80]

HfOₓ/TiOₓ/HfOₓ/TiOₓ	4/2/4/2	Pt	TiN	100 μm²	<1 mA	<1.5	>500 (DC)	NS		[88]
ZrOₓ/HfOₓ	NS/3	NS	Pt	50 × 50 μm² or 250 nm diameter	NS	NS	NS	NS		[89]
SiOₓ/HfSiON	2.5	Si	NiSi	0.12 μm²	<1 μA	<2.3	>30 (DC)	>10⁴ s @ 85 C	Both bipolar and unipolar	[90]
HfO₂	20	Pt or TiN	Au	2 mm diameter	<0.1 mA	<6	NS	NS		[91]
HfO₂:Gd	20	Pt	TiN	NS	NS	<2.7	(SET/RESET: 1/1 μs @ 1.1/1.2 V)	>10⁴ s @ 120 C		[92]
Nb₂O₅	50	Pt	TiN/Ti	0.4 mm diameter	<20 mA	<1.5	>10³ (DC)	>5 × 10⁴ s @ RT		[93]
Ta₂O₅/TaOₓ	5/20	Pt	Pt	30 × 30 μm²	<0.1 mA	<2.5	>10⁶ (SET/RESET: 10/10 μs)	NS		[94]
TiO₂-δ/La₂/₃Sr₁/₃MnO₃	100/100	LAO	In	NS	<1 mA	<5	NS	NS		[95]
Ti/TiO₂	5/20	TiN	TiN	NS	<1 mA	<1.5	NS	NS		[96]
Ti/HfO₂	10/5	TiN	TiN	50 × 50 nm²—40 × 40 μm²	<1 mA	<2	>4 × 10⁴ (SET/RESET: 0.5 ms/0.5 ms)	>3 × 10⁴ s @ 200 C		[97]
WOₓ	46	W	Pt	50 nm diameter—0.1 × 0.1 mm²	<5 mA	<1.2	>10⁷ (SET/RESET: 10 μs/10 μs)	NS		[98]

(continued)

TABLE 13.6 (continued)
Recent Reports on VCM-Based RRAM Systems

Type	Switching Region	Thickness (nm)	BE	TE	Device Size	Current	Voltage (V)	Speed Endurance	Retention	Remarks	Reference
	ZnO:Ga (H$_2$; 5%)	240	ZnO:Ga	ZnO:Ga	NS	<0.1 mA	<2	>20 (SET/RESET: 0.5 ms/0.5 ms)	NS		[99]
	Ga$_2$O$_3$–ZnO–Ga$_2$O$_3$	50/120/50	ZnO:Ga	ZnO:Ga	0.03, 0.8 or 10 mm^2	<100 mA	<15	>50 (DC)	>10^5 s		[100]
	ZrO$_2$	19	Pt	Ti	25 μm^2	<25 μA (single level)	<3	>4 × 10^3	26,400 s (150 C) 11,000 s (175 C) 470 s (200 C)		[101]
Perovskite	Pr$_{0.7}$Ca$_{0.3}$MnO$_3$	60	Pt	Multilayer Graphene	~900–20,000 μm^2	<0.1 mA	<5	>10^3 (SET/RESET: 10 μs/10 μs)	>10^4 s @ 85 C		[78]
	Pr$_{0.7}$Ca$_{0.3}$MnO$_3$	40	Pt	Ge$_2$Sb$_2$Te$_5$/Ti	150 nm diameter	<0.1 μA	<4	>10^5 (SET/RESET: 100 μs/20 μs)	>10^4 s @ 125 C	Possible of being charge injection type []	[102]
	La$_{0.7}$Ca$_{0.3}$MnO$_3$	200	Pt	Ag, Al or Ag–Al	50 μm diameter	<50 mA (10 nm Al top)	<2.5 (10 nm Al top)	>1.1 × 10^4 (SET/RESET: 5 μs/5 μs) (10 nm Al top)	>10^5 s (10 nm Al top)	Possible of being Mott insulator–transition type []	[103]

or homogenous switching device. However, in RESET process, electric field dominates over temperature gradient in ion migrations, and the reversed field polarity causes the rupture of the conducting bridge or filament. VCM mechanisms have been reviewed by several groups as listed in Table 13.2.

13.2.2.3.1   Physical Characterization

Because TCM and VCM filament switching systems share strong resemblance, the physical characterizations for TCM systems can also be applied to VCM systems. And bipolar switching is also spotted frequently in unipolar switching system. (i.e., TiO_2).

In 2006, Szot et al. have reported on the conduction spots on $SrTiO_3$ formed by thermal reduction (as SET process, to create oxygen vacancies) and oxidation (as RESET process, to annihilate oxygen vacancies) [104]. CAFM has detected the conducting spots that are of nanometer level size as shown in Figure 13.26.

Similar filament CAFM images of WO_x were reported by Shang et al. recently [105] (see Figure 13.27). They also claim that carrier flux is mainly confined at grain surface rather than grain boundary surface.

13.2.2.3.2   Switching and Conduction Mechanism

Although oxygen vacancies have been identified as having a strong correlation with the switching phenomena of VCM RRAM, the detailed mechanism is still unknown. Different models have been proposed by different groups.

Regarding the switching mechanism, McKenna et al. have made an important claim that segregation energy of neutral and positively charged oxygen vacancies is lower at the (101) grain boundary via first principles simulation as shown in Figure 13.28. This may serve as evidence that percolation paths favor grain boundaries [106].

Gao et al. have reported an ion-transport-recombination-based model to simulate switching, with the prediction of oxygen profile, switching speed, and endurance illustrated in Figure 13.29 [107].

A similar model is reported by Chang et al. with qualitative description [108]. Joule heating effect is taken into consideration in the ion migration model by Yu et al. [109].

In terms of transport mechanism, in 2006, Broqvist et al. calculated the electronic structure of HfO_2 with oxygen vacancy at different charge states as well as its formation energy [110]. Hybrid functional was used, which has prevented the self-interaction caused underestimation of bandgap (see Figure 13.30).

In 2007, Ho et al. suggested that HRS is of variable-range hopping-type transport and LRS is of a metallic nature [79]. Gao et al. proposed a universal model for carrier transport in bipolar RRAM in 2009 based on thermal-assisted hopping through the chain of oxygen vacancies [111]. Momida et al. calculated the electronic properties of isolated oxygen vacancy in amorphous Al_2O_3 under different charge conditions. The calculated DOS is illustrated in Figure 13.31 [112].

Recently, a trap-assisted tunneling model has been worked out by Yu et al. [113].

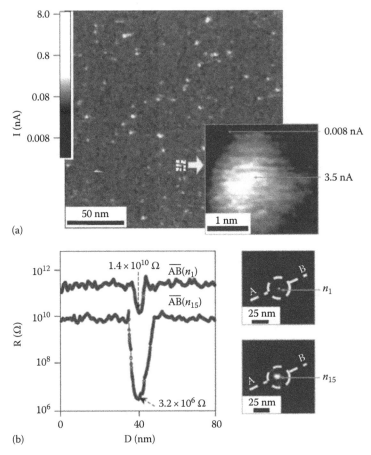

FIGURE 13.26 (a) Conductivity map of the surface of a $SrTiO_3$ single crystal as recorded by the LC-AFM. Filamentary paths with enhanced conductance are present on the surface after thermal reduction and reoxidation under ambient conditions. Inset: spot with a dimension of 1–2 nm, where the main current is concentrated in a region corresponding to the size of the core of a typical edge-type dislocation. (b) Line scan across the selected spot (D denoting distance along AB) showing the dynamic range of the resistance change as a result of the application of a negative tip voltage bias, that is, selective electroformation. Right: Conductivity maps of the selected spot before (n_1) and after electroformation (n_{15}) with an increase in diameter at the surface from 5 to 10 nm. (Reproduced from Szot K. et al., *Nat. Mater.*, 5, 312, April 2006.)

13.2.3 ELECTROCHEMICAL METALLIZATION RRAM MATERIAL SYSTEM

13.2.3.1 General Properties

ECM-type RRAM sometimes is called atomic switches. It has received more attention in the early phase of RRAM development due to its clear operation mechanism. Unlike TCM- and VCM-type RRAM, the key elements in resistive switching are atoms from electrodes, and the bulk dielectric insulating layer serves

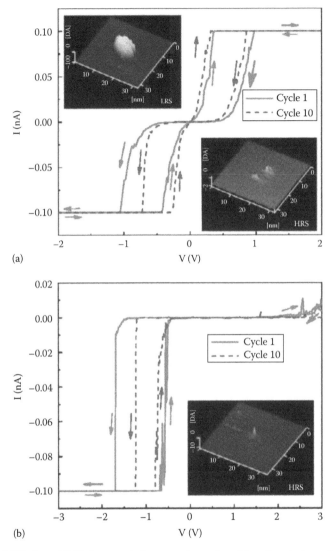

FIGURE 13.27 Local I–V characteristics at grain surface (a) and grain boundary surface (b) measured by the CAFM tip with the external biases ranging from –3 to 3 V. The current compliance is 100 pA. Upper left and lower right insets in (a) show the current mappings after negative and positive voltage sweeps, respectively. Inset in (b) shows the current mappings after negative voltage sweep. The current images were obtained under the tip bias of –0.5 V. (Reproduced from Shang, D.-S. et al., *Nanotechnology*, 22, 254008, 2011.)

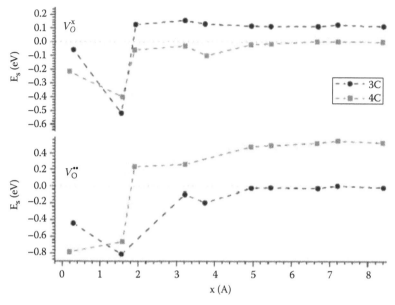

FIGURE 13.28   Segregation energies (Es) for neutral and double positively charged oxygen vacancy defects in the vicinity of the (101) twin boundary. O sites with different coordination number (i.e., number of nearest Hf neighbors) are separated for clarity. (Reproduced from McKenna, K. et al., *Appl. Phys. Lett.*, 95, 222111, 2009.)

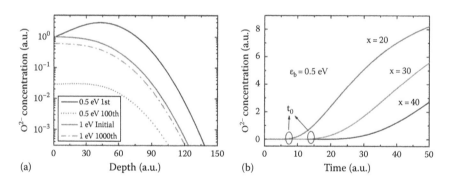

FIGURE 13.29   (a) Calculated O_{2-} concentration distribution as a function of depth from the interface with different interface barrier. (b) Calculated time evolution of O_{2-} concentration at different points. (Reproduced from Gao, B. et al., *Appl. Phys. Lett.*, 98, 232108, 2011.)

as solid electrolyte. The mechanism has projected special requirements on the selection of electrode materials. They should be easily oxidized and reduced, and they must have reasonable mobility in the corresponding electrolyte.

13.2.3.1.1   Area Dependence of Resistance

In ECM RRAM, since the metal ions form a bridge, which is the filament, LRS resistance typically is not area dependent.

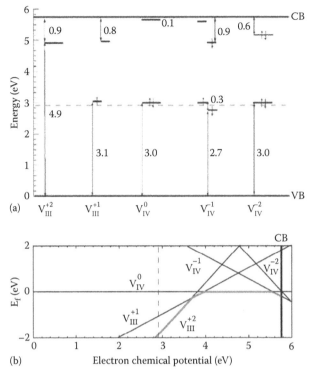

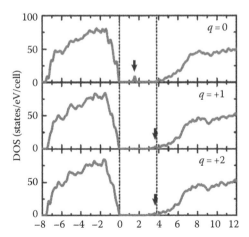

FIGURE 13.30 (a) Kohn–Sham energy levels for neutral and charged oxygen vacancies in m-HfO$_2$. The valence (VB) and conduction band (CB) edges are indicated. (b) Formation energy vs. electron chemical potential for the O vacancy in m-HfO$_2$. The midgap and CB levels are indicated by dashed and solid vertical lines, respectively. The most stable charge states are highlighted. (Reproduced from Broqvist, P. et al., *Appl. Phys. Lett.*, 89, 262904, 2006.)

FIGURE 13.31 Density of states (DOS) for the V$_{Oq}$ in the amorphous model of the most energetically favorable configuration. (Reproduced from Momida, H. et al., *Appl. Phys. Lett.*, 98, 042102, 2011.)

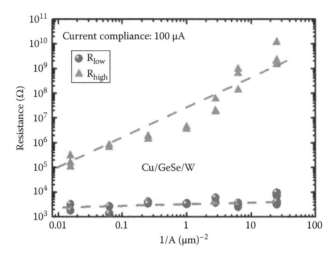

FIGURE 13.32   Area-independent LRS resistance and -dependent HRS resistance. (Reproduced from Rahaman, S.Z. et al., *Electrochem. Solid-State Lett.*, 13, H159, 2010.)

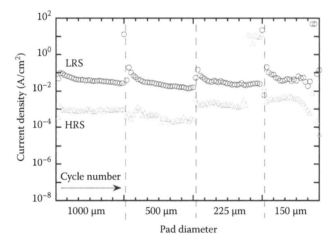

FIGURE 13.33   Area-independent LRS and HRS resistances. (Reproduced from Thomas, M. et al., From micrometric to nanometric scale switching of CuTCNQ-based nonvolatile memory structures, in *Non-Volatile Memory Technology Symposium*, pp. 1–4, 2008).

For HRS, both size-dependent and size-independent resistances have been reported. The reason may be the same with TCM and VCM cases as the current flow through ruptured filament competes with current flow through the whole cross section (see Figures 13.32 and 13.33).

13.2.3.1.2 Temperature Dependence of Resistance

For LRS, majority reports treat its temperature dependence with the same manner for metallic nanowire in low-temperature region (Figure 13.34). However,

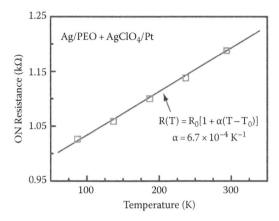

FIGURE 13.34 Metallic-like LRS in low-temperature range. (Reproduced from Wu, S. et al., *Adv. Funct. Mater.*, 21, 93, 2011.)

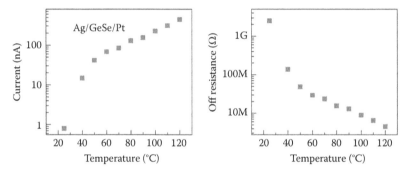

FIGURE 13.35 (a) LRS current at −2 V measured at elevated temperatures. (b) Corresponding HRS resistance at −2 V at temperatures up to 120°C. (Reproduced from Schindler, C. et al., *Z. Phys. Chem.*, 221, 1469, 2007.)

in high-temperature region, the behavior may deviate from the predictions based on metallic nanowire due to severe temperature-driven ion diffusion (see Figure 13.35).

Transport in HRS is sensitive to the choice of electrolyte. Typically, it exhibits the property of thermally activated transport.

13.2.3.1.3 Frequency Dependence of Admittance

Very few reports concern admittance (or impedance = 1/admittance) as a function of frequency in ECM RRAM. Similar behaviors with TCM or VCM RRAM cells have been observed. The circuit model could be considered as a resistor-type filament in parallel with the nonideal capacitor. For pristine device, the resistance of the filament is very large, and thus nonideal capacitor dominates the low-frequency behaviors. For LRS, since filament is of low resistance, the low-frequency response will be dominated by metal nanowire like filament (see Figure 13.36).

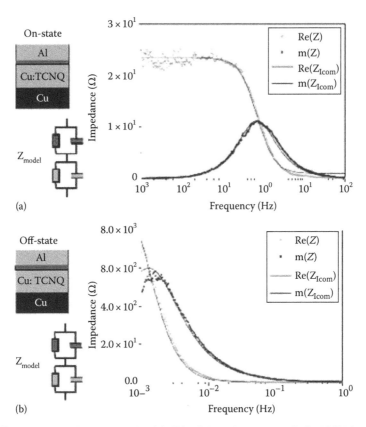

FIGURE 13.36 Impedance analysis of CuTCNQ-based memory cells in LRS (a) and HRS (b) together with the circuit model including a thin Al_2O_3 interface layer at the interface between CuTCNQ and Al top electrode. (From Mikolajick, T. et al., *Adv. Eng. Mater.*, 11, 235, 2009.)

13.2.3.2 Material Combinations

Materials showing ECM switching behaviors share some unique features that their electrodes are commonly Ag or Cu. Table 13.7 lists very recent publications on ECM-type RRAMs.

13.2.3.3 ECM Mechanism Study

Direct identification of metal atom bridge in ECM-type resistive switching system can date back to 1976 when Ag filament in $Ag/As_2S_3/Ag$ RRAM was reported by Hirose et al. [134]. The SET/RESET operations correspond to the creation/annihilation of this metallic ion bridge. Recently, with the advanced microscopy, more reports of nanoscale filament are available. Due to the length limit of this chapter, only very recent pieces of work are discussed in this section.

TABLE 13.7
Recent Reports of ECM-Based RRAM Systems

Type	Switching Region	Thickness (nm)	BE	TE	Device Size	Current	Voltage (V)	Speed Endurance	Retention	Remarks	Reference
Carbon and polymer	PI/GO-PI/PI	10/30/10	ITO	Ag	0.5 mm diameter	<1 mA	<5	>130 (DC)	>1400 s		[119]
	GO	30	Pt	Ag, Au, Cu or Ti	0.2 mm diameter	<10 mA (Cu/GO/Pt)	1.1.1 <1 (Cu/GO/Pt)	>100 (DC)	>8 × 10^4 s		[120]
	PEO+AgClO$_4$	1.1.2 125–500	Pt	Ag	50 × 50 μm^2	<0.1 mA	1.1.3 <2.5	>400 (SET/RESET:8.3 ms/8.3 ms)	>10^5 s		[117]
1.1.4 Chalcogenide glass	Amorphous or crystalline Ag$_{10}$Ge$_{15}$Te$_{75}$	1.1.5 200	Pt	Ag	0.2 mm diameter	<0.3 or 1 mA	1.1.6 <0.3 or 0.15	>1.5 × 10^5 or 3.5 × 10^5 (SET/RESET: 100 ns/100 ns)	NS		[121]
	Ge$_{0.3}$Se$_{0.7}$	1.1.7 30–120	Pt	Cu	1.1.8 50 × 50—400 × 400 μm^2	<0.5 mA	1.1.9 <1	NS	NS		[122]
	GeTe:Cu	1.1.10 100	W	Cu	2 × 2 μm^2	<50 mA	1.1.11 <1	>10^6 (SET/RESET: 200 ns/500 ns)	>5 × 10^{3}s		[123]
1.1.12 Metal carbide	CuC or CuC/BO	NS	Pt	Pt	50 × 50 μm^2	<10 mA	1.1.13 <1.5 or 2.5	>10^3 (SET/RESET: 1 μs/500 ns)	NS	1.1.14 BO = Buffer oxide	[124]

(continued)

TABLE 13.7 (continued)
Recent Reports of ECM-Based RRAM Systems

Type	Switching Region	Thickness (nm)	BE	TE	Device Size	Current	Voltage (V)	Speed Endurance	Retention	Remarks	Reference
1.1.15 Metal oxide	Al_2O_3	1.1.16 50	Si	Cu-Te	3 mm diameter	<5 µA	1.1.17 <1	>10^3 (DC)	>10^4 s @ 85 C		[125]
	Li_xCoO_2	1.1.18 100	Si	Au	0.1 × 0.4 mm^2	<0.1 mA	1.1.19 <4	>12 (DC)	NS		[126]
	CuO	1.1.20 200	ATO	Cu	NS	<10 mA	1.1.21 <5	NS	NS		[127]
	SiO_2	1.1.22 20	Pt	Cu	NS	NS	NS	NS	NS	1.1.23 Complementary cell	[128]
	Ta_2O_5	1.1.24 15	Pt	Cu	20 × 20 µm^2	<1 mA	<2	>100 (DC)	NS		[129]
	ZrO_2:Cu	40	Pt	Cu	200 × 100 nm^2	<0.1 mA	<0.8	NS	NS		[130]
Metal sulfide	Ag_2S	<120	Ag	CAFM tip	44 nm	<100 nA	<0.5	NS	>10 h		[131]
	Cu_2S	100	Cu	Ti	2 mm diameter	<1 mA	<0.4	NS	NS		[132]
	GeS	8–60	Cu	Pt-Ir	<20 nm diameter	<0.5 µA	<4	NS	NS	TE = STEM tip	[133]

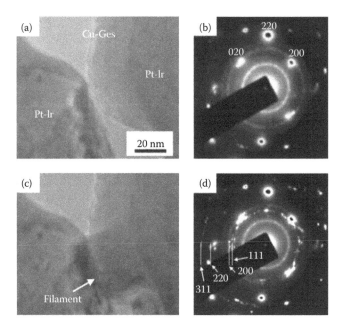

FIGURE 13.37 (a) TEM image and (b) SAD pattern before voltage application. Clear spots caused by Pt–Ir and weak Debye ring patterns were observed. The latter diffraction, which did not change during the resistance switching operation, may come from the Ge nanocrystals. The indices are those of Pt–Ir. (c) TEM image and (d) SAD pattern during voltage application of 1 V. The filament-like deposit and the appearance of fine and sharp diffraction spots forming Debye rings were recognized. They are thought to be from Cu nanocrystals, four of which corresponded with reflection indices. (Reproduced from Fujii, T. et al., *Appl. Phys. Lett.*, 98, 212104, 2011.)

13.2.3.3.1 Physical Characterization

As shown in Figure 13.37, for GeS system, Fujii et al. have reported the direct observation of Cu filament by TEM with in situ Pt–Ir tip electrode [133]. The gradual formation/annihilation process has also been included. EDX has revealed that Cu peak significantly intensifies when the voltage is applied in filament region.

In situ observation of filament formation in GeTe electrolyte has been reported by Choi et al. [135]. HRTEM image of the middle region of the channel reveals significant change of chemical composition with biasing applied, which is shown in Figure 13.38.

For GeSe system, Chen et al. reported time-dependent evolution of precipitation and percolation of Ag cations [136] (see Figure 13.39).

Similar observation was made by Rahaman et al. They reported the direct observation of Cu filaments in GeSe electrolyte by HRTEM also [114]. After cycling, the solid electrolyte is observed of formation of Cu filament, which is also confirmed by EDX (Figure 13.40).

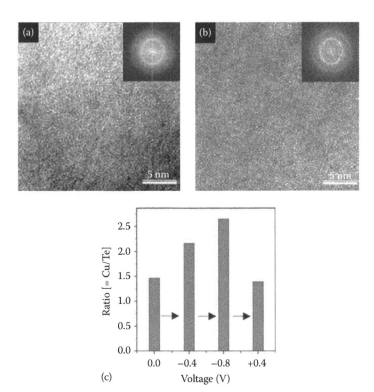

FIGURE 13.38 (a) HRTEM of the middle region in the electrolyte of the Pt–Ir/GeTe:Cu/Cu RRAM with V = 0 applied at top electrode, (b) HRTEM image of the same location with V = −0.8 V, and (c) EDS analysis of Cu/Te ratio under different biasing. (Reproduced from Choi, S.-J. et al., *Adv. Mater.*, 23, 3272, 2011.)

Guo et al. have reported an interesting visualization of Ag filaments in Ag/H$_2$O/Pt switching system [137] (see Figure 13.41). Tree-shaped Ag dendrites grow from the Pt electrode toward the Ag electrode. Once the front-most dendrite contacts the Ag electrode, the cell is switched on.

A novel way to indirectly probe the underlying electrochemical reaction is achieved by employing strain image to see the tiny change of volume associated with biasing by Takata et al. [132]. They employ periodic voltage signal to drive the RRAM and using AFM to record the periodic change of surface morphology (see Figure 13.42).

13.2.3.3.2 Switching and Conduction Mechanism

Quantitative formulation of metal cation migration in ECM-type switching memory can be found from the review by Waser et al. [20] and Valov et al. [25].

In 2009, Chen et al. have employed Monte Carlo simulation to imitate the evolution of filament growth and rupture [138]. The electric field-driven conduction bridge formation is given in Figure 13.43.

Similar efforts have been made by Feng et al. [139,140] (see Figure 13.44).

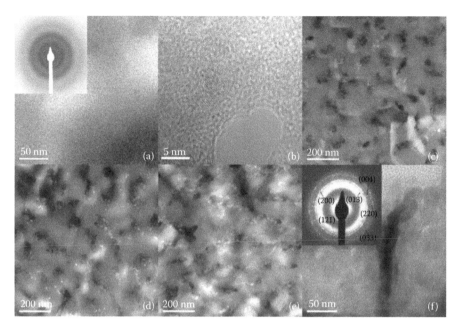

FIGURE 13.39 (a) As deposited film (b) (c) (d) (e) 3, 60, 180, 480 min of electric field, (f) zoom of (e), which shows network-type precipitation. (Reproduced from Chen, L. et al., *Appl. Phys. Lett.*, 94, 162112, 2009.)

13.3 TYPICAL RRAM STRUCTURES

13.3.1 CONVENTIONAL STRUCTURE

Most reported RRAMs employ vertical device architecture for the demonstration of resistive switching performance. This is based on two reasons. The first reason is that ultrahigh density integration desires crossbar structures, requiring layers to be vertically stacked. The other reason is that vertical stacking makes thickness control of individual layer more precise due to the advanced deposition technology such as atomistic layer deposition. The conventional RRAM cell design has also been discussed in the review by Akinaga et al. [21]. Schematic illustration is given for the popular vertical stacking systems (see Figure 13.45).

13.3.2 UNCONVENTIONAL STRUCTURES

Despite the typical MIM vertical stacking structures, a few of the unconventional device architectures have received the attention of researchers due to their unique features.

Son et al. employed a lateral structure where a NiO nanodot is sandwiched by the ends of two Au nanowires [46]. Such a system is suitable for scaling effect study (see Figure 13.46).

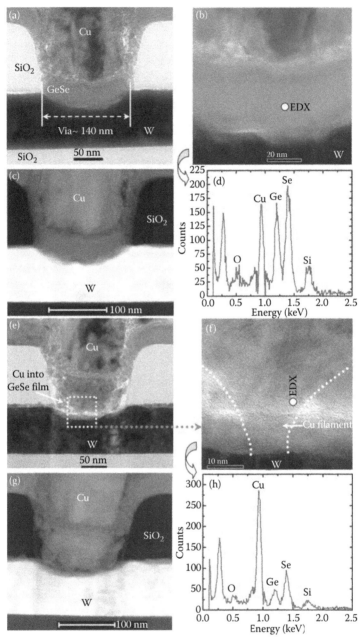

FIGURE 13.40 (a) HRTEM image of W/Ge$_{0.4}$Se$_{0.6}$/Cu memory device without external bias (i.e., fresh device), (b) HRTEM image with a small scale bar of 20 nm, (c) HAADF-STEM image of W/Ge$_{0.4}$Se$_{0.6}$/Cu memory device, (d) EDX spectrum in Ge$_{0.4}$Se$_{0.6}$ layer, (e) HRTEM image with external bias and 4×10^3 cycles (i.e., LRS), and (f) HRTEM image after cycles with small scale bar of 10 nm. A Cu filament is observed clearly. (g) HAADF-STEM image after cycles (i.e., LS) and (h) EDX spectrum in Ge$_{0.4}$Se$_{0.6}$ layer after P/E cycles. (Reproduced from Rahaman, S.Z. et al., *Electrochem. Solid-State Lett.*, 13, H159, 2010.)

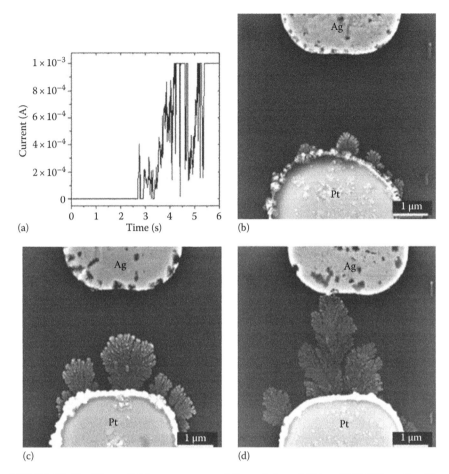

FIGURE 13.41 The switching-on process while applying −1 V to a Pt/H$_2$O/Ag cell with a Pt/Ag gap of 3 μm: (a) I–t curve, (b), (c), and (d) SEM images showing the Ag dendrite growth after applying −1 V for about 1, 2, and 4 s, respectively. (Reproduced from Guo, X. et al., *Appl. Phys. Lett.*, 91, 133513, 2007.)

Yao et al. have reported the SiO$_2$-based nanogap-type RRAM with different electrodes [141,142] (see Figure 13.47). Despite lithography, dielectric breakdown of amorphous carbon or CNT is used for the nanogap generation. Similar nanogap system for charge injection SrTiO$_3$ RRAM has been reported by Chen et al. [143].

Yao has also reported on the Pt/CNT/Pt lateral architecture. In this case, CNT is responsible for switching rather than nanogap [144] (see Figure 13.48).

Another novel structure fringing field RRAM developed by Lee et al. is for the sake of a better electric field confinement in RRAM [145]. The SiO$_2$ dielectric layer embedded by two planar electrodes is of no switching function. Instead, the encapsulation NiO is used for switching under fringing field, as illustrated in Figure 13.49.

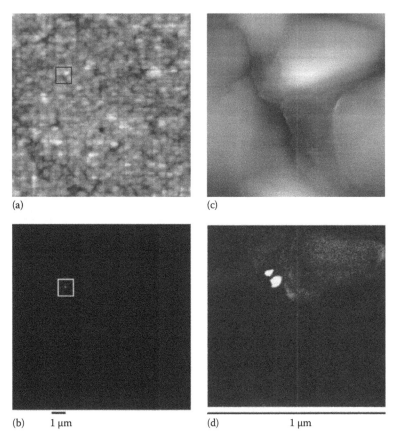

FIGURE 13.42   20-mm-square (a) topographic and (b) amplitude images taken with 0.5 V_{pp} at 60 kHz. 1-mm-square (c) topographic and (d) amplitude images of the square region in (a) and (b). (Reproduced from Takata, K. et al., *Curr. Appl. Phys.*, 11, 1364, 2011.)

13.4 TYPICAL RRAM CIRCUITS

The simple MIM structure makes RRAM suitable for high-density integration. Since the contemporary RAM is of a matrix-like architecture, each single bit element is connected to the decoders via two lines, which are word line and bit line. However, sneak current or parasitic-path-effect exists if the elements are directly incorporated into the matrix. In that case, the signal may bypass the addressed cell by flowing through the LRS cells not addressed. A possible scenario is illustrated in Figure 13.50.

To resolve sneak current issue, depending on the operation schemes, either a nonlinear passive element or active element could be added. They are the most two straightforward approaches but not the only approach. Other sophisticated treatment has been explored also. Recently, the novel concept of complementary RRAM was developed [128,146,147]. The circuit design topic has been recently concerned in the reviews of Kügeler et al. [23] and Valov et al. [25].

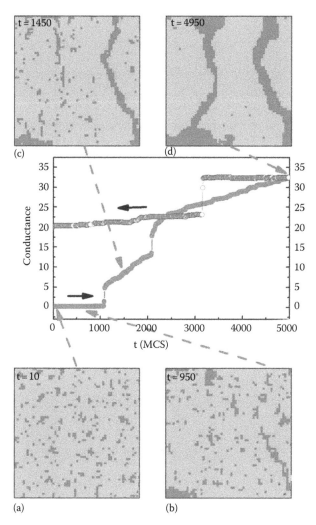

FIGURE 13.43 (a through d) Four snapshots are taken from an evolution, and the phase-separated lattice consisted of high-conductance phase (dark gray) and insulating (light gray) sites. The inset is the AGS conductance as a function of MCS. The red line (solid cycle) shows the result under the positive bias and the blue line (ring) exhibits the result under the reversed bias. (Reproduced from Chen, L. et al., *Appl. Phys. Lett.*, 95, 242106, 2009.)

13.4.1 PASSIVE MATRIX WITH NONLINEAR DEVICE

For unipolar operation scheme, a nonlinear device that is typically a diode can be added in series with the RRAM cell with schematic illustration given in Figure 13.51. Passive matrix is of more attraction than active array due to good potential of scaling. Besides, passive array also features for ease of fabrication.

In 2002, Zhuang et al. have reported the fabrication of the RRAM arrays under 1R–1D scheme [16] (see Figure 13.52).

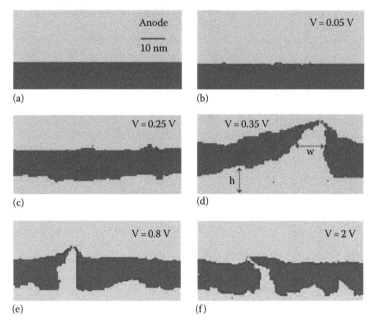

FIGURE 13.44 Filament topographies obtained at different voltage levels. (a) Initial state, (b) V = 0.05 V, (c) V = 0.25 V, (d) V = 0.35 V, (e) V = 0.8 V, and (f) V = 2 V. The arrows in (d) with labels—h and w indicate the thickness of isotropic deposition and the average width of the filament, respectively. (Reproduced from Pan, F. et al., *IEEE Electron Device Lett.*, 32, 949, 2011.)

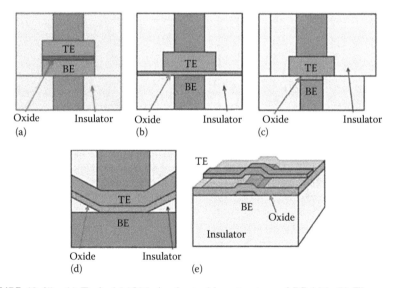

FIGURE 13.45 (a) Typical MOM simple stacking structure of RRAM. (b) The memory element on the metallic via as a bottom electrode with oxide and top electrode layers. (c) The oxidized via material for the resistance switching oxide layer. (d) The concave structure for RRAM. (e) The crossbar structure consisting of the bottom and top electrode wires and blanket oxide film. (Reproduced from Akinaga, H. et al., *Proc. IEEE*, 98, 2237, 2010.)

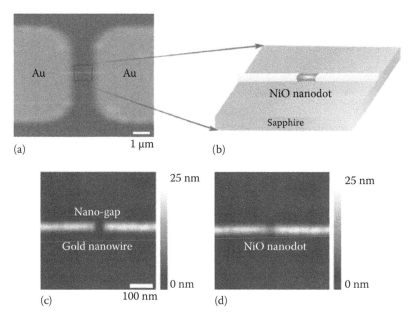

FIGURE 13.46 The Au/NiO/Au nanowire resistive switch. (a) SEM image of the Au/NiO/Au RRAM connected with two large Au electrodes. (b) The schematic drawing of the Au/NiO/Au RRAM as a replica of Figure 1d. (c) AFM image of the nanogap formed between the two DPN Au nanowires with a width of approximately 25 nm where the nanogap distance was approximately 30 nm. (d) AFM image of the Au/NiO/Au nanowire resistive switch with a DPN NiO nanodot with a diameter of approximately 30 nm. (Reproduced from Son, J.Y. et al., *Appl. Surf. Sci.*, 257, 9885, 2011.)

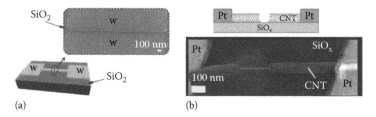

FIGURE 13.47 (a) Schematic of the W–W gap and the SEM image. (b) Upper: A laterally confined embodiment by using the two broken ends of an electrically broken CNT as effective electrodes. Bottom: SEM image of the switching CNT-SiO_x nanogap system.

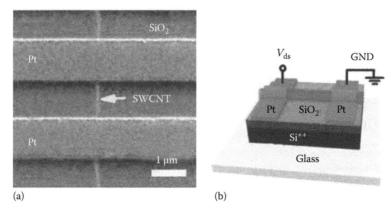

(a)   (b)

FIGURE 13.48 (a) An SEM image of a typical SWCNT device. (b) Schematic of the two-terminal configuration for electrical measurements. (Reproduced from Yao, J. et al., *ACS Nano*, 3, 4122, 2009.)

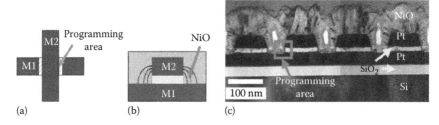

(a)   (b)   (c)

FIGURE 13.49 (a) Top view in the crossbar structure memory array. (b) Cross-section view. (c) TEM cross-section image. (Reproduced from Lee, B. et al., *IEEE Trans. Electron Devices*, 58, 3270, 2011.)

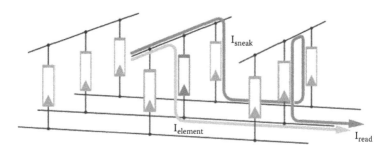

FIGURE 13.50 Possible scenario of sneaking path. Cell in dark gray is of HRS and addressed. Cells in light gray are of LRS and not addressed. Only one sneak current path is depicted. (Reproduced from Linn, E. et al., *Nat. Mater.*, 9, 403, 2010.)

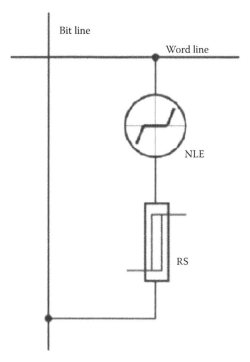

FIGURE 13.51 Schematics of a storage node in a passive array with nonlinear element (NLE) and resistive switching (RS) cell. (Reproduced from Waser, R. et al., *Adv. Mater.*, 21, 2632, 2009.)

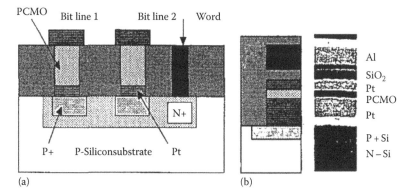

FIGURE 13.52 (a) Schematics of 1R–1D RRAM architecture. (b) TEM microphotograph of a fabricated IR-ID memory cell. (Reproduced from Zhuang, W.W. et al., Novel colossal magnetoresistive thin film nonvolatile resistance random access memory (RRAM), in *IEDM Technical Digest*, pp. 193–196, 2002.)

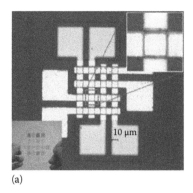
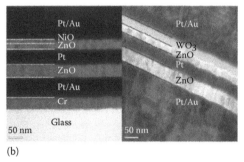

(a)   (b)

FIGURE 13.53 (a) A 4×4 crossbar array ZnO RRAM device stacked with heterostructure diodes. The lower inset is an image of the entire structure of the device. (b) Cross-sectional TEM images of the stacked contact area. (Reproduced from Jung, W.S. et al., *Appl. Phys. Lett.*, 98, 233505, 2011.)

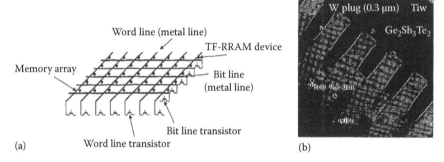

(a)   (b)

FIGURE 13.54 (a) Schematic of the architecture of self-rectifying RRAM memory array. The word lines and the bit lines are selected by word-line transistors and bit-line transistors, respectively. The TF-RRAM device serves both as the memory element and as the access element. (b) Bright-field TEM picture of the mini-array. The diameter of W plugs is 0.3 μm and the pitch of the mini-array is 0.6 μm. (Reproduced from Chen, Y.-C. et al., An access-transistor-free (0T/1R) nonvolatile resistance random access memory (RRAM) using a novel threshold switching, self-rectifying chalcogenide device, in *IEDM Technical Digest*, pp. 37.4.1–37.4.4, 2003.)

Recently, Seo et al. have reported a 4 × 4 bit crossbar array of ZnO unipolar RRAM with selection diode integrated [55] (see Figure 13.53).

Some RRAM systems are of self-rectifying characteristics; they do not desire external nonlinear devices such as diodes. A 4-kbit memory matrix based on self-rectifying RRAM cells has been reported by Chen et al. [148] (see Figure 13.54).

13.4.2 ACTIVE MATRIX

Similar to DRAM, active matrix can be implemented on RRAM. In this case, the access to the individual storage element is controlled by the active element such as

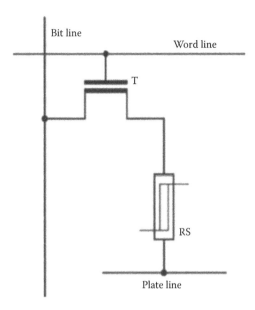

FIGURE 13.55 Schematics of a storage node in an active array with transistor (T) and resistive switching (RS) cell. (Reproduced from Waser, R. et al., *Adv. Mater.*, 21, 2632, 2009.)

transistors (see Figure 13.55). Up to now, majority of large-scale RRAM chip is demonstrated based on active matrix due to mature technology and reliable performance.

The fabrication of crossbar active array of RRAM can be dated back to 2002 when Zhuang et al. reported the 64-bit RRAM with 1R–1T configurations [16] (see Figure 13.56).

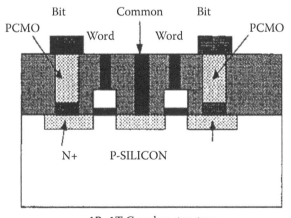

1R–1T Crossbar structure

FIGURE 13.56 Schematics 1R–1T RRAM architectures of crossbar structures. (Reproduced from Zhuang, W.W. et al., Novel colossal magnetoresistive thin film nonvolatile resistance random access memory (RRAM), in *IEDM Technical Digest*, pp. 193–196, 2002.)

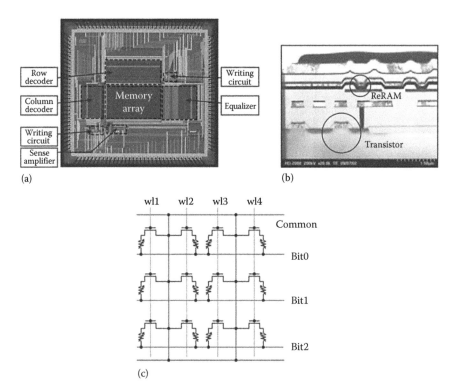

FIGURE 13.57 (a) Microphotograph of 128-kbit RRAM chip. (b) Cross-sectional view of 1T–1R memory cell structure. (c) The architecture of the 1T–1R memory array. (Reproduced from Akinaga, H. et al., *Proc. IEEE*, 98, 2237, 2010.)

Akinaga et al. have reported on a 128-kbit RRAM chipset under 1R–1T scheme in their review [21] (see Figure 13.57). The memory array is demonstrated on a TiN/CoOx/Ta cell.

Recently, successful fabrication of 1Mb RRAM crossbar has been demonstrated by Wang et al. [149] (see Figure 13.58). The single RRAM cell employs a 40 nm size TaN/Cu_xSi_yO/Cu RRAM structure corresponding to the 22 nm node of via.

13.4.3 COMPLEMENTARY RRAM

Complementary RRAM is proposed by Linn et al. and Rosezin et al. [128,146,147]. This novel concept employs two bipolar RRAM cells that are connected back to back. Unlike passive array with nonlinear selector device or active array with transistor, it's feasible to integrate complementary RRAM with the simplest crossbar passive array directly. As proposed, the complementary RRAM consists of four states and I–V characteristics illustrated in Figure 13.59.

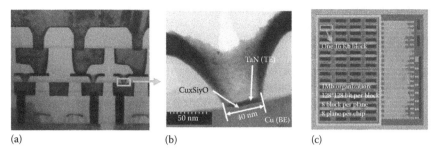

(a) (b) (c)

FIGURE 13.58 TEM cross section of (a) 1T–1R cell in BL direction in 1Mb test chip. The photo shows metal level up to M3 while the chip has totally six metal layers. (b) M1 Copper (BE)/CuSi$_x$O$_y$/TaN(TE) RRAM structure. The active area is shrunk to 40 nm successfully, corresponding to 22 nm node via; (c) micrograph and organization of 1 Mb test chip (in red box) with test features. (Reproduced from Wang, Y. et al., Logic-based mega-bit Cu$_x$Si$_y$O emRRAM with excellent scalability down to 22 nm node for post-emFLASH SOC Era, in *Proceedings of the IEEE International Memory Workshop*, pp. 1–2, 2011.)

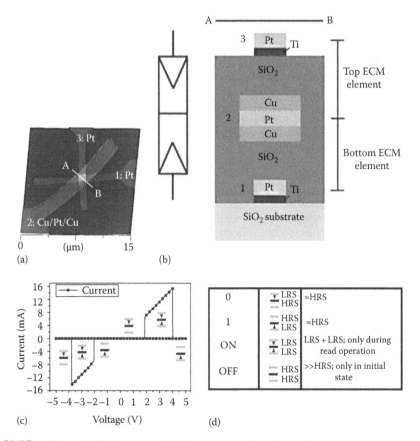

FIGURE 13.59 (a) AFM image and (b) the corresponding sketch of a cross section along the line AB. (c) I–V characteristics, the symbols show states of the complementary RRAM, and (d) states table. (Reproduced from Rosezin, R. et al., *IEEE Electron Device Lett.*, 32, 191, 2011.)

13.5 CONCLUSION

Recent progress in nanoscale RRAM is reviewed in terms of material systems, device architecture, and circuit integration. Particularly, recent development in physical characterization and theoretical formulation of the switching phenomena of different material systems is discussed.

As one of the most promising candidates of next-generation nonvolatile memory, RRAM has received a lot of attention and has been demonstrated with excellent performances. Rapid advancement of fabrication technology and physical characterization will further optimize the performance of RRAM and decipher the underlying switching conduction mechanisms. Atomistic-level formulation of the resistive switching will be developed.

REFERENCES

1. T. W. Hickmott, Low-frequency negative resistance in thin anodic oxide films, *J. Appl. Phys.*, 33, 2669–2682, 1962.
2. D. V. Geppert, A new negative-resistance device, *Proc. IEEE*, 51, 223–223, 1963.
3. H. T. Mann, Electrical properties of thin polymer films. Part I. Thickness 500–2500 Å, *J. Appl. Phys.*, 35, 2173, 1964.
4. S. Pakswer, Negative resistance in thin anodic oxide films, *J. Appl. Phys.*, 34, 711, 1963.
5. K. L. Chopra, Current-controlled negative resistance in thin niobium oxide films, *Proc. IEEE*, 51, 941–942, 1963.
6. W. R. Beam et al., Further properties and a suggested model for niobium oxide negative resistance, *Proc. IEEE*, 52, 300–301, 1964.
7. T. W. Hickmott, Impurity conduction and negative resistance in thin oxide films, *J. Appl. Phys.*, 35, 2118, 1964.
8. T. Mukerjee et al., Negative resistance in Al–Al_2O_3–Se–Au sandwich cells, *J. Appl. Phys.*, 35, 2270, 1964.
9. J. F. Gibbons et al., Switching properties of thin NiO films, *Solid-State Electron.*, 7, 785–790, 1964.
10. K. L. Chopra, Avalanche-induced negative resistance in thin oxide films, *J. Appl. Phys.*, 36, 184, 1965.
11. W. R. Hiatt et al., Bistable switching in niobium oxide diodes, *Appl. Phys. Lett.*, 6, 106, 1965.
12. G. Dearnaley et al., Electrical phenomena in amorphous oxide films, *Rep. Prog. Phys.*, 33, 1129, 1970.
13. R. F. Rutz et al., An AlN switchable memory resistor capable of a 20-MHz cycling rate and 500-picosecond switching time, *IBM J. Res. Dev.*, 17, 61–65, 1973.
14. H.-D. Kim et al., Stable bipolar resistive switching characteristics and resistive switching mechanisms observed in aluminum nitride-based ReRAM devices, *IEEE Trans. Electron Devices*, 58, 3566–3573, 2011.
15. A. Beck et al., Reproducible switching effect in thin oxide films for memory applications, *Appl. Phys. Lett.*, 77, 139, 2000.
16. W. W. Zhuang et al., Novel colossal magnetoresistive thin film nonvolatile resistance random access memory (RRAM), in *IEDM Technical Digest*, 2002, pp. 193–196.
17. R. Waser et al., Nanoionics-based resistive switching memories, *Nat. Mater.*, 6, 833–840, 2007.
18. A. Sawa, Resistive switching in transition metal oxides, *Mater. Today*, 11, 28–36, 2008.

19. A. Ignatiev et al., Resistance switching in oxide thin films, *Phase Transit.*, 81, 791–806, 2008.
20. R. Waser et al., Redox-based resistive switching memories—Nanoionic mechanisms, prospects, and challenges, *Adv. Mater.*, 21, 2632–2663, 2009.
21. H. Akinaga et al., Resistive random access memory (ReRAM) based on metal oxides, *Proc. IEEE*, 98, 2237–2251, 2010.
22. Y. V. Pershin et al., Memory effects in complex materials and nanoscale systems, *Adv. Phys.*, 60, 145–227, 2011.
23. C. Kügeler et al., Materials, technologies, and circuit concepts for nanocrossbar-based bipolar RRAM, *Appl. Phys. A*, 102, 791–809, 2011.
24. K. Szot et al., TiO_2—A prototypical memristive material, *Nanotechnology*, 22, 254001, June 24, 2011.
25. I. Valov et al., Electrochemical metallization memories—Fundamentals, applications, prospects, *Nanotechnology*, 22, 289502, 2011.
26. K. M. Kim et al., Nanofilamentary resistive switching in binary oxide system; a review on the present status and outlook, *Nanotechnology*, 22, 254002, June 24, 2011.
27. D. Ielmini et al., Thermochemical resistive switching: Materials, mechanisms, and scaling projections, *Phase Transit.*, 84, 570–602, 2011.
28. L. Courtade et al., Improvement of resistance switching characteristics in NiO films obtained from controlled Ni oxidation, in *Non-Volatile Memory Technology Symposium*, 2007, pp. 1–4.
29. D. Ielmini, Size-dependent switching and reliability of NiO RRAMs, *ECS Trans.*, 33, 323–331, 2010.
30. X. A. Tran et al., A high-yield HfO_x-based unipolar resistive RAM employing Ni electrode compatible with Si-Diode selector for crossbar integration, *IEEE Electron Device Lett.*, 32, 396–398, 2011.
31. D. S. Jeong et al., Impedance spectroscopy of TiO_2 thin films showing resistive switching, *Appl. Phys. Lett.*, 89, 082909, 2006.
32. F. Di et al., Unipolar resistive switching properties of diamondlike carbon-based RRAM devices, *IEEE Electron Device Lett.*, 32, 803–805, 2011.
33. K. S. Vasu et al., Nonvolatile unipolar resistive switching in ultrathin films of graphene and carbon nanotubes, *Solid State Commun.*, 151, 1084–1087, 2011.
34. P. Y. Lai et al., Ultrahigh on/off-current ratio for resistive memory devices with Poly(N-Vinylcarbazole)/Poly(3,4-Ethylenedioxythiophene)-Poly(Styrenesulfonate) stacking bilayer, *IEEE Electron Device Lett.*, 32, 387–389, 2011.
35. R. Rosezin et al., Electroforming and resistance switching characteristics of silver-doped MSQ with inert electrodes, *IEEE Trans. Nanotechnol.*, 10, 338–343, 2011.
36. S. Hong et al., Unipolar resistive switching mechanism speculated from irreversible low resistance state of Cu_2O films, *Appl. Phys. Lett.*, 99, 052105, 2011.
37. N. Raghavan et al., Filamentation mechanism of resistive switching in fully silicided high-κ gate stacks, *IEEE Electron Device Lett.*, 32, 455–457, 2011.
38. T.-H. Hou et al., Evolution of RESET current and filament morphology in low-power HfO_2 unipolar resistive switching memory, *Appl. Phys. Lett.*, 98, 103511, 2011.
39. Y. Y. Chen et al., Switching by Ni filaments in a HfO_2 matrix: A new pathway to improved unipolar switching RRAM, in *Proceedings of the IEEE International Memory Workshop*, 2011, pp. 1–4.
40. X. A. Tran et al., Ni electrode unipolar resistive RAM performance enhancement by AlO_y incorporation into HfO_x switching dielectrics, *IEEE Electron Device Lett.*, 32, 1290–1292, 2011.
41. Z. Q. Wang et al., Flexible resistive switching memory device based on amorphous InGaZnO film with excellent mechanical endurance, *IEEE Electron Device Lett.*, 32, 1442–1444, 2011.

42. K. Jung et al., Unipolar resistive switching in insulating niobium oxide film and probing electroforming induced metallic components, *J. Appl. Phys.*, 109, 054511, 2011.

43. C. Dumas et al., Resistive switching characteristics of NiO films deposited on top of W or Cu pillar bottom electrodes, *Thin Solid Films*, 519, 3798–3803, 2011.

44. J. Kim et al., Effect of W impurity on resistance switching characteristics of NiO_x films, *Curr. Appl. Phys.*, 11, e70–e74, 2011.

45. Q.-Q. Sun et al., Controllable filament with electric field engineering for resistive switching uniformity, *IEEE Electron Device Lett.*, 32, 1167–1169, 2011.

46. J. Y. Son et al., Nanoscale resistive random access memory consisting of a NiO nanodot and Au nanowires formed by dip-pen nanolithography, *Appl. Surf. Sci.*, 257, 9885–9887, 2011.

47. R. Huang et al., Resistive switching of silicon-rich-oxide featuring high compatibility with CMOS technology for 3D stackable and embedded applications, *Appl. Phys. A*, 102, 927–931, 2011.

48. C.-Y. Liu et al., Different resistive switching characteristics of a $Cu/SiO_2/Pt$ structure, *Jpn. J. Appl. Phys.*, 50, 091101, 2011.

49. S. Almeida et al., Resistive switching of SnO_2 thin films on glass substrates, *Integr. Ferroelectr.*, 126, 117–124, 2011.

50. H. K. Yoo et al., Conversion from unipolar to bipolar resistance switching by inserting Ta_2O_5 layer in $Pt/TaO_x/Pt$ cells, *Appl. Phys. Lett.*, 98, 183507, 2011.

51. L. F. Liu et al., Unipolar resistive switching and mechanism in Gd-doped-TiO_2-based resistive switching memory devices, *Semicond. Sci. Technol.*, 26, 115009, 2011.

52. G. H. Kim et al., Improved endurance of resistive switching TiO_2 thin film by hourglass shaped Magnéli filaments, *Appl. Phys. Lett.*, 98, 262901, 2011.

53. Y. H. Kang et al., Thickness dependence of the resistive switching behavior of nonvolatile memory device structures based on undoped ZnO films, *Solid State Commun.*, 151, 1739–1742, 2011.

54. M. H. Tang et al., Resistive switching behavior of La-doped ZnO films for nonvolatile memory applications, *Solid-State Electron.*, 63, 100–104, 2011.

55. W. S. Jung et al., A ZnO cross-bar array resistive random access memory stacked with heterostructure diodes for eliminating the sneak current effect, *Appl. Phys. Lett.*, 98, 233505, 2011.

56. Q. Xu et al., Bipolar and unipolar resistive switching in $Zn_{0.98}Cu_{0.02}O$ films, *J. Phys. D Appl. Phys.*, 44, 335104, 2011.

57. Y. Li et al., Reset instability in Cu/ZrO_2:Cu/Pt RRAM device, *IEEE Electron Device Lett.*, 32, 363–365, 2011.

58. S. B. Lee et al., Time-dependent current-voltage curves during the forming process in unipolar resistance switching, *Appl. Phys. Lett.*, 98, 053503, 2011.

59. S.-L. Li et al., Unipolar resistive switching in high-resistivity $Pr_{0.7}Ca_{0.3}MnO_3$ junctions, *Appl. Phys. A*, 103, 21–26, 2011.

60. X. Sun et al., Coexistence of the bipolar and unipolar resistive switching behaviours in $Au/SrTiO_3/Pt$ cells, *J. Phys. D Appl. Phys.*, 44, 125404, 2011.

61. M. H. Tang et al., Bipolar and unipolar resistive switching behaviors of sol–gel-derived $SrTiO_3$ thin films with different compliance currents, *Semicond. Sci. Technol.*, 26, 075019, 2011.

62. J. Wu et al., Current–voltage characteristics of sol–gel derived $SrZrO_3$ thin films for resistive memory applications, *J. Alloys Compd.*, 509, 2050–2053, 2011.

63. Z. Yan et al., High-performance programmable memory devices based on co-doped $BaTiO_3$, *Adv. Mater.*, 23, 1351–1355, March 18, 2011.

64. D. Ielmini et al., Physical models of size-dependent nanofilament formation and rupture in NiO resistive switching memories, *Nanotechnology*, 22, 254022, June 24, 2011.

65. I. K. Yoo et al., Fractal dimension of conducting paths in nickel oxide (NiO) thin films during resistance switching, *IEEE Trans. Nanotechnol.*, 9, 131–133, 2010.

66. X. Li et al., Resistive switching in NiSi gate metal-oxide-semiconductor transistors, *Appl. Phys. Lett.*, 97, 202904, 2010.

67. P. Calka et al., Origin of resistivity change in NiO thin films studied by hard x-ray photoelectron spectroscopy, *J. Appl. Phys.*, 109, 124507, 2011.

68. H. Kondo et al., The observation of conduction spot on NiO resistance random access memory, *Jpn. J. Appl. Phys.*, 50, 081101, 2011.

69. D.-H. Kwon et al., Atomic structure of conducting nanofilaments in TiO_2 resistive switching memory, *Nat. Nano.*, 5, 148–153, 2010.

70. U. Russo et al., Filament conduction and reset mechanism in NiO-based resistive-switching memory (RRAM) devices, *IEEE Trans. Electron Devices*, 56, 186–192, 2009.

71. M. Bocquet et al., Self-consistent physical modeling of set/reset operations in unipolar resistive-switching memories, *Appl. Phys. Lett.*, 98, 263507, 2011.

72. L. Zhao et al., Dynamic modeling and atomistic simulations of SET and RESET operations in TiO_2-based unipolar resistive memory, *IEEE Electron Device Lett.*, 32, 677–679, 2011.

73. F. S.-S. Chien et al., Disproportionation and comproportionation reactions of resistive switching in polycrystalline NiO_x films, *Appl. Phys. Lett.*, 98, 153513, 2011.

74. S.-G. Park et al., Impact of oxygen vacancy ordering on the formation of a conductive filament in TiO_2 for resistive switching memory, *IEEE Electron Device Lett.*, 32, 197–199, 2011.

75. B. Magyari-Köpe et al., Resistive switching mechanisms in random access memory devices incorporating transition metal oxides: TiO_2, NiO and $Pr_{0.7}Ca_{0.3}MnO_3$, *Nanotechnology*, 22, 254029, June 24 2011.

76. H. Y. Lee et al., Low power and high speed bipolar switching with a thin reactive Ti buffer layer in robust HfO2 based RRAM, in *IEDM Technical Digest*, 2008, pp. 1–4.

77. Y. L. Song et al., Low reset current in stacked AlO_x/WO_x resistive switching memory, *IEEE Electron Device Lett.*, 32, 1439–1441, 2011.

78. W. Lee et al., Nonvolatile resistive switching in $Pr_{0.7}Ca_{0.3}MnO_3$ devices using multilayer graphene electrodes, *Appl. Phys. Lett.*, 98, 032105, 2011.

79. C. Ho et al., A highly reliable self-aligned graded oxide WO_x resistance memory: Conduction mechanisms and reliability, in *VLSI Symposium Technology Digest*, 2007, pp. 228–229.

80. C. H. Cheng et al., Ultralow switching energy $Ni/GeO_x/HfON/TaN$ RRAM, *IEEE Electron Device Lett.*, 32, 366–368, 2011.

81. P. Gonon et al., Resistance switching in HfO_2 metal-insulator-metal devices, *J. Appl. Phys.*, 107, 074507, 2010.

82. I. Kim et al., Low temperature solution-processed graphene oxide/$Pr_{0.7}Ca_{0.3}MnO_3$ based resistive-memory device, *Appl. Phys. Lett.*, 99, 042101, 2011.

83. H.-D. Kim et al., Transparent resistive switching memory using ITO/AlN/ITO capacitors, *IEEE Electron Device Lett.*, 32, 1125–1127, 2011.

84. Q. Lu et al., Reproducible resistive-switching behavior in copper-nitride thin film prepared by plasma-immersion ion implantation, *Phys. Status Solidi (a)*, 208, 874–877, 2011.

85. L. Chen et al., Enhancement of resistive switching characteristics in Al_2O_3-based RRAM with embedded ruthenium nanocrystals, *IEEE Electron Device Lett.*, 32, 794–796, 2011.

86. Y.-E. Syu et al., Redox reaction switching mechanism in RRAM device with $Pt/CoSiO_X\backslash TiN$ structure, *IEEE Electron Device Lett.*, 32, 545–547, 2011.

87. C. H. Cheng et al., Stacked $GeO/SrTiO_x$ resistive memory with ultralow resistance currents, *Appl. Phys. Lett.*, 98, 052905, 2011.

88. Z. Fang et al., $HfO_x/TiO_x/HfO_x/TiO_x$ multilayer-based forming-free RRAM devices with excellent uniformity, *IEEE Electron Device Lett.*, 32, 566–568, 2011.

89. D. Lee et al., Noise-analysis-based model of filamentary switching ReRAM with ZrO_x/HfO_x stacks, *IEEE Electron Device Lett.*, 32, 964–966, 2011.

90. N. Raghavan et al., Very low reset current for an RRAM device achieved in the oxygen-vacancy-controlled regime, *IEEE Electron Device Lett.*, 32, 716–718, 2011.

91. C. Vallée et al., Plasma treatment of HfO_2-based metal–insulator–metal resistive memories, *J. Vac. Sci. Technol. A*, 29, 041512, 2011.

92. H. Zhang et al., Gd-doping effect on performance of HfO_2 based resistive switching memory devices using implantation approach, *Appl. Phys. Lett.*, 98, 042105, 2011.

93. L. Chen et al., Bipolar resistive switching characteristics of atomic layer deposited Nb_2O_5 thin films for nonvolatile memory application, *Curr. Appl. Phys.*, 11, 849–852, 2011.

94. C. B. Lee et al., Highly uniform switching of tantalum embedded amorphous oxide using self-compliance bipolar resistive Switching, *IEEE Electron Device Lett.*, 32, 399–401, 2011.

95. Y. S. Chen et al., Understanding the intermediate initial state in $TiO_{2-\delta}/La_{2/3}Sr_{1/3}MnO_3$ stack-based bipolar resistive switching devices, *Appl. Phys. Lett.*, 99, 072113, 2011.

96. J.-K. Lee et al., Extraction of trap location and energy from random telegraph noise in amorphous TiO_x resistance random access memories, *Appl. Phys. Lett.*, 98, 143502, 2011.

97. Y.-S. Chen et al., Good endurance and memory window for Ti/HfO_x pillar RRAM at 50-nm scale by optimal encapsulation layer, *IEEE Electron Device Lett.*, 32, 390–392, 2011.

98. S. Kim et al., Effect of scaling WO_x-based RRAMs on their resistive switching characteristics, *IEEE Electron Device Lett.*, 32, 671–673, 2011.

99. K. Kinoshita et al., Flexible and transparent ReRAM with GZO memory layer and GZO-electrodes on large PEN sheet, *Solid-State Electron.*, 58, 48–53, 2011.

100. K. Zheng et al., An indium-free transparent resistive switching random access memory, *IEEE Electron Device Lett.*, 32, 797–799, 2011.

101. M.-C. Wu et al., Low-power and highly reliable multilevel operation in ZrO_2 1T1R RRAM, *IEEE Electron Device Lett.*, 32, 1026–1028, 2011.

102. M. Siddik et al., Thermally assisted resistive switching in $Pr_{0.7}Ca_{0.3}MnO_3/Ti/Ge_2Sb_2Te_5$ stack for nonvolatile memory applications, *Appl. Phys. Lett.*, 99, 063501, 2011.

103. R. Yang et al., Improvement of resistance switching properties for metal/$La_{0.7}Ca_{0.3}MnO_3$/Pt devices, *Phys. Status Solidi (a)*, 208, 1041–1046, 2011.

104. K. Szot et al., Switching the electrical resistance of individual dislocations in single-crystalline $SrTiO_3$, *Nat. Mater.*, 5, 312–320, April 2006.

105. D.-S. Shang et al., Local resistance switching at grain and grain boundary surfaces of polycrystalline tungsten oxide films, *Nanotechnology*, 22, 254008, June 24, 2011.

106. K. McKenna et al., The interaction of oxygen vacancies with grain boundaries in monoclinic HfO_2, *Appl. Phys. Lett.*, 95, 222111–222113, 2009.

107. B. Gao et al., A physical model for bipolar oxide-based resistive switching memory based on ion-transport-recombination effect, *Appl. Phys. Lett.*, 98, 232108, 2011.

108. H.-L. Chang et al., Physical mechanism of HfO_2-based bipolar resistive random access memory, in *Proceedings of the VLSI-TSA*, 2011, pp. 1–2.

109. S. Yu et al., Investigating the switching dynamics and multilevel capability of bipolar metal oxide resistive switching memory, *Appl. Phys. Lett.*, 98, 103514, 2011.

110. P. Broqvist et al., Oxygen vacancy in monoclinic HfO_2: A consistent interpretation of trap assisted conduction, direct electron injection, and optical absorption experiments, *Appl. Phys. Lett.*, 89, 262904, 2006.

111. B. Gao et al., Unified physical model of bipolar oxide-based resistive switching memory, *IEEE Electron Device Lett.*, 30, 1326–1328, 2009.
112. H. Momida et al., Effect of vacancy-type oxygen deficiency on electronic structure in amorphous alumina, *Appl. Phys. Lett.*, 98, 042102, 2011.
113. S. Yu et al., Conduction mechanism of TiN/HfO$_x$/Pt resistive switching memory: A trap-assisted-tunneling model, *Appl. Phys. Lett.*, 99, 063507, 2011.
114. S. Z. Rahaman et al., Bipolar resistive switching memory using Cu metallic filament in Ge$_{0.4}$Se$_{0.6}$ solid electrolyte, *Electrochem. Solid-State Lett.*, 13, H159, 2010.
115. M. Thomas et al., From micrometric to nanometric scale switching of CuTCNQ-based non-volatile memory structures, in *Non-Volatile Memory Technology Symposium*, 2008, pp. 1–4.
116. C. Schindler et al., Resistive switching in Ge$_{0.3}$Se$_{0.7}$ films by means of copper ion migration, *Z. Phys. Chem.*, 221, 1469–1478, 2007.
117. S. Wu et al., A polymer-electrolyte-based atomic switch, *Adv. Funct. Mater.*, 21, 93–99, 2011.
118. T. Mikolajick et al., Nonvolatile memory concepts based on resistive switching in inorganic materials, *Adv. Eng. Mater.*, 11, 235–240, 2009.
119. C. Wu et al., Highly reproducible memory effect of organic multilevel resistive-switch device utilizing graphene oxide sheets/polyimide hybrid nanocomposite, *Appl. Phys. Lett.*, 99, 042108, 2011.
120. F. Zhuge et al., Mechanism of nonvolatile resistive switching in graphene oxide thin films, *Carbon*, 49, 3796–3802, 2011.
121. H. Xu et al., Phase change behavior in Ag$_{10}$Ge$_{15}$Te$_{75}$ and the electrolytic resistive switching in both amorphous and crystalline Ag$_{10}$Ge$_{15}$Te$_{75}$ films, *Electrochem. Solid-State Lett.*, 14, H99, 2011.
122. R. Soni et al., On the stochastic nature of resistive switching in Cu doped Ge$_{0.3}$Se$_{0.7}$ based memory devices, *J. Appl. Phys.*, 110, 054509, 2011.
123. S.-J. Choi et al., Multibit operation of Cu/Cu-GeTe/W resistive memory device controlled by pulse voltage magnitude and width, *IEEE Electron Device Lett.*, 32, 375–377, 2011.
124. S. Kim et al., Forming-free CuC-buffer oxide resistive switching behavior with improved resistance ratio, *Electrochem. Solid-State Lett.*, 14, H322, 2011.
125. L. Goux et al., Influence of the Cu-Te composition and microstructure on the resistive switching of Cu-Te/Al$_2$O$_3$/Si cells, *Appl. Phys. Lett.*, 99, 053502, 2011.
126. A. Moradpour et al., Resistive switching phenomena in Li$_x$CoO$_2$ thin films, *Adv. Mater.*, 23, 4141–4145, September 22, 2011.
127. Y. Li et al., Top electrode effects on resistive switching behavior in CuO thin films, *Appl. Phys. A*, 104, 1069–1073, 2011.
128. R. Rosezin et al., Integrated complementary resistive switches for passive high-density nanocrossbar arrays, *IEEE Electron Device Lett.*, 32, 191–193, 2011.
129. T. Tohru et al., Temperature effects on the switching kinetics of a Cu–Ta$_2$O$_5$-based atomic switch, *Nanotechnology*, 22, 254013, 2011.
130. S. Long et al., Resistive switching mechanism of Ag/ZrO$_2$:Cu/Pt memory cell, *Appl. Phys. A*, 102, 915–919, 2011.
131. D. Wang et al., Fabrication and characterization of extended arrays of Ag$_2$S/Ag nanodot resistive switches, *Appl. Phys. Lett.*, 98, 243109, 2011.
132. K. Takata et al., Strain imaging of a Cu$_2$S switching device, *Curr. Appl. Phys.*, 11, 1364–1367, 2011.
133. T. Fujii et al., In situ transmission electron microscopy analysis of conductive filament during solid electrolyte resistance switching, *Appl. Phys. Lett.*, 98, 212104, 2011.
134. Y. Hirose et al., Polarity-dependent memory switching and behavior of Ag dendrite in Ag-photodoped amorphous As$_2$S$_3$ films, *J. Appl. Phys.*, 47, 2767, 1976.

135. S.-J. Choi et al., In situ observation of voltage-induced multilevel resistive switching in solid electrolyte memory, *Adv. Mater.*, 23, 3272–3277, August 2, 2011.

136. L. Chen et al., Electrical field induced precipitation reaction and percolation in $Ag_{30}Ge_{17}Se_{53}$ amorphous electrolyte films, *Appl. Phys. Lett.*, 94, 162112, 2009.

137. X. Guo et al., Understanding the switching-off mechanism in Ag^+ migration based resistively switching model systems, *Appl. Phys. Lett.*, 91, 133513, 2007.

138. L. Chen et al., Monte Carlo simulation of the percolation in $Ag_{30}Ge_{17}Se_{53}$ amorphous electrolyte films, *Appl. Phys. Lett.*, 95, 242106, 2009.

139. F. Pan et al., A comprehensive simulation study on metal conducting filament formation in resistive switching memories, in *Proceedings of the IEEE International Memory Workshop*, 2011, pp. 1–4.

140. F. Pan et al., A detailed study of the forming stage of an electrochemical resistive switching memory by KMC simulation, *IEEE Electron Device Lett.*, 32, 949–951, 2011.

141. J. Yao et al., Resistive switching in nanogap systems on SiO_2 substrates, *Small,* 5, 2910–2915, December 2009.

142. J. Yao et al., Resistive switches and memories from silicon oxide, *Nano Lett.*, 10, 4105–4110, October 13, 2010.

143. X. G. Chen et al., Comprehensive study of the resistance switching in $SrTiO_3$ and Nb-doped $SrTiO_3$, *Appl. Phys. Lett.*, 98, 122102, 2011.

144. J. Yao et al., Two-terminal nonvolatile memories based on single-walled carbon nanotubes, *ACS Nano*, 3, 4122–4126, December 22, 2009.

145. B. Lee et al., Fabrication and characterization of nanoscale NiO resistance change memory (RRAM) cells with confined conduction paths, *IEEE Trans. Electron Devices*, 58, 3270–3275, 2011.

146. E. Linn et al., Complementary resistive switches for passive nanocrossbar memories, *Nat. Mater.*, 9, 403–406, 2010.

147. R. Rosezin et al., Crossbar logic using bipolar and complementary resistive switches, *IEEE Electron Device Lett.*, 32, 710–712, 2011.

148. Y.-C. Chen et al., An access-transistor-free (0T/1R) non-volatile resistance random access memory (RRAM) using a novel threshold switching, self-rectifying chalcogenide device, in *IEDM Technical Digest*, 2003, pp. 37.4.1–37.4.4.

149. Y. Wang et al., Logic-based mega-bit Cu_xSi_yO emRRAM with excellent scalability down to 22 nm node for post-emFLASH SOC Era, in *Proceedings of the IEEE International Memory Workshop*, 2011, pp. 1–2.

Index